SAN JOAQUIN DELTA COLLEGE LIBRARY
1 0200 2190465 4

D0769120

Organic Reactions

Organic Reactions

VOLUME 17

JOHN WILEY & SONS, INC.

NEW YORK • LONDON • SYDNEY • TORONTO

Library of Congress Catalog Card Number: 42-20265
SBN 471 19615 0
PRINTED IN THE UNITED STATES OF AMERICA

PREFACE TO THE SERIES

In the course of nearly every program of research in organic chemistry the investigator finds it necessary to use several of the better-known synthetic reactions. To discover the optimum conditions for the application of even the most familiar one to a compound not previously subjected to the reaction often requires an extensive search of the literature; even then a series of experiments may be necessary. When the results of the investigation are published, the synthesis, which may have required months of work, is usually described without comment. The background of knowledge and experience gained in the literature search and experimentation is thus lost to those who subsequently have occasion to apply the general method. The student of preparative organic chemistry faces similar difficulties. The textbook and laboratory manuals furnish numerous examples of the application of various syntheses, but only rarely do they convey an accurate conception of the scope and usefulness of the processes.

For many years American organic chemists have discussed these problems. The plan of compiling critical discussions of the more important reactions thus was evolved. The volumes of *Organic Reactions* are collections of chapters each devoted to a single reaction, or a definite phase of a reaction, of wide applicability. The authors have had experience with the processes surveyed. The subjects are presented from the preparative viewpoint, and particular attention is given to limitations, interfering influences, effects of structure, and the selection of experimental techniques. Each chapter includes several detailed procedures illustrating the significant modifications of the method. Most of these procedures have been found satisfactory by the author or one of the editors, but unlike those in *Organic Syntheses* they have not been subjected to careful testing in two or more laboratories. When all known examples of the reactions are not mentioned in the text, tables are given to list compounds which have been prepared by or subjected to the reaction. Every effort has been made to include in the tables all such compounds and references; however, because of the very nature of the reactions discussed and their frequent use as one of the several steps of syntheses in which not all of the intermediates have been isolated, some instances may well have been missed. Nevertheless, the investigator will be able to use the tables and their accompanying bibliographies in place of most or all of the literature search so often required.

Because of the systematic arrangement of the material in the chapters and the entries in the tables, users of the books will be able to find information desired by reference to the table of contents of the appropriate chapter. In the interest of economy the entries in the indices have been kept to a minimum, and, in particular, the compounds listed in the tables are not repeated in the indices.

The success of this publication, which will appear periodically, depends upon the cooperation of organic chemists and their willingness to devote time and effort to the preparation of the chapters. They have manifested their interest already by the almost unanimous acceptance of invitations to contribute to the work. The editors will welcome their continued interest and their suggestions for improvements in *Organic Reactions*.

CONTENTS

CHAPTER 1

THE SYNTHESIS OF SUBSTITUTED FERROCENES AND OTHER π-CYCLOPENTADIENYL-TRANSITION METAL COMPOUNDS

Donald E. Bublitz
Western Division Research Laboratories, The Dow Chemical Company

Kenneth L. Rinehart, Jr.,
University of Illinois

CONTENTS

INTRODUCTION

The discovery of dicyclopentadienyliron (ferrocene) in 1951[1, 2] has led to the development of an entirely new area in the field of organometallic chemistry, that of the π-metallohydrocarbons. Representative compounds are the π-metalloarenes like dibenzenechromium, the π-metallopseudoarenes like ferrocene, and the π-metalloolefin complexes like butadieneiron tricarbonyl.

Numerous reviews[3–43] of this fascinating new branch of organometallic

[1] T. J. Kealy and P. L. Pauson, *Nature*, **168**, 1039 (1951).

[2] S. A. Miller, J. A. Tebboth, and J. F. Tremaine, *J. Chem. Soc.*, 632 (1952).

[3] S. Swaminathan and S. Ranganathan, *Current Sci.* (*India*), **25**, 6 (1956).

[4] Shiro Senoh, *Kagaku No Ryoiki*, **7**, 167 (1953) [*C.A.*, **48**, 13639 (1954)].

[5] P. L. Pauson, in H. Zeiss, *Organometallic Chemistry*, pp. 346–379, Reinhold, New York, 1960.

[6] P. L. Pauson, in D. Ginsburg, *Non-Benzenoid Aromatic Compounds*, pp. 107–140, Interscience, New York, 1959.

[7] M. Rausch, M. Vogel, and H. Rosenberg, *J. Chem. Ed.*, **34**, 268 (1957).

[8] K. Plesske, *Angew. Chem.*, **74**, 301, 347 (1962); *Angew. Chem.* (*Intern. Ed.*), **1**, 312, 394 (1964).

[9] A. N. Nesmeyanov, *Proc. Roy. Soc.* (*London*), *Ser. A.*, **246**, 495 (1958).

[10] A. N. Nesmeyanov and E. G. Perevalova, *Usp. Khim.*, **27**, 3 (1958) [*C.A.*, **52**, 14579 (1958)].

[11] A. N. Nesmeyanov, *Zh. Vses. Khim. Obshchestva im. D. I. Mendeleeva*, **7**, 249 (1962) [*C.A.*, **58**, 4596 (1963)].

[12] M. Dub, *Organometallic Compounds Literature Survey 1937–1958*, Vol. 1, Springer-Verlag, 1961.

[13] A. N. Nesmeyanov and E. G. Perevalova, *Khim. Nauka i Promy.*, **3**, 146 (1958) [*C.A.*, **52**, 20108 (1958)].

[14] P. L. Pauson, *Quart. Rev.* (*London*), **9**, 391 (1955).

[15] E. O. Fischer, *Angew. Chem.*, **67**, 475 (1955).

[16] E. O. Fischer and H. P. Fritz, *Advan. Inorg. Chem. Radiochem.*, **1**, 55 (1959).

[17] M. D. Rausch, *J. Chem. Ed.*, **37**, 568 (1960).

[18] K. Schlögl, *Oesterr. Chemiker-Z.*, **59**, 93 (1958) [*C.A.*, **52**, 18358 (1958)].

[19] M. E. Dyatkina, *Usp. Khim.*, **27**, 57 (1958) [*C.A.*, **52**, 14579 (1958)].

[20] H. Zeiss, in H. Zeiss, *Organometallic Chemistry*, pp. 380–425, Reinhold, New York, 1960.

[21] E. O. Fischer and H. P. Fritz, *Angew. Chem.*, **73**, 353 (1961).

[22] G. Wilkinson and F. A. Cotton, *Progr. Inorg. Chem.*, **1**, 1 (1959).

[23] P. L. Pauson, *Endeavour*, 175 (1962).

[24] M. L. H. Green, *Angew. Chem.*, **72**, 719 (1960).

[25] D. Braun, *Angew. Chem.*, **73**, 197 (1961).

[26] M. D. Rausch, *Can. J. Chem.*, **41**, 1289 (1963).

[27] M. D. Rausch, *Advan. Chem. Ser.*, **37**, 56 (1963).

[28] W. F. Little, *Survey of Progress in Chemistry*, Vol. 1, Academic, New York, 1963.

[29] J. M. Birmingham, *Chem. Eng. Progress*, **58**, 74 (1962).

[30] M. A. Bennett, *Chem. Rev.*, **62**, 611 (1962).

chemistry have appeared since that year. Some cover mainly the organic chemistry of ferrocene;[5–13, 26–28] others cover general preparative methods, physical properties, some organic chemistry of ferrocene and related metal cyclopentadienyls;[14–19, 29, 35, 40] still others deal largely with metal arene and metal olefin complexes (refs. 20–23, 30–34, 37, 41), with hydride complexes of transition metals,[24] with macromolecular organometallic compounds,[25] or with infrared and Raman spectral studies.[36] The rapid progress in π-metallohydrocarbon chemistry has in large measure been responsible for the founding of a new review series[38] devoted exclusively to organometallic chemistry; and a book has recently appeared dealing largely with the chemistry of ferrocene.[39]

In view of this already overwhelming body of π-metallohydrocarbon literature, we have chosen arbitrarily to limit our treatment in three dimensions: to include only compounds containing cyclopentadienyl rings, of these to include only those which have been shown to undergo aromatic substitution reactions, and within this group to discuss only synthetic aspects of their chemistry. In effect, our choice very nearly restricts the discussion to syntheses of substituted ferrocenes.

Although this chapter is concerned with the synthetic aspects of aromatic cyclopentadienyl compounds, a few brief comments on bonding and geometry in ferrocene are appropriate.[44–50] The bonding picture generally accepted is that which ascribes stability of the π-metallohydrocarbons to overlap of one or more d orbitals of the metal atom with the π-electron cloud of the hydrocarbon ligand. Detailed reviews of descriptions of the

[31] R. Pettit and G. F. Emerson, *Advan. Organometal. Chem.*, **1,** 1 (1964).

[32] R. Pettit, G. F. Emerson, and J. Mahler, *J. Chem. Ed.*, **40,** 175 (1963).

[33] H. D. Kaesz, *J. Chem. Ed.*, **40,** 159 (1963).

[34] E. O. Fischer and H. Werner, *Angew. Chem.* (*Intern. Ed.*), **2,** 80 (1963).

[35] G. E. Coates, *Organometallic Compounds*, p. 286, Wiley, New York, 1960.

[36] H. P. Fritz, *Advan. Organometal. Chem.*, **1,** 239 (1964).

[37] R. G. Guy and B. L. Shaw, *Advan. Inorg. Chem. Radiochem.*, **4,** 77 (1962).

[38] *Advan. Organometal. Chem.*, F. G. A. Stone and R. West, eds., Academic, 1964.

[39] M. Rosenblum, *Chemistry of the Iron Group Metallocenes: Ferrocene, Ruthenocene, and Osmocene*, Wiley, 1965.

[40] J. Birmingham, *Advan. Organometal. Chem.*, **2,** 365 (1964).

[41] M. L. H. Green and P. L. I. Nagy, *Advan. Organometal. Chem.*, **2,** 325 (1964).

[42] A. N. Nesmeyanov and E. G. Perevalova, *Ann. N.Y. Acad. Sci.*, **125,** 67 (1965).

[43] A. N. Nesmeyanov, *Bull. Soc. Chim. France*, 1229 (1965).

[44] J. D. Dunitz and L. E. Orgel, *Nature*, **171,** 121 (1953); *J. Chem. Phys.*, **23,** 954 (1955).

[45] W. Moffitt, *J. Am. Chem. Soc.*, **76,** 3386 (1954).

[46] E. M. Shustorovich and M. E. Dyatkina, *Dokl. Akad. Nauk SSSR.*, **128,** 1234 (1959) : **133,** 141 (1960) [*C.A.*, **54,** 23845 (1960)].

[47] E. O. Fischer and W. Pfab, *Z. Naturforsch.*, **7b,** 377 (1952).

[48] E. O. Fischer and H. Leipfinger, *Z. Naturforsch.*, **10b,** 353 (1955).

[49] E. O. Fischer, *Rec. Trav. Chim.*, **75,** 629 (1956).

[50] J. D. Dunitz, L. E. Orgel, and A. Rich, *Acta Cryst.*, **9,** 373 (1956).

electronic structure of ferrocene and other π-metallohydrocarbons are available.[22, 39]

All the electronic descriptions take into account the remarkable geometry of these compounds established by x-ray and electron diffraction studies.[39, 44, 47, 51] Ferrocene, for example, was shown early to have a "sandwich" or "Doppelkegel" structure, in which the two cyclopentadienyl rings lie parallel to one another and the iron atom is buried in the π-electron cloud between them. Two conformations are possible for this "sandwich"—prismatic (**1**) and antiprismatic (**2**). In crystals, ferrocene

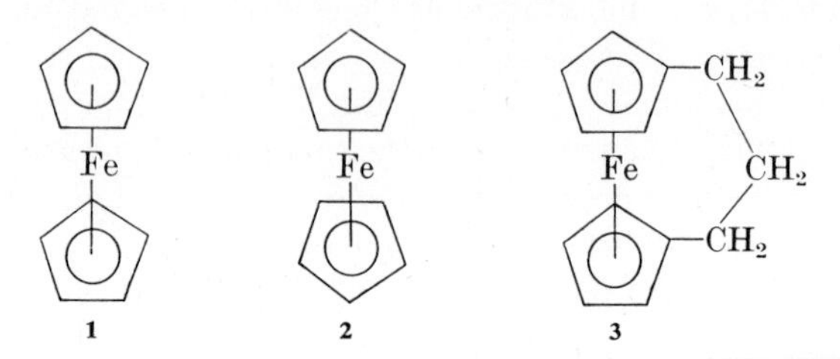

1 **2** **3**

itself exists in the antiprismatic or staggered form (**2**), with the carbon atoms of one ring between those of the other. However, in solution the rings are free to rotate (except where tied together by a bridge as in **3**), since the barrier to rotation of the rings of ferrocene is negligible.[52] This means that the sterochemistry of ferrocene in the solid state is determined largely by crystal lattice forces; in agreement with this postulate, related compounds have been determined to be in various conformations:[39] ruthenocene and osmocene in the eclipsed conformation, diindenyliron and di-*p*-chlorophenylcyclopentadienyliron in the skew conformation.

Moreover, the number of isomers of a substituted ferrocene corresponds precisely to the number of non-equivalent positions predicted on the basis of freely rotating rings. It is, in fact, perhaps simplest to regard ferrocene as being like an ethane in which the two central carbon atoms have five substituents instead of three. One can then draw ferrocene by looking down the central bond, as in **4**. It is apparent that, with freedom of ring rotation, only one heteroannularly disubstituted ferrocene (with

4 **5** **6** **7**

[51] E. A. Seibold and L. E. Sutton, *J. Chem. Phys.*, **23**, 1967 (1955).
[52] Y. T. Struchkov, *J. Gen. Chem.*, **27**, 2093 (1957).

substituents on different rings) is possible, but that ferrocenes with at least two different substituents on the same ring (homoannularly disubstituted) like **5** or **6** can exist in optically active forms (as has been demonstrated),[53–60] and that ferrocenes with at least two different substituents on both rings (like **7**) can exist as geometrical isomers (as has also been demonstrated),[61, 62] better described as *meso* and *racemic* than as *cis* and *trans*.

Although one can name substituted ferrocenes as derviatives of dicyclopentadienyliron, it is more convenient to name them as derivatives of ferrocene; this is the system used throughout this chapter. The rings are numbered separately, the less substituted ring being indicated by primed numbers (1′, 2′, etc.). In bridged ferrocenes the positions bearing the bridge are assigned the numbers 1 and 1′. For continuity of representation (important with bridged ferrocenes) we employ the prismatic symbol **1** throughout, recognizing that, in most compounds, ring rotation is possible. An example of this numbering system is found in 1,1′-trimethyleneferrocene **(3)**.

Except for their structures, probably the most striking feature of the π-metallohydrocarbons is the ability of many of them to undergo aromatic substitution. Early investigations[63, 64] indicated that ferrocene was a new type of aromatic system which readily underwent Friedel-Crafts acylation, and observations of its ability to undergo reactions typical of an aromatic system were extended to include alkylation, arylation, metalation, and sulfonation. In ferrocene the iron atom can be regarded as formally in the Fe^{II} state and, as such, to be oxidizable. For this reason, attempts to nitrate or halogenate ferrocene directly (with nitric acid or bromine) led only to oxidation of the iron nucleus to the Fe^{III} state, in which the molecule bears a net positive charge and is resistant to further electrophilic attack. The product of this oxidation, called the ferricenium ion **(8)**, is water-soluble and is blue-green, whereas ferrocene itself is soluble in organic solvents but not water and is orange. The oxidation of ferrocene to ferricenium ion is a reversible process, and

[53] L. Westman and K. L. Rinehart, Jr., *Acta Chem. Scand.*, **16,** 1199 (1962).
[54] J. B. Thomson, *Tetrahedron Letters*, No. 6, 26 (1959).
[55] K. Schlögl and M. Fried, *Monatsh. Chem.*, **95,** 558 (1964).
[56] K. Schlögl, M. Fried, and H. Falk, *Monatsh. Chem.*, **95,** 576 (1964).
[57] K. Schlögl and M. Fried, *Tetrahedron Letters*, 1473 (1963).
[58] K. Schlögl and H. Falk, *Angew. Chem.*, **76,** 570 (1964).
[59] H. Falk and K. Schlögl, *Monatsh Chem.*, **96,** 1081 (1965).
[60] H. Falk and K. Schlögl, *Monatsh. Chem.*, **96,** 1065 (1965).
[61] R. E. Benson and R. V. Lindsey, *J. Am. Chem. Soc.*, **79,** 5471 (1957).
[62] K. L. Rinehart, Jr., and K. L. Motz, *Chem. Ind.* (*London*), 1150 (1957).
[63] R. B. Woodward, M. Rosenblum, and M. C. Whiting, *J. Am. Chem. Soc.*, **74,** 3458 (1952).
[64] G. Wilkinson, M. Rosenblum, M. C. Whiting, and R. B. Woodward, *J. Am. Chem. Soc.*, **74,** 2125 (1952).

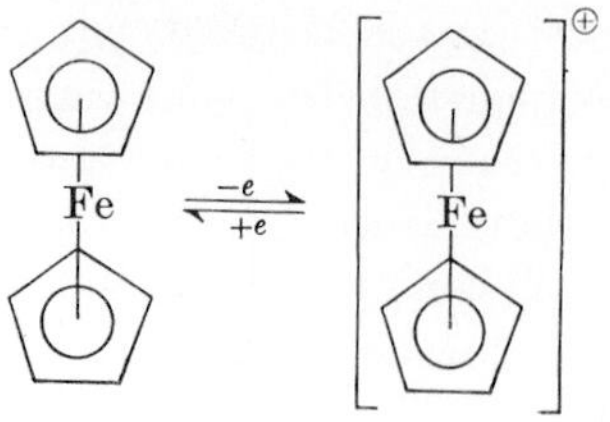

$$\mathrm{(C_5H_5)_2Fe} \ \underset{+e}{\overset{-e}{\rightleftharpoons}} \ \left[\mathrm{(C_5H_5)_2Fe}\right]^{\oplus}$$

1 8

numerous studies on the controlled oxidation of ferrocene and its derivatives have been reported.[65–73]

Although many other cyclopentadienyl compounds like ferrocene have been discovered and, by their very existence, may be somewhat aromatic in nature, most, such as nickelocene and cobaltocene, are not capable of undergoing typical aromatic substitution reactions because they are oxidized or undergo addition reactions more readily.

The compounds discussed in this review have all been shown to undergo aromatic substitution reactions which have led to numerous derivatives of most of the parent structures. The compounds treated include ferrocene, ruthenocene, osmocene, cyclopentadienylmanganese tricarbonyl, cyclopentadienylrhenium tricarbonyl, cyclopentadienyltechnetium tricarbonyl, cyclopentadienylvanadium tetracarbonyl, and cyclopentadienylchromium nitrosyl dicarbonyl, listed in approximately descending order according to the degree to which they have been studied.

FERROCENE

Synthesis of Ferrocene and Substituted Ferrocenes from Cyclopentadienes

Numerous synthetic procedures for the preparation of ferrocene itself have appeared since the compound was first identified; they are

[65] G. L. K. Hoh, W. E. McEwen, and J. Kleinberg, *J. Am. Chem. Soc.*, **83**, 3949 (1961).

[66] J. G. Mason and M. Rosenblum, *J. Am. Chem. Soc.*, **82**, 4206 (1960).

[67] S. P. Gubin and E. G. Perevalova, *Dokl. Akad. Nauk SSSR*, **143**, 1351 (1962) [*C.A.*, **57**, 9575 (1962)].

[68] T. Kuwana, D. E. Bublitz, and G. Hoh, *J. Am. Chem. Soc.*, **82**, 5811 (1960).

[69] J. Komenda and J. Tirouflet, *Compt. Rend.*, **254**, 3093 (1962).

[70] J. Komenda, *Chem. Zvesti*, **18**, 378 (1964) [*C.A.*, **61**, 11622 (1964)].

[71] E. G. Perevalova, S. P. Gubin, S. A. Smirnova, and A. N. Nesmeyanov, *Dokl. Akad. Nauk SSSR*, **155**, 857 (1964) [*C.A.*, **61**, 1509 (1964)].

[72] A. N. Nesmeyanov, E. G. Perevalova, L. P. Yur'eva, and S. P. Gubin, *Izv. Akad. Nauk SSSR, Ser. Khim.*, 909 (1965) [*C.A.*, **63**, 5507 (1965)].

[73] D. M. Knight and R. C. Schlitt, *Anal. Chem.*, **37**, 470 (1965).

summarized in Table I. Most of them involve the treatment of cyclopentadiene with a base to give the resonance-stabilized cyclopentadienyl anion,

$$2\,C_5H_6 + 2\,\text{Base} \rightleftharpoons 2\,\text{Base}\cdot H^+ + 2\,C_5H_5^- \xrightarrow{FeCl_2} (C_5H_5)_2Fe\ (\mathbf{1}) + 2\,Cl^-$$

which then reacts with an anhydrous iron salt to give ferrocene. Bases employed range in strength from the weak base diethylamine[74] (which is added simultaneously with ferrous chloride so that the small amount of cyclopentadienyl anion formed is removed by immediate reaction with ferrous iron) to metallic sodium[75] and Grignard reagents.[76] For the laboratory preparation of ferrocene itself the diethylamine procedure is one of the best.[74, 77, 78] Another useful procedure involves the reaction of cyclopentadiene with metallic sodium in tetrahydrofuran or 1,2-dimethoxyethane and subsequent treatment with ferrous chloride.[75] Ferric chloride is reduced to the ferrous state by the stronger bases and so may also be used, though yields are necessarily reduced.

On a commercial scale, routes proceeding directly from cyclopentadiene and iron oxides[79] or iron carbonyls[80] would probably be preferred, and ferrocene is presently available from a number of commercial sources.

For the preparation of substituted ferrocenes two basically different approaches involving "direct" or "indirect" routes can be employed. The great majority of the substituted ferrocenes reported have been prepared by the "indirect" route in which the preformed ferrocene molecule serves as a starting point. Substituents are then introduced and the products are subsequently converted to the desired material. A few of the more important examples of these indirect syntheses are discussed on pp. 13–29, and others are found under the individual classes of compounds on pp. 29–63.

The remainder of this section is devoted to "direct" syntheses of substituted ferrocenes, those in which the compound is prepared from an appropriately substituted cyclopentadiene and an iron compound in

[74] G. Wilkinson, *Org. Syn.*, **36,** 34 (1956).

[75] G. Wilkinson, *Org. Syn.*, **36,** 31 (1956).

[76] E. B. Sokolova, M. P. Shebanova, and V. A. Zhichkina, *Zh. Obshch. Khim.*, **30,** 2040 (1960) [*C.A.*, **55,** 6457 (1961)].

[77] R. L. Pruett and E. L. Morehouse, *Advan. Chem. Ser.*, **23,** 368 (1958).

[78] K. Jones and M. F. Lappert, *J. Organometal. Chem.*, **3,** 295 (1965).

[79] E. I. du Pont de Nemours and Co., Brit. pat. 764,058 [*C.A.*, **52,** 5480 (1958)].

[80] A. Tartter, Ger. pat. 1,101,418 [*C.A.*, **56,** 10193 (1962)].

syntheses similar to those used for the preparation of ferrocene itself. Only the stronger bases—Grignard or organolithium reagents, metallic sodium—appear to give good yields of ferrocenes from substituted cyclopentadienes. Substituted ferrocenes prepared by direct synthesis are shown in Table II. Simple alkyl, aryl, and silyl cyclopentadienes have been most frequently employed as ferrocene precursors, but even highly substituted ferrocenes like *sym*-octaphenylferrocene have been prepared by this route.

Direct synthesis has been less frequently employed for the preparation of more functionally substituted compounds, but the symmetrical diacetyl, dicarbomethoxy, and bis(phenylazo) derivatives have been prepared from the corresponding cyclopentadienes. A dihydroxyferrocene has also been prepared from the enolate of 3-methyl-2-cyclopentenone and has been isolated as its oxyacetic acid **(9a)** and benzoate **(9b)** derivatives.[61,82] Both of the derivatives are capable of existing as *racemic* and *meso* isomers; the *meso* forms are shown.

CH_3 $\xrightarrow{NaNH_2}$ CH_3 $O^{\ominus}$ $\xrightarrow[2.\ RCl]{1.\ FeCl_2}$ CH_3 OR Fe CH_3 OR

9a, b

a: R = CH_2CO_2H
b: R = COC_6H_5

Functional groups are sometimes reduced in the formation of ferrocenes. For example, tetrabenzylferrocenes **(10)** are synthesized when 1,2-diaroylcyclopentadienes are treated with lithium aluminum hydride followed by ferrous chloride.[81]

COAr COAr $\xrightarrow[2.\ FeCl_2]{1.\ LiAlH_4}$ CH_2Ar CH_2Ar Fe CH_2Ar CH_2Ar

10

Fused-ring compounds (listed in Table II-C) have been prepared from precursor indenes, which may be regarded as substituted cyclopentadienes and, as such, give benzoferrocenes, though in low yields. Attempts to prepare di- and tetra-benzoferrocenes from fluorenes have been

[81] W. F. Little and R. C. Koestler, *J. Org. Chem.*, **26**, 3245 (1961).
[82] R. E. Benson, U.S. pat. 2,849,470 [*C.A.*, **53**, 406 (1959)].

unsuccessful, but the synthesis of bis(*as*-indacenyl)iron (**11**) from the *as*-indacenyl dianion has been accomplished.[83] By contrast, reaction of the pentalene dianion yielded the diolefin **12** rather than a binuclear ferrocene.[84]

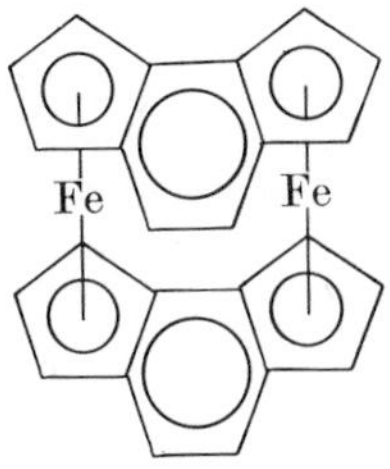
Fe
Fe

11

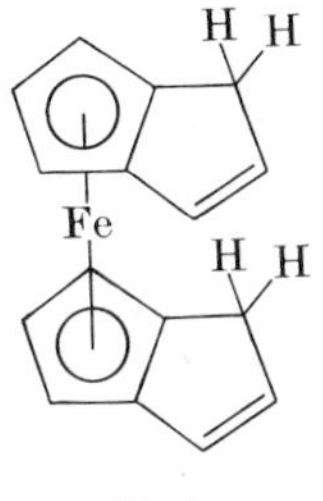
H H
Fe
H H

racemic and *meso*

12

A few ferrocene derivatives of formula **13**, containing bridges from one ring to another (listed in Table II-D), have been synthesized from precursors containing two cyclopentadiene rings linked by alkane or siloxane chains.

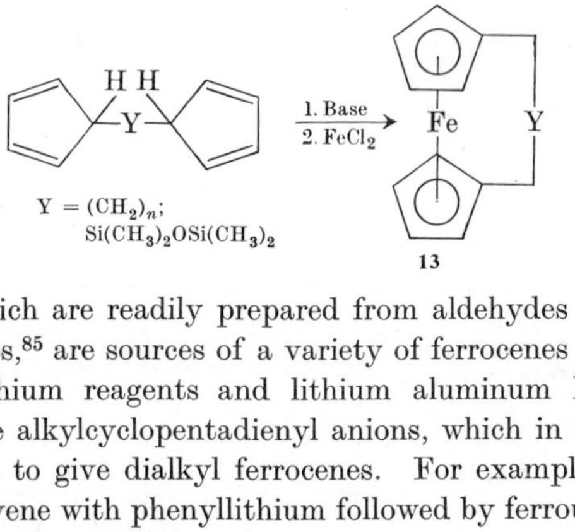
H H
Y
1. Base
2. FeCl2
Fe
Y
Y = (CH2)n;
Si(CH3)2OSi(CH3)2

13

Fulvenes, which are readily prepared from aldehydes or ketones and cyclopentadienes,[85] are sources of a variety of ferrocenes (see Table III). Both organolithium reagents and lithium aluminum hydride add to fulvenes to give alkylcyclopentadienyl anions, which in turn react with ferrous chloride to give dialkyl ferrocenes. For example, treatment of 6,6-diphenylfulvene with phenyllithium followed by ferrous chloride gave 1,1′-ditritylferrocene (**14**),[86] while treatment of azulene with lithium

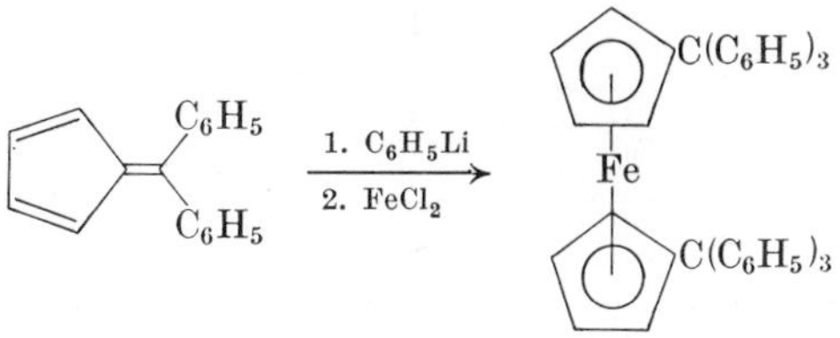
C6H5
C6H5
1. C6H5Li
2. FeCl2
C(C6H5)3
Fe
C(C6H5)3

14

[83] T. J. Katz and J. Shulman, *J. Am. Chem. Soc.*, **86,** 3169 (1964).

[84] T. J. Katz and M. Rosenberger, *J. Am. Chem. Soc.*, **85,** 2030 (1963).

[85] J. H. Day, *Chem. Rev.*, **53,** 167 (1953).

[86] R. C. Koestler and W. F. Little, *Chem. Ind.* (*London*), 1589 (1958).

aluminum hydride, then ferrous chloride, gave the fused-ring compound **(15)** in 17% yield.[87, 88] The syntheses of 1,1′-bis(dimethylaminomethyl)-

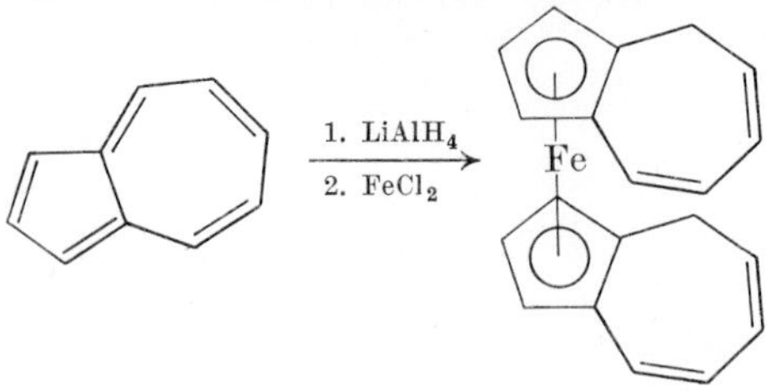

1. $LiAlH_4$
2. $FeCl_2$

Fe

racemic and *meso*

15

ferrocene and 1,1′-bis-(α-dimethylaminoethyl)ferrocene **(16a** and **16b)** have been accomplished in the same way.[90]

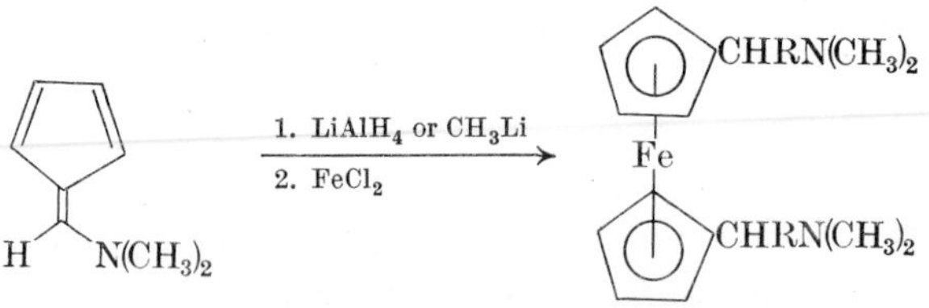

H $N(CH_3)_2$

1. $LiAlH_4$ or CH_3Li
2. $FeCl_2$

$CHRN(CH_3)_2$
Fe
$CHRN(CH_3)_2$

16a,b
a: R = H
b: R = CH_3

In a reaction related to the fulvene reactions discussed above, treatment of diazocyclopentadiene with phenyllithium, then with ferrous chloride, gave 1,1′-bis(phenylazo)ferrocene **(17)** (see Table II).

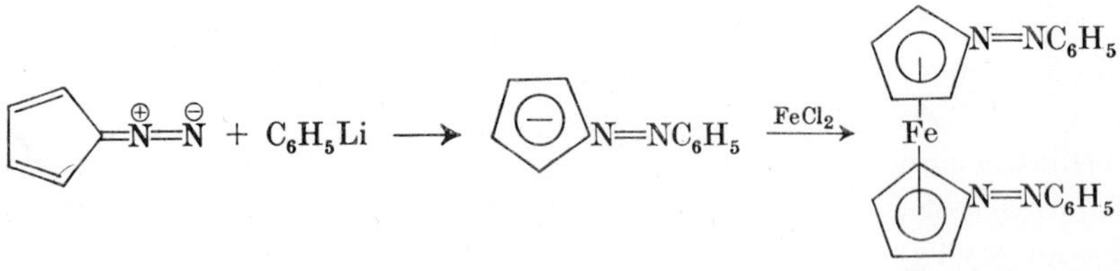

$\overset{\oplus}{N}=\overset{\ominus}{N}$ + C_6H_5Li → $N{=}NC_6H_5$ $\xrightarrow{FeCl_2}$ Fe, $N{=}NC_6H_5$, $N{=}NC_6H_5$

17

A second reaction of fulvenes occurs with strong bases which are not nucleophiles. Treatment of 6,6-dimethylfulvene with sodium amide gives a cyclopentadienyl anion which, when treated with ferrous chloride, gives 1,1′-diisopropenylferrocene **(18)**.[87, 88] In yet a third reaction some

[87] G. R. Knox and P. L. Pauson, *Proc. Chem. Soc.*, 289 (1958).

[88] G. R. Knox and P. L. Pauson, *J. Chem. Soc.*, 4610 (1961).

[89] K. L. Rinehart, Jr., A. K. Frerichs, P. A. Kittle, L. F. Westman, D. H. Gustafson, R. L. Pruett, and J. E. McMahon, *J. Am. Chem. Soc.* **82**, 4111 (1960).

[90] G. R. Knox, J. D. Munro, P. L. Pauson, G. H. Smith, and W. E. Watts, *J. Chem. Soc.*, 4619 (1961).

fulvenes react with sodium in tetrahydrofuran to give 1,2-dicyclopentadienylethanes which, when treated with ferrous chloride, give bridged ferrocenes (**19**),[89] although the reaction to give 1,1′-dialkenylferrocenes (like **18**) may also occur with this reagent.

$$\mathbf{18} \xleftarrow[\text{2. } FeCl_2]{\text{1. } NaNH_2} C_5H_4{=}C(CH_3)_2 \xrightarrow[\text{2. } FeCl_2]{\text{1. Na, THF}} \mathbf{19}$$

Unsymmetrically substituted ferrocenes usually are prepared by substitution on ferrocene ("indirect" synthesis), but some have been prepared in a direct manner by reaction of equimolar amounts of a substituted cyclopentadienyl reagent and an unsubstituted cyclopentadienyl reagent with an iron source (see Table IV). The yields are low or unspecified, and indirect methods are far superior except for the preparation of phenylazoferrocene and benzoferrocene. An extension of the process employing a mixture of cyclopentadienes has provided a unique synthesis of 1,1′-bis(dimethylaminomethyl)ferrocene (**16a**) which cannot be prepared by substitution on ferrocene. This involves cleavage of dimethylaminomethylferrocene (**20**) with lithium in ethylamine, followed by treatment of the resulting cyclopentadienyl anions with ferrous chloride to give ferrocene (14%), dimethylaminomethylferrocene (36%), and 1,1′-bis-(dimethylaminomethyl)ferrocene (15%).[91, 92]

$$\underset{\mathbf{20}}{C_5H_5FeC_5H_4CH_2N(CH_3)_2} \xrightarrow[C_2H_5NH_2]{Li} C_5H_4^{-}CH_2N(CH_3)_2 + C_5H_5^{-} \xrightarrow{FeCl_2}$$

$$\underset{\mathbf{20}}{C_5H_5FeC_5H_4CH_2N(CH_3)_2} + \underset{\mathbf{1}}{C_5H_5FeC_5H_5}$$

$$+ \underset{\mathbf{16a}}{(CH_3)_2NCH_2C_5H_4FeC_5H_4CH_2N(CH_3)_2}$$

Useful Intermediates in the Synthesis of Substituted Ferrocenes

In the indirect synthesis of substituted ferrocenes a few key intermediates have proved especially useful. In general, these are acylated, dimethylaminomethylated, or metalated ferrocenes, easily prepared from ferrocene but sufficiently versatile and reactive to be convertible to many

[91] J. M. Osgerby and P. L. Pauson, *J. Chem. Soc.*, 4604 (1961).
[92] P. L. Pauson, G. R. Knox, J. D. Munro, and J. M. Osgerby, *Angew. Chem.*, **72**, 37 (1960).

other derivatives. In this section the preparation of some of these useful intermediates is discussed in detail.

These intermediates have proved to be more useful for the synthesis of monosubstituted ferrocene derivatives than for disubstituted ferrocenes. Thus, while monoacyl ferrocenes have been used for the preparation of a host of substituted ferrocenes, comparatively little use has been made of diacyl ferrocenes in such conversions. Similarly, chloromercuriferrocene has been shown to be very useful for preparing iodoferrocene, diferrocenylmercury, etc., but little has been done with 1,1′-bis(chloromercuri)-ferrocene.

The lack of widespread use of disubstituted intermediates may be due in part to the fact that many symmetrically disubstituted ferrocenes are conveniently prepared from substituted cyclopentadienes and a base or by the route involving fulvene compounds (see pp. 11–13 and Tables II and III).

Alkali Metal Derivatives

Lithioferrocenes. *Monolithioferrocene.* Ferrocene is metalated by *n*-butyllithium to give a mixture of mono- and di-lithiated products (**21** and **22**). Attempts to circumvent the formation of the mixtures of lithiation products have been made by various authors with some success.[93–95] Utilization of a mixed solvent system (ether-tetrahydrofuran) with varying ratios of *n*-butyllithium to ferrocene produced mono- and di-lithiated materials in varying ratios,[94] and use of 1.04 molar equivalents of *n*-butyllithium in the solvent system hexane-ether was reported to give only the monolithio product.[96]

Fe $\xrightarrow{n\text{-}C_4H_9Li}$ Fe (Li) + Fe (Li)(Li)

21 **22**

Other workers have reported the preparation of pure ferrocenyllithium by methods utilizing substituted ferrocenes as the starting material. Diferrocenylmercury (**23a**) and 2.8 molar equivalents of *n*-butyllithium

[93] R. A. Benkeser, D. Goggin, and G. Schroll, *J. Am. Chem. Soc.*, **76**, 4025 (1954).
[94] D. W. Mayo, P. D. Shaw, and M. Rausch, *Chem. Ind.* (*London*), 1388 (1957).
[95] D. Seyferth and J. F. Helling, *Chem. Ind.* (*London*), 1568 (1961).
[96] S. I. Goldberg, L. H. Keith, and T. S. Prokopov, *J. Org. Chem.*, **28**, 850 (1963).

gave pure monolithioferrocene in 80% yield,[97] while monochloromercuriferrocene **(23b)** and ethyllithium gave only monolithioferrocene in 64–96% yield.[95, 98, 99] The yields in both cases were based on carbonation to the acid.[98] In both of these reactions, however, one must reckon with the presence of excess alkyl- or aryl-lithium reagent as a contaminant.

23a, b
a: X = $C_5H_4FeC_5H_5$
b: X = Cl

Lithiation of 1,1′-dimethylferrocene leads to a mixture of 2- and 3-lithiated isomers which are converted to the corresponding acids by carbonation.[53] While this metalation leads to a mixture of isomers, lithiation of diphenylferrocenylcarbinol[100] or dimethylaminomethylferrocene[101, 102] has been reported to give, after carbonation, the corresponding 2-substituted ferrocenoic acids as the sole products, presumably the result of coordination of the metal with the carbinol or amine.

Although the lithio derivatives of ferrocene are never isolated, they have been extremely useful intermediates in the preparation of many ferrocene derivatives (see Table V).

Dilithioferrocene. 1,1′-Dilithioferrocene **(22)** free of monolithioferrocene has been prepared only by treatment of 1,1′-bis(chloromercuri)-ferrocene with ethyllithium,[98] never by direct lithiation of ferrocene. Studies of the direct lithiation products show that, even with a 25-fold molar excess of butyllithium over ferrocene, 23% of the monolithiated product is obtained.[94] Nevertheless, dilithioferrocene is an important intermediate in the preparation of many 1,1′-disubstituted ferrocene derivatives (Table V).

Sodioferrocenes. Mono- and di-sodio derivatives of ferrocene, which

[97] M. D. Rausch, *Inorg. Chem.*, **1**, 414 (1962).

[98] D. Seyferth, H. P. Hoffman, R. Burton, and J. F. Helling, *Inorg. Chem.*, **1**, 227 (1962).

[99] H. Rosenberg and R. U. Schenk, 145th National A.C.S. Meeting, New York, September 9–13, 1963; *cf. Abstracts*, p. 76Q.

[100] R. A. Benkeser, W. P. Fitzgerald, and M. S. Melzer, *J. Org. Chem.*, **26**, 2569 (1961).

[101] D. W. Slocum, B. W. Rockett, and C. R. Hauser, *Chem. Ind.* (*London*), 1831 (1964).

[102] D. W. Slocum, B. W. Rockett, and C. R. Hauser, *J. Am. Chem. Soc.*, **87**, 1241 (1965).

have been studied less than the lithio derivatives, are formed by exchange reactions involving phenylsodium[103, 105] or amylsodium.[98, 106, 107] When the product of the exchange reaction is treated with carbon dioxide,[98] trialkylchlorosilanes,[98, 103] or triphenylbromogermane,[98] mixtures of the corresponding mono- and 1,1′-di-substituted products are obtained. In early work, when the reaction product was carbonated, the sole product identified was ferrocene-1,1′-dicarboxylic acid,[104] but it has subsequently been shown that the monocarboxylic acid is also formed.[98]

Ferroceneboronic Acids

Ferroceneboronic Acid. One derivative prepared from lithioferrocene worthy of special note is ferroceneboronic acid **(24)**, which is an important intermediate in the preparation of chloro-, bromo-, amino-, and, particularly, hydroxy-ferrocenes. Treatment of a mixture of lithiated ferrocenes with tri-*n*-butyl borate, followed by hydrolysis, leads to a mixture of ferroceneboronic acid and 1,1′-ferrocenediboronic acid **(25)**,[108, 110] separable either by fractional acidification of a mixed salt solution to give first the less soluble dibasic acid then ferroceneboronic acid,[108, 109] or by selective extraction[110] of the more soluble ferroceneboronic acid by diethyl ether. The latter method of separation is clearly

$$C_5H_4Li\text{-}Fe\text{-}C_5H_5 + C_5H_4Li\text{-}Fe\text{-}C_5H_4Li \xrightarrow[2.\ H_2O]{1.\ (n\text{-}C_4H_9O)_3B} \underset{\mathbf{24}}{C_5H_4B(OH)_2\text{-}Fe\text{-}C_5H_5} + \underset{\mathbf{25}}{C_5H_4B(OH)_2\text{-}Fe\text{-}C_5H_4B(OH)_2}$$

less tedious and seems preferable. Replacement of the boronic acid group by a variety of others has been reported for both ferroceneboronic

[103] S. I. Goldberg, D. W. Mayo, M. Vogel, H. Rosenberg, and M. Rausch, *J. Org. Chem.*, **24,** 824 (1959).

[104] A. N. Nesmeyanov, E. G. Perevalova, and Z. A. Beinoravichute, *Dokl. Akad. Nauk SSSR*, **112,** 439 (1957) [*C.A.*, **51,** 13855 (1957)].

[105] M. Okawara, Y. Takemoto, H. Kitaoka, E. Haruki, and I. Imoto, *Kogyo Kagaku Zasshi*, **65,** 685 (1962) [*C.A.*, **58,** 577 (1963)].

[106] P. J. Graham, U.S. pat. 2,835,686 [*C.A.*, **52,** 16366 (1958)].

[107] R. A. Benkeser, Y. Nagai, and J. Hooz, *J. Am. Chem. Soc.*, **86,** 3742 (1964).

[108] A. N. Nesmeyanov, V. A. Sazonova, and V. N. Drozd, *Dokl. Akad. Nauk SSSR*, **126,** 1004 (1959) [*C.A.*, **54,** 6673 (1960)].

[109] A. N. Nesmeyanov, V. A. Sazonova, and V. N. Drozd, *Chem. Ber.*, **93,** 2717 (1960).

[110] H. Schechter and J. F. Helling, *J. Org. Chem.* **26,** 1034 (1961).

acid[108, 109, 111–113] and substituted ferroceneboronic acids,[114–116] (Table VI-A, B) and is illustrated for ferroceneboronic acid in Fig. 1.

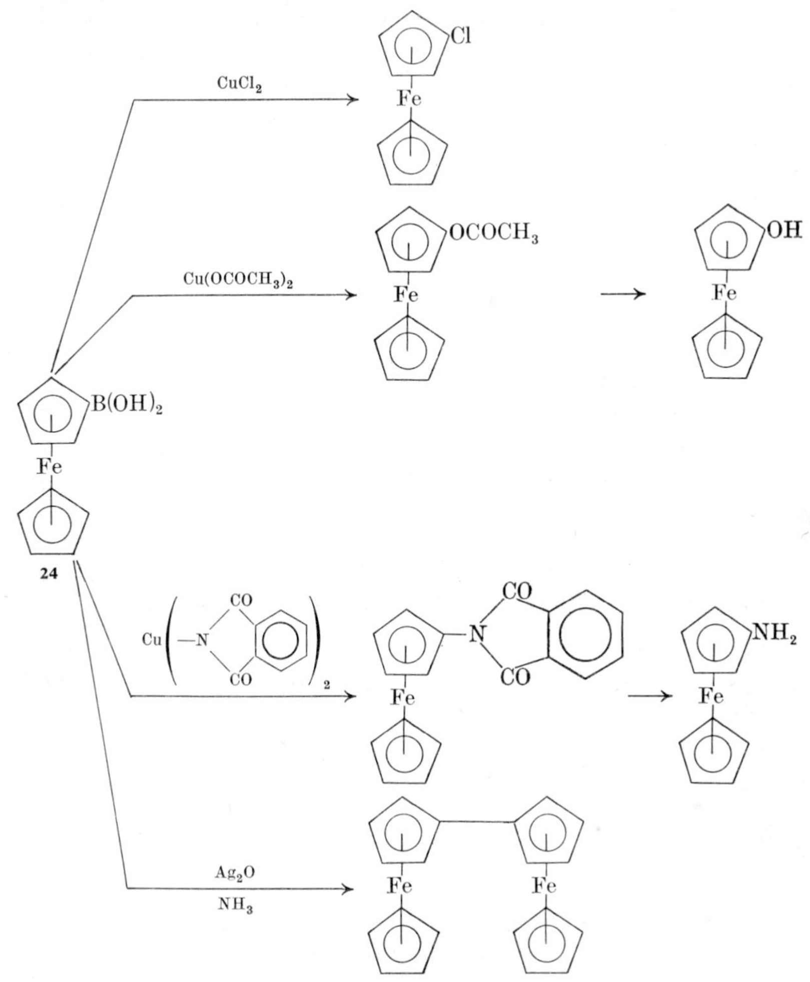

FIG. 1. Some replacement reactions of ferroceneboronic acid.

[111] A. N. Nesmeyanov, V. A. Sazonova, A. V. Gerasimenko, and V. G. Medvedeva, *Izv. Akad. Nauk SSSR, Otd. Khim. Nauk*, 2073 (1962) [*C.A.*, **58**, 9133 (1963)].

[112] A. N. Nesmeyanov, V. A. Sazonova, and V. N. Drozd, *Dokl. Akad. Nauk SSSR*, **129**, 1060 (1959) [*C.A.* **53**, 16111 (1959)].

[113] A. N. Nesmeyanov, V. A. Sazonova, and V. N. Drozd, *Tetrahedron Letters*, No. 17, 13 (1959).

[114] A. N. Nesmeyanov, V. A. Sazonova, V. N. Drozd, and L. A. Nikonova, *Dokl. Akad. Nauk. SSSR*, **131**, 1088 (1960) [*C.A.*, **54**, 21025 (1960)].

[115] A. N. Nesmeyanov, V. A. Sazonova, V. N. Drozd, and L. A. Nikonova, *Dokl. Akad. Nauk SSSR*, **133**, 126 (1960) [*C.A.*, **54**, 24616 (1960)].

[116] A. N. Nesmeyanov, V. A. Sazonova, and V. I. Romanenko, *Dokl. Akad. Nauk SSSR*, **157**, 922 (1964) [*C.A.*, **61**, 13343 (1964)].

1,1′-Ferrocenediboronic Acid. 1,1′-Ferrocenediboronic acid **(25)**, prepared in addition to the corresponding monoacid by treatment of the mixed lithioferrocenes with tri-*n*-butyl borate,[109, 110] has served as an intermediate for numerous disubstituted ferrocenes (Table VI-C).

Mercurated Ferrocenes

Monomercurated Ferrocenes. When ferrocene is treated with mercuric acetate in ether-methanol or ether-ethanol, a mixture of acetoxymercuri- and 1,1′-bis(acetoxymercuri)-ferrocene is formed.[106, 117, 118] The ratio of the mercuration products can be varied somewhat,[118] but one cannot be prepared to the exclusion of the other. Although the mixed acetoxymercuri products have been isolated,[106] the usual procedure involves conversion of the products to the corresponding chloromercuri compounds **(23b** and **26)** by treatment with potassium chloride[117, 118] or lithium chloride.[118–120] The chloromercuri derivatives are readily

Fe $\xrightarrow[\text{2. LiCl}]{\text{1. } Hg(OCOCH_3)_2}$ Fe–HgCl (**23b**) + ClHg–Fe–HgCl (**26**)

separable because of the solubility of the mono-substituted product **(23b)** in hot 1-butanol[117–119] or methylene chloride.[120] The most detailed directions also appear to give the best yield and offer an alternative procedure using sodium acetate and mercuric chloride.[119] Other methods, involving displacement of a functional group of a substituted ferrocene, are also available for the preparation of chloromercuriferrocene, but they appear to offer no advantage. These are presented in Table VII together with direct substitution results.

p-Nitrophenylferrocene forms a dimercuration product of undesignated structure;[121] this is the only reported mercuration of a substituted ferrocene.

[117] A. N. Nesmeyanov, E. G. Perevalova, R. V. Golovnya, and O. A. Nesmeyanova, *Dokl. Akad. Nauk SSSR*, **97**, 459 (1954) [*C.A.*, **49**, 9633 (1955)].

[118] M. Rausch, M. Vogel, and H. Rosenberg, *J. Org. Chem.*, **22**, 900 (1957).

[119] M. D. Rausch, L. C. Klemann, A. Siegel, R. F. Kovar, and J. H. Maldines, Submitted to *Organometal. Syn.*

[120] R. W. Fish and M. Rosenblum, *J. Org. Chem.*, **30**, 1253 (1965).

[121] A. N. Nesmeyanov, E. G. Perevalova, R. V. Golovnya, N. A. Simukova, and O.V. Starovskii, *Izv. Akad. Nauk SSSR, Otd. Khim. Nauk*, 638 (1957) [*C.A.*, **51**, 15422 (1957)].

Chloromercuriferrocene (**23b**) has been converted to many other ferrocene derivatives as illustrated in Fig. 2. A more complete list of conversions is found in Table VIII-A.

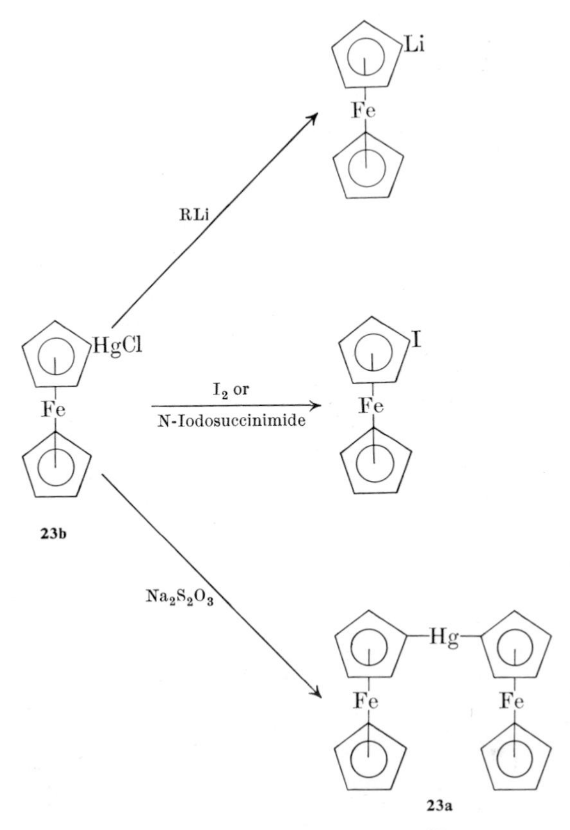

FIG. 2. Some reactions of chloromercuriferrocene.

Diferrocenylmercury **(23a)**, prepared by reduction of chloromercuriferrocene,[117, 118] has been shown to undergo displacement reactions like those of chloromercuriferrocene. Thus heating in the presence of silver[97, 122] or palladium black[123] gives biferrocenyl, while reaction with *n*-butyllithium provides another route to pure monolithioferrocene.[97, 98] These and other displacements are listed in Table VIII-B and are illustrated in Fig. 3.

[122] M. D. Rausch, *J. Am. Chem. Soc.*, **82,** 2080 (1960).

[123] O. A. Nesmeyanova and E. G. Perevalova, *Dokl. Akad. Nauk SSSR*, **126,** 1007 (1959) [*C.A.*, **54,** 1478 (1960)].

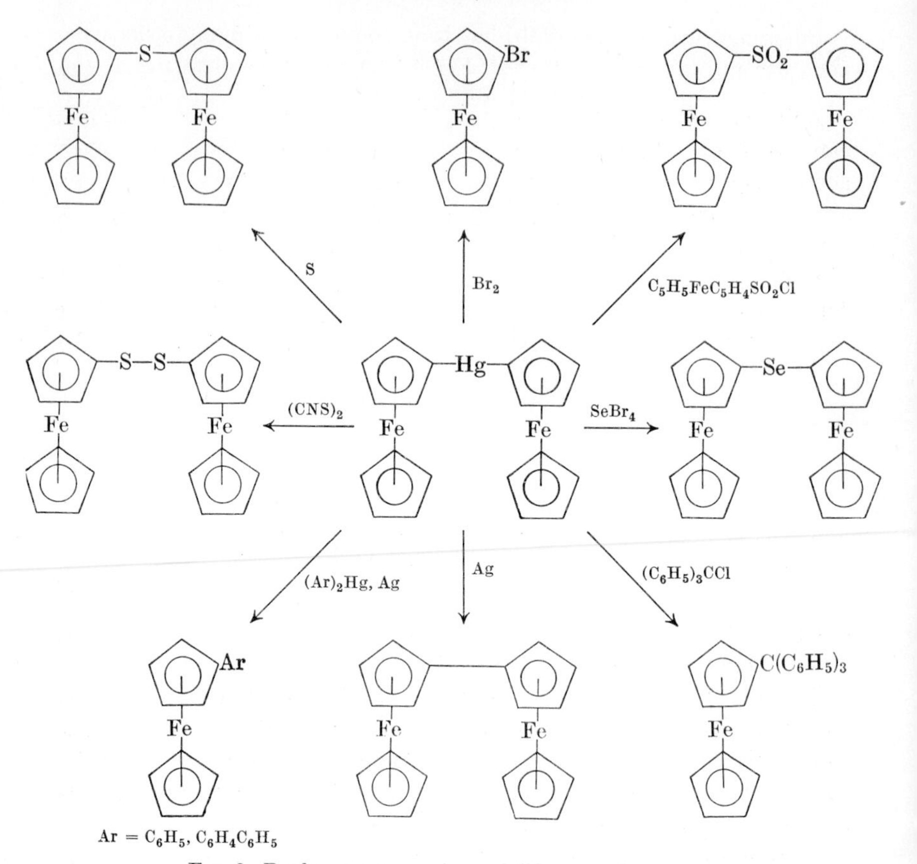

FIG. 3. Replacement reactions of diferrocenylmercury.

Dimercurated Ferrocenes. As described above, mercuration of ferrocene by mercuric acetate followed by treatment with potassium chloride or lithium chloride gives both mono- and di-substituted products,[117–119] which are readily separable because of the solubility of the monosubstituted product in hot 1-butanol[117–119] or methylene chloride.[120] Reactions of 1,1′-bis(chloromercuri)ferrocene, listed in Table VIII-C, generally parallel those of chloromercuriferrocene. Of special interest is the reaction of 1,1′-bis(chloromercuri)ferrocene with sodium iodide in ethanol to give polyferrocenylmercury.[124] Heating this polyferrocenylmercury compound with silver at 300° gives biferrocenyl and polyferrocenyl.

Dimercuration of *p*-nitrophenylferrocene has been reported but the product was of undesignated structure.[121]

[124] M. D. Rausch, *J. Org. Chem.*, **28**, 3337 (1963).

N,N-Dialkylaminomethyl Ferrocenes (Table IX)

Aminomethylation of ferrocene to N,N-dimethylaminomethylferrocene can be effected by reaction with formaldehyde and dimethylamine in acetic acid or by using N,N,N′,N′-tetramethyldiaminomethane as the aminomethylating agent.[125] The yield is improved by using phosphoric acid in addition to acetic acid.[126] A standard preparation using added phosphoric acid gives yields of 68–81% of the amine as its methiodide.[127] The phosphoric acid modification has been used successfully for the aminomethylation of phenylferrocene[128] and methylferrocene.[129]

$$\text{Ferrocene} \xrightarrow[\text{CH}_3\text{CO}_2\text{H},\ \text{H}_3\text{PO}_4]{(\text{CH}_3)_2\text{NCH}_2\text{N}(\text{CH}_3)_2} \text{C}_5\text{H}_5\text{FeC}_5\text{H}_4\text{CH}_2\text{N}(\text{CH}_3)_2$$

The most valuable reaction of dimethylaminomethylferrocene is that with methyl iodide to give (ferrocenylmethyl)trimethylammonium iodide **(27)**,[130] which undergoes S_N2 displacement of trimethylamine by many nucleophiles—alkoxide and cyanide anions, Grignard reagents, amines, and a variety of carbanionic reagents (Table X). The quaternary methiodide treated with sodium cyclopentadienide and then with ferrous chloride gives 1,1′-bis(ferrocenylmethyl)ferrocene **(28)**.[131]

Rearrangement of (ferrocenylmethyl)trimethylammonium iodide can also occur, however, and treatment with potassium amide in liquid ammonia has been shown to give β-(dimethylamino)ethylferrocene **(29)**.[132]

Recently a large number of ferrocenylmethylamine quaternary ammonium salts containing different N-alkyl substituents has been prepared. Subsequent reactions of these quaternary salts have been

[125] J. K. Lindsay and C. R. Hauser, *J. Org. Chem.*, **22,** 355 (1957).

[126] J. M. Osgerby and P. L. Pauson, *J. Chem. Soc.*, 656 (1958).

[127] D. Lednicer and C. R. Hauser, *Org. Syn.*, **40,** 31 (1960).

[128] A. N. Nesmeyanov, E. G. Perevalova, S. P. Gubin, T. V. Nikitina, A. A. Ponomarenko, and L. S. Shilovtseva, *Dokl. Akad. Nauk SSSR,* **139,** 888 (1961) [*C.A.*, **56,** 1477 (1962)].

[129] A. N. Nesmeyanov, E. G. Perevalova, L. S. Shilovtseva, and Yu. A. Ustynyuk, *Dokl. Akad. Nauk SSSR,* **124,** 331 (1959) [*C.A.*, **53,** 11332 (1959)].

[130] C. R. Hauser and J. K. Lindsay, *J. Org. Chem.*, **21,** 382 (1956).

[131] P. L. Pauson and W. E. Watts, *J. Chem. Soc.*, 3880 (1962).

[132] C. R. Hauser, J. K. Lindsay, and D. Lednicer, *J. Org. Chem.*, **23,** 358 (1958).

reported to parallel in most cases those of the simpler ferrocenylmethyltrimethylammonium compound.[133–138]

CH_2 Fe Fe Fe CH_2

28

1. C_5H_5Na 2. $FeCl_2$

$CH_2\overset{\oplus}{N}(CH_3)_3I^{\ominus}$

Fe

27

KNH NH_3

$CH_2CH_2N(CH_3)_2$

Fe

29

Methylferrocene has been shown to give 19% of a disubstituted aminomethylation product of undesignated structure in addition to the monosubstituted material.[129] This appears to be the only reported case of disubstitution in the aminomethylation reaction. Bis-(aminomethylation)

[133] E. G. Perevalova, Yu. A. Ustynyuk, and A. N. Nesmeyanov, *Izv. Akad. Nauk SSSR, Otd. Khim. Nauk*, 1045 (1963) [*C.A.*, **59**, 7557 (1963)].

[134] E. G. Perevalova, Yu. A. Ustynyuk L. A. Ustynyuk, and A. N. Nesmeyanov, *Izv. Akad. Nauk SSSR, Ser. Khim.*, 1977 (1963) [*C.A.*, **60**, 6865 (1964)].

[135] Yu. A. Ustynyuk and E. G. Perevalova, *Izv. Akad. Nauk SSSR, Ser. Khim.*, 62 (1964) [*C.A.*, **60**, 9310 (1964)].

[136] E. G. Perevalova and Yu. A. Ustynyuk, *Izv. Akad. Nauk SSSR, Ser. Khim.*, 1776 (1963) [*C.A.*, **60**, 5549 (1964)].

[137] E. G. Perevalova, Yu. A. Ustynyuk, and A. N. Nesmeyanov, *Izv. Akad. Nauk SSSR, Otd. Khim. Nauk*, 1036 (1963) [*C.A.*, **59**, 7557 (1963)].

[138] A. N. Nesmeyanov, E. G. Perevalova, L. I. Leonteva and Yu. A. Ustynyuk, *Izv. Akad. Nauk SSSR, Ser. Khim.*, 1696 (1965) [*C.A.*, **63**, 18146 (1965)].

of ferrocene itself does not occur under the normal reaction conditions.

As discussed on pp. 12–13, 1,1′-bis(dimethylaminoalkyl)ferrocenes have been prepared by the direct route—through cleavage of the monosubstituted material by lithium in ethylamine followed by recombination of the cyclopentadienes in the presence of ferrous chloride[91, 92] and by treatment of 6-dimethylaminofulvene with either lithium aluminum hydride or methyllithium, followed by ferrous chloride. However, conversion of bis(dimethylaminoalkyl)ferrocenes to quaternary salts for replacement reactions of the salts have not been reported.

Ferrocenecarboxaldehyde

Ferrocenecarboxaldehyde (**30**) has been prepared in one step from ferrocene by two methods. The first consists of treatment of ferrocene with phosphorus oxychloride and N-methylformanilide[139–146] or dimethylformamide;[139] reported yields of aldehyde are as high as 80%.[140] The second method consists of the reaction of ferrocene and 1,1-dichloromethyl ethyl ether in the presence of aluminum chloride; the aldehyde was formed in about 50% yield.[131] The former reaction has been more frequently used and is better suited for general preparative purposes. Synthesis of the aldehyde has also been accomplished commercially by the Sommelet reaction of ferrocenylmethyltrimethylammonium iodide.[147] Preparative routes to ferrocenecarboxaldehyde are listed in Table XI.

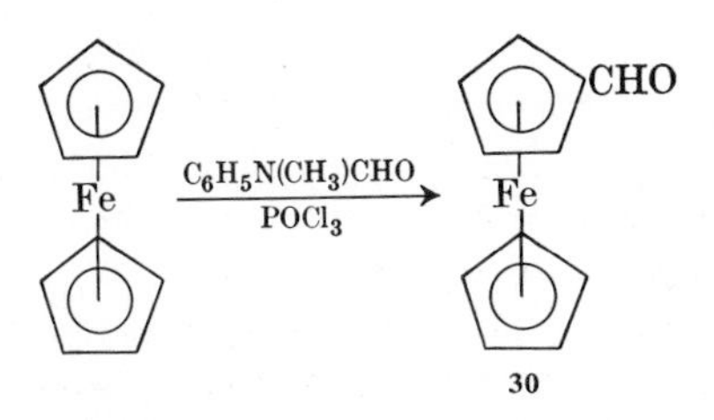

[139] P. J. Graham, R. V. Lindsey, G. W. Parshall, M. L. Peterson, and G. M. Whitman, *J. Am. Chem. Soc.*, **79**, 3416 (1957).

[140] M. Rosenblum, A. K. Banerjee, N. Danieli, R. W. Fish, and V. Schlatter, *J. Am. Chem. Soc.*, **85**, 316 (1963).

[141] G. D. Broadhead, J. M. Osgerby, and P. L. Pauson, *J. Chem. Soc.*, 650 (1958).

[142] K. Schlögl, *Monatsh. Chem.*, **88**, 601 (1957).

[143] M. Rosenblum, *Chem. Ind.* (*London*), 72 (1957).

[144] P. J. Graham, U.S. pat. 2,849,469 [*C.A.*, **53**, 4298 (1959)].

[145] P. J. Graham, U.S. pat. 2,988,564 [*C.A.*, **56**, 12948 (1962)].

[146] C. Jutz, *Tetrahedron Letters*, No. 21, 1 (1959).

[147] P. Pratter, Research Organic Chemical Co., Sun Valley, California, personal communication.

The formylation of certain ferrocene derivatives by means of N-methylformanilide and phosphorus oxychloride has been accomplished.[55, 148–152]

The two reports of the diformylation of ferrocene by N-methylformanilide in the presence of phosphorus oxychloride which have appeared describe crude material only.[144, 145] 1,1′-Ferrocenedicarboxaldehyde has been prepared by manganese dioxide oxidation of 1,1′-bis(α-hydroxymethyl)ferrocene.[91]

Ferrocenecarboxaldehyde undergoes the usual reactions of an aromatic aldehyde and has been utilized for the synthesis of many substituted ferrocenes. Especially useful are its condensations with active methylene compounds (refs. 91, 126, 139–142, 149, 150, 153–162). Reactions of the aldehyde are illustrated in Fig. 4 (p. 25), and useful products derived from it are listed in Table XII. Similar reactions have been reported for the dialdehyde[91] and for substituted ferrocenecarboxaldehydes.[149, 150, 161]

Acyl Ferrocenes and Related Compounds

Acyl ferrocenes are prepared by Friedel-Crafts acylation employing acid halides or anhydrides in the presence of the usual catalysts, e.g., aluminum chloride, boron trifluoride, or phosphoric acid. They were the first substituted ferrocenes prepared and are among the most useful intermediates in the preparation of other ferrocene derivatives. However, the vastly greater reactivity of ferrocenes toward electrophilic aromatic substitution, as compared with the corresponding benzenoid derivatives, poses the problem of control of multisubstitution in acylation. This is particularly manifest in a tendency toward heteroannular disubstitution, since the second ring, unacylated, is only partially deactivated by the acyl group on the substituted ring. Usually the amount of disubstitution can be controlled by the proper choice of catalyst and ratio of reagents.

[148] K. Schlögl and H. Seiler, *Tetrahedron Letters*, No. 7, 4 (1960).
[149] K. Schlögl and M. Peterlik, *Monatsh. Chem.*, **93**, 1328 (1962).
[150] K. Schlögl, M. Peterlik, and H. Seiler, *Monatsh. Chem.*, **93**, 1309 (1962).
[151] G. Tainturier and J. Tirouflet, *Compt. Rend.*, **258**, 5666 (1964).
[152] G. Tainturier and J. Tirouflet, *Bull. Soc. Chim. France*, 2739 (1964).
[153] J. Tirouflet and J. Boichard, *Compt. Rend.*, **250**, 1861 (1960).
[154] C. R. Hauser and J. K. Lindsay, *J. Org. Chem.*, **22**, 906 (1957).
[155] G. D. Broadhead, J. M. Osgerby, and P. L. Pauson, *Chem. Ind.* (*London*), 209 (1957).
[156] B. Loev and M. Flores, *J. Org. Chem.*, **26**, 3595 (1961).
[157] I. K. Barben, *J. Chem. Soc.*, 1827 (1961).
[158] J. Boichard and J. Tirouflet, *Compt. Rend.*, **251**, 1394 (1960).
[159] K. Sonogashira and N. Hagihara. *J. Chem. Soc. Japan*, **66**, 1090 (1963) [*C.A.* **62**, 7794 (1965)].
[160] H. Egger and K. Schlögl. *J. Organometal. Chem.*, **2**, 398 (1964).
[161] K. Schlögl and M. Peterlik, *Tetrahedron Letters*, 573 (1962).
[162] P. daRe and E. Sainesi. *Experientia*, **21**, 648 (1965).

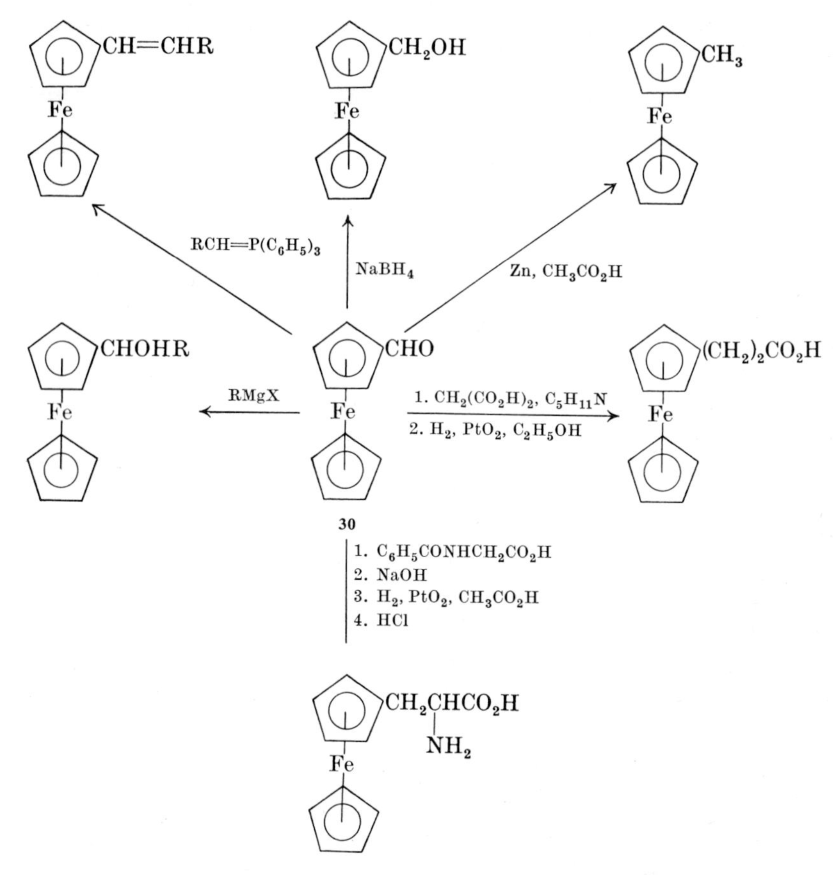

FIG. 4. Useful synthetic reactions of ferrocenecarboxaldehyde.

Monoacylated Ferrocenes. Monoacylation of ferrocenes can be accomplished readily by three procedures: with an acid anhydride-boron trifluoride complex in methylene chloride;[163] with an acid anhydride in liquid hydrogen fluoride;[164] and with the acid anhydride as solvent in the presence of phosphoric acid.[139] The simplest reaction, the last of the three, has found application in the commercial synthesis of acetylferrocene **(31)** (p. 26).[147] The amount of anhydride required obviously eliminates the last method when the anhydride is expensive, and all three methods are limited by the availability of the desired acid anhydride. The boron trifluoride procedure has been used by various workers for the acylation of substituted ferrocenes and appears to be quite practical.

[163] C. R. Hauser and J. K. Lindsay, *J. Org. Chem*,. **22,** 482 (1957).
[164] V. Weinmayr, *J. Am. Chem. Soc.*, **77,** 3009 (1955).

However, other workers have found the reaction products to be pure diacylated materials[165] or mixtures of mono- and di-acylated products.[166] To obtain only the monosubstituted material, only 1 molar equivalent of acetic anhydride must be used.[167] A method of lesser importance involves the addition of Grignard reagents to cyanoferrocene,[168] though this method may assume greater importance in light of the ease with which cyanoferrocene can be prepared (see p. 40).[169, 170]

$$\text{Ferrocene} \xrightarrow[\text{BF}_3,\ \text{HF, or } \text{H}_3\text{PO}_4]{(\text{CH}_3\text{CO})_2\text{O}} \text{Acetylferrocene (COCH}_3\text{)}$$

31

Acylation of ferrocene by means of an acid chloride and aluminum chloride leads to significant amounts of disubstitution unless considerable care is taken to control the amount of aluminum chloride present. Ferrocene can form a cationic complex with aluminum chloride and hydrogen chloride that is not acylated under the usual reaction conditions.[171] Any aluminum chloride over a stoichiometric amount will form a non-reactive complex with ferrocene and the hydrogen chloride liberated in the acylation reaction; this removal of ferrocene from the reaction is responsible for the formation of diacyl products.

Monoacyl ferrocenes that have been prepared by these methods are listed in Table XIII.

Diacylated Ferrocenes. *Heteroannular Diacyl Ferrocenes.* Although several procedures are available for the preparation of monoacyl ferrocenes, only two have been used for the diacylation of ferrocene. The first method involves the acid anhydride, aluminum chloride, and ferrocene. In general a molar ratio of 2:1 of acylating agent over ferrocene is sufficient, but observation of some monoacylation has been reported. The second method involves acylation by an excess of acid chloride-aluminum chloride; this procedure is used for the commercial synthesis of

[165] G. R. Buell, personal communication.

[166] K. L. Rinehart Jr., D. E. Bublitz, and D. H. Gustafson *J. Am. Chem. Soc.*, **85**, 970 (1963).

[167] D. E. Bublitz and K. L. Rinehart Jr. unpublished observations.

[168] A. N. Nesmeyanov, E. G. Perevalova, L. P. Yur'eva, and L. I. Denisovich, *Izv. Akad. Nauk SSSR, Otd. Khim. Nauk*, 2241 (1962) [*C.A.*, **58,** 12597 (1963)]

[169] E. G. Perevalova, L. P. Yur'eva, Yu. I. Baukov, *Dokl. Akad. Nauk SSSR*, **135,** 1402 (1960) [*C.A.*, **55,** 12378 (1961)].

[170] A. N. Nesmeyanov, E. G. Perevalova, and L. P. Juryeva, *Chem. Ber.*, **93,** 2729 (1960).

[171] M. Rosenblum, J. O. Santer, and W. G. Howells, *J. Am. Chem. Soc.*, **85,** 1450 (1963).

1,1′-diacetylferrocene **(32)**.[147] Diacyl and diaroyl ferrocenes prepared by these procedures are presented in Table XIV.

Fe + Excess $(CH_3CO)_2O$, $AlCl_3$ or Excess CH_3COCl, $AlCl_3$ → $COCH_3$ Fe $COCH_3$

32

Homoannular Diacyl Ferrocenes. During the diacetylation of ferrocene a small amount of 1,2-diacetylferrocene (1–2%) is formed in addition to the predominant 1,1′-diacetyl product.[172, 173]

No other simple 1,2-diacylferrocene has been reported from acylation—a reflection of the directive effect of the acyl group.

Diacyl groupings which are part of a second, fused carbocycle are known, however, and are discussed on pp. 49, 51.

Acylation of Substituted Ferrocenes. The acylation of substituted ferrocenes has been extensively studied, especially with regard to directive effects of substituents. A detailed discussion of these effects, given elsewhere,[39] is outside the scope of this review. Suffice it to generalize here that acylation of ferrocenes substituted with electron-donating substituents like alkyl groups usually gives all possible isomers, the homoannular 3-acyl isomer being favored, while acylation of ferrocenes substituted with electron-withdrawing substituents like acyl groups gives almost exclusively the heteroannular 1′-acyl isomer. A singular exception is the reported isolation of a tetraacetyl ferrocene of undesignated structure from the reaction of ferrocene with acetic anhydride in the presence of trifluoroacetic acid at 200–350°.[174]

Acylation reactions of alkyl (refs. 62, 151, 152, 166, 172, 175–184), acyl,[141, 178] aryl,[185–187]halo,[188, 189] acetamido,[189–191] carbomethoxy,[192, 193] alkylcarbamido ($C_5H_5FeC_5H_4NHCO_2R$),[189] and bridged ferrocenes[140, 166, 176] have been reported.

[172] M. Rosenblum and R. B. Woodward, *J. Am. Chem. Soc.*, **80,** 5443 (1958).

[173] J. H. Richards and T. J. Curphey, *Chem. Ind.* (*London*), 1456 (1956).

[174] W. M. Sweeney, U.S. pat. 2,852,542 [*C.A.*, **53,** 4297 (1959)].

[175] A. N. Nesmeyanov, E. G. Perevalova, Z. A. Beinoravichute, and I. G. Malygina, *Dokl. Akad. Nauk SSSR,* **120,** 1263 (1958) [*C.A.*, **53,** 1293 (1959)].

[176] E. A. Hill and J. H. Richards, *J. Am. Chem. Soc.*, **83,** 4216 (1961).

[177] J. P. Monin, G. Tainturier, R. Dabard, and J. Tirouflet, *Bull. Soc. Chim. France,* 667 (1963).

[178] A. N. Nesmeyanov and N. A. Volkenau, *Dokl. Akad. Nauk. SSSR,* **111,** 605 (1956) [*C.A.*, **51,** 9599 (1957).

[179] L. A. Day, Brit. pat. 864,198 [*C.A.*, **55,** 17647 (1961)].

[180] S. Birtwell, Brit. pat. 861,834 [*C.A.*, **55,** 16565 (1961)].

The diacetylation of dimethylferrocene has been described.[62, 175] In one study four of the six possible isomeric heteroannular diacetyl-1,1′-dimethylferrocenes were characterized.[62]

Compared to the large number of heteroannular diacyl- and diaroylferrocenes in which both groups are identical, there are relatively few in which these groups are different. Those reported are presented in Table XIV-B.

Transformations of Acyl Ferrocenes. The carbonyl groups of acyl ferrocenes and substituted acyl ferrocenes, like those of ferrocenecarboxaldehydes, undergo the same reactions as benzenoid compounds of similar structures. Both mono- and di-acyl ferrocenes are thus quite useful as synthetic intermediates en route to other ferrocenes.

Acyl ferrocenes are particularly important in the synthesis of alkyl ferrocenes (see Table XVI), of alcohols,[140, 154, 195–209] and of alkenyl ferrocenes (see Table XVIII) from the alcohols. They also undergo the

[181] T. Leigh, Brit. pat. 819,108 [*C.A.*, **54**, 7732 (1960)].

[182] K. L. Rinehart, Jr., K. L. Motz, and S. Moon, *J. Am. Chem. Soc.*, **79**, 2749 (1957).

[183] Y. Nagai, J. Hooz, and R. A. Benkeser, *Bull. Chem. Soc. Japan*, **37**, 53 (1964).

[184] J. Tirouflet, G. Tainturier, and R. Dabard, *Bull. Soc. Chim. France*, 2403 (1963).

[185] M. Rosenblum and W. G. Howells, *J. Am. Chem. Soc.*, **84**, 1167 (1962).

[186] M. Rosenblum, *J. Am. Chem. Soc.*, **81**, 4530 (1959).

[187] D. E. Bublitz, W. E. McEwen, and J. Kleinberg, *J. Am. Chem. Soc.*, **84**, 1845 (1962).

[188] A. N. Nesmeyanov, V. A. Sazonova, and V. N. Drozd, *Dokl. Akad. Nauk SSSR*, **137**, 102 (1961) [*C.A.*, **55**, 21081 (1961)].

[189] D. W. Håll and J. H. Richards, *J. Org. Chem.*, **28**, 1549 (1963).

[190] R. E. Bozak and K. L. Rinehart, Jr., unpublished results.

[191] A. N. Nesmeyanov, V. N. Drozd, and V. A. Sazonova, *Izv. Akad. Nauk SSSR, Ser., Khim.*, 1205 (1965) [*C.A.*, **63**, 13313 (1965)].

[192] N. A. Nesmeyanov and O. A. Reutov, *Dokl. Akad. Nauk SSSR*, **115**, 518 (1957) [*C.A.*, **52**, 5393 (1958)].

[193] E. G. Perevalova, M. D. Reshetova, K. I. Grandberg, and A. N. Nesmeyanov, *Izv. Akad. Nauk SSSR, Ser. Khim.*, 1901 (1964) [*C.A.*, **62**, 2792 (1965)].

[194] M. Furdik, S. Toma, M. Dzurilla, and J. Suchy, *Chem. Zvesti*, **18**, 607 (1964) [*C.A.*, **61**, 14710 (1964)].

[195] K. Schlögl and H. Egger, *Monatsh. Chem.*, **94**, 376 (1963).

[196] J. W. Huffman and D. J. Rabb, *J. Org. Chem.*, **26**, 3588 (1961).

[197] R. Dabard and B. Gautheron, *Compt. Rend.*, **254**, 2014 (1962).

[198] N. Sugiyama, H. Suzuki, Y. Shioura, and T. Teitei, *Bull. Chem. Soc. Japan*, **35**, 767 (1962).

[199] A. N. Nesmeyanov, N. S. Kochetkova, V. D. Vil'chevskaya, U. N. Sheinker, L. B. Senyavina, and M. I. Struchkova, *Izv. Akad. Nauk SSSR, Otd. Khim. Nauk*, 1990 (1962) [*C.A.*, **58**, 9133 (1963)].

[200] N. Weliky and E. S. Gould, *J. Am. Chem. Soc.*, **79**, 2742 (1957).

[201] K. Schlögl and H. Pelousek, *Ann.* **651**, 1 (1962).

[202] K. Schlögl and A. Mohar, *Monatsh. Chem.*, **92**, 219 (1961).

[203] C. R. Hauser, R. L. Pruett, and T. A. Mashburn, Jr., *J. Org. Chem.*, **26**, 1800 (1961).

[204] F. S. Arimoto and A. C. Haven, Jr., *J. Am. Chem. Soc.*, **77**, 6295 (1955).

[205] K. L. Rinehart, Jr., R. J. Curby, Jr., D. H. Gustafson, K. G. Harbison, R. E. Bozak, and D. E. Bublitz, *J. Am. Chem. Soc.*, **84**, 3263 (1962).

[206] M. Rausch, M. Vogel, and H. Rosenberg, *J. Org. Chem.*, **22**, 903 (1957).

[207] R. Riemschneider and D. Helm, *Chem. Ber.*, **89**, 155 (1956).

[208] D. S. Trifan and R. Bacskai, *Tetrahedron Letters*, No. 13, 1 (1960).

[209] W. F. Little and R. Eisenthal, *J. Org. Chem.*, **26**, 3609 (1961).

usual ketone condensation (refs. 131, 142, 153, 163, 166, 205, 210–220) and alkylation[212] reactions, and react with Grignard reagents,[55, 65, 207, 221–224] other organometallic reagents,[131, 168, 202, 223] and Wittig reagents.[225] Several of these reactions are illustrated in Fig. 5 (p. 30) for acetylferrocene, and a number of useful compounds obtained from acyl ferrocenes are listed in Table XV.

The behavior of the diacyl ferrocenes generally parallels that of the monoacyl derivatives. Many have been reduced to the corresponding carbinols[226–228] and to the corresponding alkyl ferrocenes (Table XVI), and a number have been shown to undergo the usual ketone condensation reactions,[212, 214, 216, 229–231] addition of Grignard reagents,[207, 221] the Willgerodt reaction,[211] and formation of enol acetates.[233] Substituted acyl ferrocenes behave similarly provided that the other substituents are unreactive under the reaction conditions.

Syntheses of Substituted Ferrocenes

Alkyl Ferrocenes

Two general methods are available for the synthesis of mono- or multialkyl ferrocenes. The better one involves the use of appropriate acyl

[210] M. D. Rausch and L. E. Coleman, *J. Org. Chem.*, **23,** 107 (1958).

[211] K. L. Rinehart, Jr., R. J. Curby, Jr., and P. E. Sokol, *J. Am. Chem. Soc.*, **79,** 3420 (1957).

[212] C. R. Hauser and T. A. Mashburn, Jr., *J. Org. Chem.*, **26,** 1795 (1961).

[213] M. Furdik, P. Elecko, S. Toma, and J. Suchy, *Chem. Zvesti*, **15,** 501 (1960) [*C.A.*, **55** 16508 (1961)]

[214] M. Furdik, S. Toma, and J. Suchy, *Chem. Zvesti*, **16,** 449 (1962) [*C.A.*, **58,** 11398 (1963)].

[215] C. J. Pederson and V. Weinmayr, U.S. pat. 2,875,223 [*C.A.*, **53,** 16149 (1959)].

[216] C. E. Cain, T. A. Mashburn, Jr., and C. R. Hauser, *J. Org. Chem.*, **26,** 1030 (1961).

[217] V. Weinmayr, *Naturwiss.* **45,** 311 (1958).

[218] L. Wolf and H. Hennig, *Z. Chem.*, **3,** 469 (1963).

[219] B. Gautheron and R. Dabard, *Bull. Soc. Chim. France*, 2009 (1963).

[220] J. Boichard, J. P. Monin, and J. Tirouflet, *Bull. Soc. Chim. France*, 851 (1963).

[221] R. Riemschneider and D. Helm, *Ann.* **646,** 10 (1961).

[222] Haun-Li Wu, E. B. Sokolova, L. A. Leites and A. D. Petrov, *Izv. Akad. Nauk SSSR, Otd. Khim. Nauk*, 887 (1962) [*C.A.*, **57,** 12532 (1962)].

[223] Haun-Li Wu, E. B. Sokolova, I. I. Chlenov, and A. D. Petrov, *Dokl. Akad. Nauk SSSR*, **137,** 111 (1961) [*C.A.*, **55,** 19885 (1961)].

[224] J. Boichard and J. Tirouflet, *Compt. Rend.*, **253,** 1337 (1961).

[225] P. L. Pauson and W. E. Watts, *J. Chem. Soc.*, 2990 (1963).

[226] T. A. Mashburn, Jr., and C. R. Hauser, *J. Org. Chem.*, **26,** 1671 (1961).

[227] E. C. Winslow and E. W. Brewster, *J. Org. Chem.*, **26,** 2982 (1961).

[228] K. Yamakawa, H. Ochi, and K. Arakawa, *Chem. Pharm. Bull.* (*Tokyo*), **11,** 905 (1963) [*C.A.*, **59,** 8787 (1963)].

[229] T. A. Mashburn, Jr., C. E. Cain, and C. R. Hauser, *J. Org. Chem.*, **25,** 1982 (1960).

[230] M. Furdik, S. Toma, and J. Suchy, *Chem. Zvesti*, **15,** 547 (1961) [*C.A.*, **56,** 7355 (1962)].

[231] C. R. Hauser and C. E. Cain, *J. Org. Chem.*, **23,** 1142 (1958).

[232] C. E. Cain, T. A. Mashburn, Jr., and C. R. Hauser, *J. Org. Chem.*, **26,** 1030 (1961).

[233] R. L. Pruett, U.S. pat. 2,947,769 [*C.A.*, **55,** 565 (1961)].

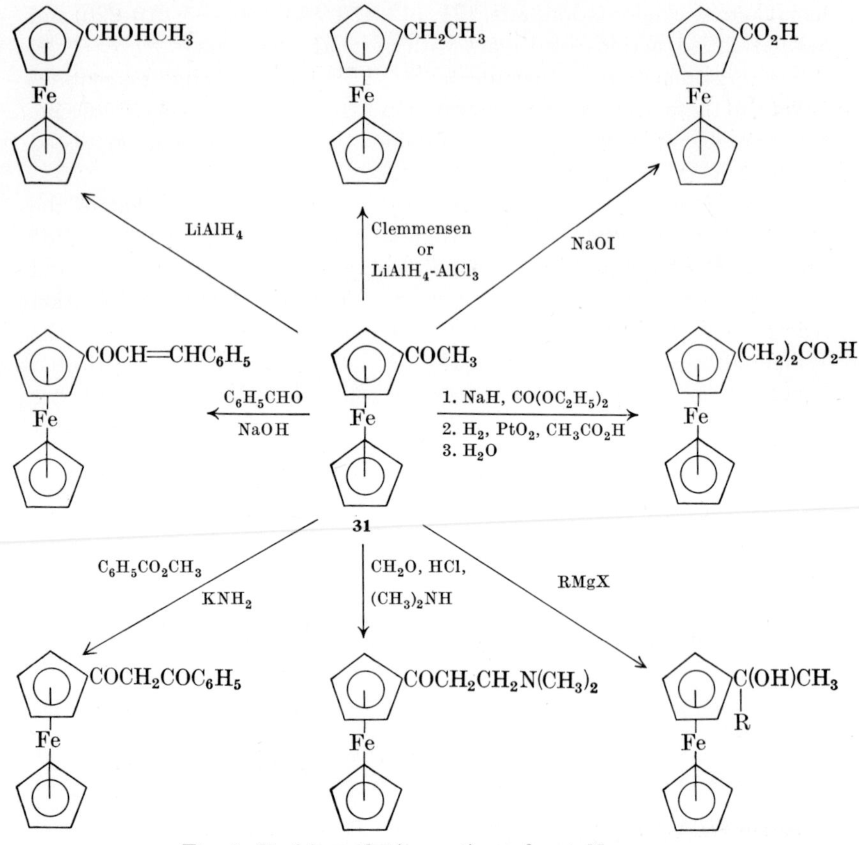

FIG. 5. Useful synthetic reactions of acetylferrocene.

ferrocenes (see pp. 24–29), which may be reduced by Clemmensen reduction, lithium aluminum hydride-aluminum chloride treatment, or catalytic hydrogenation to the corresponding alkyl ferrocenes. Other alkyl ferrocenes are obtained from acyl ferrocenes by the addition of Grignard reagents, followed by reduction of the carbinol product. By these procedures monoalkyl ferrocenes may be obtained from the monoacyl compounds and multi-alkyl ferrocenes from acylation of alkyl or acyl ferrocenes and reduction of the products. Alkyl ferrocenes prepared from acyl ferrocenes and also from alkylation (see below) are listed in Tables XVI and XVII.

Most dialkyl ferrocenes can be prepared by acylation of a monoalkyl or monoacyl ferrocene followed by chromatographic separation of the isomers and reduction to the product sought. The groups may be the same or different and, although the method can be tedious and cumbersome, it is the preferred route to pure isomers. Acylation of acyl ferrocenes or the diacylation of ferrocene leads almost entirely to 1,1′-disubstitution products, while acylation of alkyl ferrocenes usually gives all three isomeric products; thus the route involving acylation of alkyl ferrocenes is favored for homoannular isomers.

The second general method for the preparation of mono- or multi-alkyl ferrocenes is direct alkylation of ferrocene with an appropriate Friedel-Crafts agent. Separation of the monosubstituted product from the higher-substituted materials concurrently formed is accomplished by distillation at reduced pressure or by preparative vapor phase chromatography.[167, 234]

R, Fe; 1. R′COCl, $AlCl_3$; 2. Separate; R, COR′, Fe + R′CO, R, Fe + R, Fe, COR′

↓ ↓ ↓

R, CH_2R', Fe; $R'CH_2$, R, Fe; R, Fe, CH_2R'

Although direct alkylation is intrinsically desirable, the separation of products involved is bothersome and the yield of a single product is never good. In addition to the mixture of mono- and multi-substitution products, mixtures of di- and higher-substituted materials are obtained and the separation of mixtures of dialkylated ferrocenes by the usual physical methods is difficult. Thus unresolved mixtures of di-, tri-, and poly-ethylferrocenes,[235–237] di- and poly-isopropylferrocenes,[235, 236] dimethylferrocenes,[234] and tri- and tetra-(*t*-butyl)ferrocenes,[236, 238] have been reported, together with isomeric tri- and tetra-ethylferrocenes.[239]

[234] R. A. Benkeser and J. L. Bach, *J. Am. Chem. Soc.*, **86,** 890 (1964).

[235] A. N. Nesmeyanov and N. S. Kochetkova, *Dokl. Akad. Nauk SSSR*, **114,** 800 (1957) [*C.A.*, **52,** 3794 (1958)].

[236] A. N. Nesmeyanov and N. S. Kochetkova, *Izv. Akad. Nauk SSSR, Otd. Khim. Nauk* 242 (1958) [*C.A.*, **52,** 12852 (1958)].

[237] A. N. Nesmeyanov and N. S. Kochetkova, *Dokl. Akad. Nauk SSSR*, **109,** 543 (1956) [*C.A.*, **51,** 5057 (1957)].

[238] T. Leigh, Brit. pat. 828,965 [*C.A.*, **54,** 15402 (1960)].

[239] D. E. Bublitz, *Can. J. Chem.*, **42,** 2381 (1964).

Because of the difficulty of separating mixtures of multi-alkyl ferrocenes, the acylation route is better; formation of mono- or di-substituted ferrocenes can be easily controlled and their reduction to the corresponding alkyl compounds proceeds in high yield. These products can in turn be further acylated and reduced if necessary. Even nonaethyl- and decaethyl-ferrocene[149, 161] have been prepared by repeated alternate acetylation and reduction. This technique is the one currently employed commercially for the synthesis of alkyl ferrocenes.[147]

A third method is the most convenient for preparation of many symmetrical 1,1′-dialkylated and certain tetraalkylated ferrocenes; this, the direct method, proceeds from an appropriately substituted cyclopentadiene and an iron source and has been covered earlier (see Tables II and III). Benzylferrocene is apparently the only monoalkyl ferrocene which has been prepared by the direct route.[240]

Acylation and metalation of dialkyl ferrocenes have been investigated.[53, 175, 177, 182, 241] Oxidation of dialkyl ferrocenes with manganese dioxide gives mono- and di-acyl products (aldehydes and ketones) plus some acid.[242]

A few alkyl ferrocenes have been prepared by reaction of (ferrocenylmethyl)trimethylammonium iodide with various Grignard reagents.[243] This reaction was mentioned on p. 21 (see also Table X).

Methylferrocene **(33)** has been obtained by reduction of the aldehyde and by reduction of a number of compounds like (ferrocenylmethyl)trimethylammonium iodide for which analogous starting materials are unavailable for other alkyl ferrocenes. These preparations are listed, with those of other alkyl ferrocenes, in Table XVI.

The formation of methylferrocene by reduction of ferrocenoic acid **(34)** or its methyl ester with lithium aluminum hydride,[244] has been attributed to the presence of the aluminum halide used in the preparation of the hydride.[167] When commercial lithium aluminum hydride is used, reduction stops at ferrocenylmethanol and proceeds no further, even with excess reagent.[167] When lithium aluminum hydride-aluminum chloride is the reducing agent, the product is methylferrocene.[167, 234]

An unusual route involving dimerization of an appropriately substituted ferrocene in strong acid or reaction of ferrocene and an appropriate aldehyde in acid medium is available for the preparation of symmetrical

[240] B. F. Hallam and P. L. Pauson, *J. Chem. Soc.*, 3030 (1956).

[241] J. Tirouflet, J. Monin, G. Tainturier, and R. Dabard, *Compt. Rend.*, **256**, 433 (1963).

[242] K. L. Rinehart, Jr., A. F. Ellis, C. J. Michejda, and P. A. Kittle, *J. Am. Chem. Soc.*, **82**, 4112 (1960).

[243] A. N. Nesmeyanov, E. G. Perevalova, and L. S. Shilovtseva, *Izv. Akad. Nauk SSSR, Otd. Khim. Nauk*, 1982 (1961) [*C.A.*, **56**, 10185 (1962)].

[244] A. N. Nesmeyanov, E. G. Perevalova, L. S. Shilovtseva, and Z. A. Beinoravichute, *Dokl. Akad. Nauk SSSR*, **121**, 117 (1958) [*C.A.*, **53**, 323 (1959)].

$C_5H_5FeC_5H_4CH_3$ (33) $\xrightarrow{MnO_2}$ $C_5H_5FeC_5H_4CHO$ + $C_5H_5FeC_5H_4CO_2H$ (34)

$C_5H_5FeC_5H_4CH_2\overset{\oplus}{N}(CH_3)_3I^{\ominus}$ $\xrightarrow{RMgX}$ $C_5H_5FeC_5H_4CH_2R$

$C_5H_5FeC_5H_4CH_2\overset{\oplus}{N}(CH_3)_3I^{\ominus}$ $\xrightarrow{Na(Hg)}$ $C_5H_5FeC_5H_4CH_3$ (33)

diferrocenylethanes (**40**). Thus, when ferrocene is treated with formaldehyde or benzaldehyde[164, 245–248] in the presence of hydrogen fluoride or sulfuric acid, the product is 1,2-diferrocenylethane (**40**, $R_1 = R_2 = H$) or 1,2-diferrocenyl-1,2-diphenylethane (**40**, $R_1 = H$, $R_2 = C_6H_5$), respectively. Similarly, dimerization of ferrocenylcarbinol (**35**)[249] or isopropenylferrocene (**37**)[250] in sulfuric acid leads to 1,2-diferrocenylethane or to 2,3-diferrocenyl-2,3-dimethylbutane (**40**, $R_1 = R_2 = CH_3$), respectively. 1,2-Diferrocenyl-1,2-diphenylethane is similarly formed in either the acid-catalyzed decomposition of ferrocenylphenylcarbinylazide (**36**)[251] or solvolyses involving carbonium ion centers adjacent to the ferrocene nucleus.[252–254] (See pp. 34 and 39.)

The postulated mechanism[249] for this group of reactions involves the radical ion **38b**. Generation of the intermediate cation **38a** is extremely

[245] K. L. Rinehart, Jr., C. J. Michejda, and P. A. Kittle, *J. Am. Chem. Soc.*, **81,** 3162 (1959).

[246] A. N. Nesmeyanov and I. I. Kritskaya, *Izv. Akad. Nauk SSSR, Otd. Khim. Nauk*, 352 (1962) [*C.A.*, **57,** 11230 (1962)].

[247] A. N. Nesmeyanov and I. I. Kritskaya, *Izv. Akad. Nauk SSSR, Otd. Khim Nauk*, 253 (1956) [*C.A.*, **50,** 13886 (1956)].

[248] V. Weinmayr, U.S. pat. 2,694,721 [*C.A.*, **49,** 15955 (1955)].

[249] K. L. Rinehart, Jr., C. J. Michejda, and P. A. Kittle, *Angew. Chem.*, **72,** 38 (1960).

[250] K. L. Rinehart, Jr., P. A. Kittle, and A. F. Ellis, *J. Am. Chem. Soc.*, **82,** 2082 (1960).

[251] A. Berger, W. E. McEwen, and J. Kleinberg, *J. Am. Chem. Soc.*, **83,** 2274 (1961).

[252] A. N. Nesmeyanov, and I. I. Kritskaya, *Izv. Akad. Nauk SSSR, Ser. Khim.*, 2160 (1964) [*C.A.*, **62,** 10458 (1965)].

[253] A. N. Nesmeyanov, V. A. Sazonova, V. N. Drozd, and N. A. Rodionova, *Dokl. Akad. Nauk SSSR*, **160,** 355 (1965) [*C.A.*, **62,** 14725 (1965)].

[254] A. N. Nesmeyanov, V. A. Sazonova, V. N. Drozd, N. A. Rodionova, and G. I. Zudkova, *Izv. Akad. Nauk SSSR, Ser. Khim.*, 2061 (1965) [*C.A.*, **62,** 14725 (1965)].

(*a*) $\xrightarrow[\text{H}_2\text{SO}_4 \text{ or HF}]{\text{R}_1\text{CHO}}$

Route *a*: R_1 = H or C_6H_5; R_2 = H

$H^{\oplus}$, $-H_2O$

(*b*) $\xrightarrow[-\text{H}_2\text{O}]{\text{H}^{\oplus}}$

35 38a 38b

Route *b*: $R_1 = R_2$ = H

(*c*) $H^{\oplus}$, $-HN_3$

36 39

Route *c*: $R_1 = C_6H_5$; R_2 = H

$H^{\oplus}$

$+2e$

(*d*)

37 40

Route *d*: $R_1 = R_2 = CH_3$

facile, and even weak acids like acetic and hydrazoic acids have been shown to protonate vinylferrocene.[255,256]

Following the reaction, the dipositive cation **39** is reduced and the uncharged compound is isolated. Conclusive proof for the existence of

[255] G. R. Buell, W. E. McEwen, and J. Kleinberg, *J. Am. Chem. Soc.*, **84**, 40 (1962).
[256] G. R. Buell, W. E. McEwen, and J. Kleinberg, *Tetrahedron Letters*, No. 5, 16 (1959).

the radical ion has not yet been reported, and an alternative reaction pathway has recently been proposed.[257]

A number of ferrocenes substituted by various hydrocarbon groups have been isolated in optically active form. The syntheses involved resolution of racemic amines and alcohols followed by removal of the functional groups.[56]

Alkenyl Ferrocenes

Of the many ferrocenyl substituted olefins in which the ferrocene nucleus is attached to one of the ethylenic carbon atoms, none has been prepared by substitution on ferrocene itself. Most alkenyl ferrocenes can be prepared from one of the key intermediates discussed earlier, in a normal synthetic route to the desired compound (Table XVIII). The most common route involves dehydration over alumina of an α-hydroxyalkyl ferrocene (refs. 65, 195, 202, 204, 222, 259), but other elimination reactions,[132, 198, 260] reactions of acyl compounds with phosphorus ylides[91, 225, 261] and condensation reactions of ferrocenecarboxaldehyde[154, 158] also have been used. The reaction of acyl ferrocenes with phosphorus oxychloride in dimethylformamide produces 1-chloro-1-ferrocenyl alkenes which are useful intermediates in the synthesis of ferrocenyl acetylenes and ferrocenyl allenes.[262] The syntheses of a large number of ferrocenyl polyenes by these methods together with the spectral properties of the products have recently been reported.[159, 263–265] An elimination reaction worthy of special note, since it leads to a 1,2-disubstituted ferrocene **(42)**, is shown in the accompanying equation.[260] The quaternary methiodide **41** was prepared from the corresponding

$\oplus$ N CH_3 CH_3 Fe $I^\ominus$ $\xrightarrow[\text{Liq. } NH_3]{KNH_2}$ $CH_2N(CH_3)_2$ $CH{=}CH_2$ Fe

41 **42**

[257] M. Cais and A. Eisenstadt, *J. Org. Chem.*, **30**, 1148 (1965).
[258] J. B. Thomson, *Chem. Ind.* (*London*), 1122 (1959).
[259] K. Schlögl and A. Mohar, *Naturwiss.*, **48**, 376 (1961).
[260] D. Lednicer and C. R. Hauser, *J. Org. Chem.*, **24**, 43 (1959).
[261] G. Drefahl, G. Plötner, and I. Winnefeld, *Chem. Ber.*, **95**, 2788 (1962).
[262] K. Schlögl and W. Steyer, *Monatsh. Chem.*, **96**, 1520 (1965).
[263] K. Schlögl and H. Egger, *Ann. Chem.*, **676**, 76 (1964).
[264] K. Schlögl and H. Egger, *Ann. Chem.*, **676**, 88 (1964).
[265] A. Nakamura, P. J. Kim, and N. Hagihara, *J. Organometal. Chem.*, **3**, 355 (1965).

heterocyclic product derived from β-ferrocenylethylamine, formic acid, and formaldehyde.

Synthesis of certain dialkenyl ferrocenes has been accomplished by direct synthesis of the appropriately substituted iron compound (see p. 13 and Table III). The only other 1,1′-dialkenyl ferrocenes which have been reported were prepared by dehydration of the corresponding diols. They are found in the latter portion of Table XVIII.

Ferrocenyl Acetylenes

Ferrocenyl acetylenes are prepared by standard routes. Ferrocenylacetylene itself has been prepared by dehydrohalogenation of 1,2-dibromoethylferrocene,[266] of β-chlorovinylferrocene, and of β,β-dichloroethylferrocene with sodium amide in liquid ammonia,[195] in yields of 29%, 56%, and 30%, respectively. Ferrocenylacetylene is best prepared by the dehydrohalogenation of the reaction product obtained when acetylferrocene is treated with phosphorus oxychloride in dimethylformamide (75% overall).[262]

The disubstituted acetylenes **44** have been prepared in excellent yields (85–93%) from the corresponding phosphorus ylides **(43)** as shown.[225, 267]

$$\underset{\textbf{43}}{C_5H_5FeC_5H_4\underset{\substack{|\\ COR}}{C}{=}P(C_6H_5)_3} \xrightarrow[\text{in vacuum}]{\text{Heat}} \underset{\textbf{44}}{C_5H_5FeC_5H_4C{\equiv}CR}$$

$R = C_6H_5$, $C_5H_4FeC_5H_5$, or $CO_2C_2H_5$

Phenylferrocenylacetylene is also formed (in 53% yield) from bromoferrocene and potassium phenylacetylide.[268] Dehydrohalogenation of 1-chloro-1-ferrocenyl alkenes, obtained from the respective acyl ferrocene and phosphorus oxychloride in dimethylformamide, provides a versatile route to ferrocenyl acetylenes and certain ferrocenyl alkenes.[262] Numerous acetylenes have been prepared, however, in which the ferrocene group is not directly attached to an acetylenic carbon atom.[202, 225, 269] The compounds were prepared by typical reactions upon one of the ferrocene intermediates described earlier.

Dimerization of ferrocenylacetylene with cupric acetate in methanolic pyridine gave 1,4-diferrocenyl-1,3-butadiyne in 90% yield.[195] Treatment of a mixture of ferrocenylacetylene and phenylacetylene under identical

[266] R. A. Benkeser and W. P. Fitzgerald, Jr., *J. Org. Chem.*, **26**, 4179 (1961).

[267] H. Egger and K. Schlögl, *Monatsh. Chem.*, **95**, 1750 (1964).

[268] A. N. Nesmeyanov, V. A. Sazonova, and V. N. Drozd, *Doklady Akad. Nauk SSSR*, **154**, 158 (1964) [*C.A.*, **60**, 9309 (1964)].

[269] K. Schlögl and A. Mohar, *Monatsh. Chem.*, **93**, 861 (1962).

conditions gave a 54% yield of 1-ferrocenyl-4-phenyl-1,3-butadiyne.[195] Other reactions involving ferrocenyl acetylenes have been reported.[267]

The synthesis of 1,1′-diethynylferrocene has been accomplished by dehydrohalogenation of 1,1′-bis-(α-chlorovinyl)ferrocene with sodium amide in liquid ammonia.[262]

Aryl Ferrocenes, Including Biferrocenyl and Terferrocenyl

Arylation of ferrocene was first accomplished by its reaction with aryl diazonium compounds, *p*-nitrophenylferrocene being obtained in 64% yield by this method.[117] Although this reaction is the most common method for obtaining aryl ferrocenes and is discussed below in detail, three other routes are known.

The direct synthesis of phenylferrocene from phenylcyclopentadiene and cyclopentadiene in the presence of ferrous chloride and a base[270] has been discussed above (see Table IV).

The second method involves the demercuration of a mixture of a diaryl mercury compound and diferrocenylmercury in the presence of metallic silver.[97] This method is complicated by the need of both diferrocenylmercury and a diaryl mercury compound.

Lastly, a novel method has been reported for the preparation of pure monoaryl ferrocenes from bromoferrocene and potassium tetraarylboron compounds in the presence of cupric bromide.[268] The inaccessibility of tetraarylboron salts, however, severely limits the scope.

The principal route to aryl ferrocenes, arylation by means of a diazonium salt, has been shown to lead to di- and even higher-substitution products.[117, 271] Mechanistic studies of the arylation of ferrocene have been reported,[171, 186, 272, 273] and initial formation of a ferrocene diazonium complex **(45)** seems likely. Yields in the range 40–60% have been

Fe $\xrightarrow{ArN_2^{\oplus}}$ [Fe—N_2Ar]$^{\oplus}$ ⟶ Fe(Ar) + N_2 + $H^{\oplus}$

45

[270] P. L. Pauson, *J. Am. Chem. Soc.*, **76**, 2187 (1954).
[271] V. Weinmayr, *J. Am. Chem. Soc.*, **77**, 3012 (1955).
[272] M. Rosenblum, W. G. Howells, A. K. Banerjee, and C. Bennett, *J. Am. Chem. Soc.*, **84**, 2726 (1962).
[273] A. L. J. Beckwith and R. J. Leydon, *J. Am. Chem. Soc.*, **86**, 952 (1964).

realized for the monoarylated compound,[274–276] but the yields of diarylated materials are usually below 40%. Some tri- and tetra-arylated ferrocenes of undesignated structures have been reported.[271, 274] In general the arylations are carried out in water[271, 272, 274, 277–279] or acetic acid.[66, 186, 272, 275] When the arylation is carried out in the presence of a halogenated solvent such as methylene chloride or chloroform, moderate amounts of products are formed which are derived from radical attack upon the solvent. Thus, when ferrocene is treated with 2-methylbenezenediazonium chloride in chloroform, a 25% yield of ferrocenoic acid is realized.[280]

Arylations of methylferrocene,[276] ethylferrocene,[276] *p*-tolylferrocene,[276] and *p*-anisylferrocene[171] have been reported. These reactions give mixtures in which the 1,1′-disubstituted products predominate. The known diarylated compounds containing different aryl groups are the isomeric products obtained from *p*-anisylferrocene and *p*-nitrobenzenediazonium chloride.[171] Compounds prepared by arylation and demercuration are listed in Table XIX.

The usual product of the reaction of ferrocene with aryldiazonium salts is the corresponding aryl ferrocene, and no azo coupling is observed. However, the reaction between 1,1′-diacetylferrocene and any of several aromatic diazonium salts leads to products free of iron shown to have the structure **46**.[121, 281, 282]

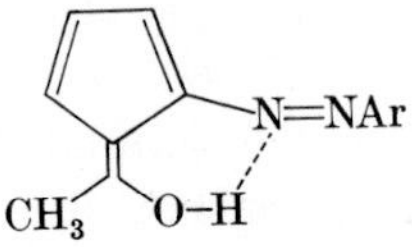

N=NAr

CH_3 O–H

46

$C_5H_5FeC_5H_4C_5H_4FeC_5H_5$

47

4,4′-Diferrocenylbiphenyl and the corresponding 3,3′ -isomer are formed in 15% and 58% yields, respectively, when 4- and 3-ferrocenylbromobenzene are treated with magnesium in diglyme at 100° and then allowed to react with carbon dioxide.[283]

[274] A. N. Nesmeyanov, E. G. Perevalova, and R. V. Golovnya, *Dokl. Akad. Nauk SSSR*, **99,** 539 (1954) [*C.A.*, **49,** 15918 (1955)].

[275] G. D. Broadhead and P. L. Pauson, *J. Chem. Soc.*, 367 (1955).

[276] E. G. Perevalova, N. A. Simukova, T. V. Nikitina, P. D. Reshetov, and A. N. Nesmeyanov, *Izv. Akad. Nauk SSSR, Otd. Khim. Nauk*, 77 (1961) [*C.A.*, **55,** 17645 (1961)].

[277] S. I. Goldberg, *J. Org. Chem.*, **25,** 482 (1960).

[278] A. N. Nesmeyanov, E. G. Perevalova, R. V. Golovnya, and L. S. Shilovtseva, *Dokl. Akad. Nauk SSSR*, **102,** 535 (1955) [*C.A.*, **50,** 4925 (1956)].

[279] W. F Little and A. K. Clark, *J. Org. Chem.*, **25,** 1979 (1960).

[280] W. F. Little, K. N. Lynn, and R. Williams, *J. Am. Chem. Soc.*, **85,** 3055 (1963).

[281] A. N. Nesmeyanov, E. G. Perevalova, N. A. Simukova, Yu. N. Sheinker, M. D. Reshetova, *Dokl. Akad. Nauk SSSR*, **133,** 851 (1960) [*C.A.*, **54,** 24790 (1960)].

[282] R. E. Bozak, and K. L Rinehart, Jr , *J. Am. Chem. Soc.*, **84,** 1589 (1962).

[283] W. F. Little, A. K. Clark, G. S. Benner, and C. Noe, *J. Org. Chem.*, **29,** 713 (1964).

Biferrocenyl (**47**) has not been obtained by the diazotization of aminoferrocene, but it has been prepared by other routes. It was observed first as a by-product in the preparation of tri-(*n*-hexyl)silylferrocene,[284] and its structure was established later.[122, 285]

Better coupling reactions which have been used for the synthesis of biferrocenyl are presented in Table XX. These reactions include demercuration of diferrocenylmercury (Fig. 3),[97, 122, 123] Ullmann coupling of haloferrocenes,[122, 285–288] coupling of ferroceneboronic acid by ammoniacal silver oxide (Fig. 1),[108, 109] and coupling of lithioferrocene by cobalt bromide.[289, 290] 1,1′-Terferrocenyl (1,1′-diferrocenylferrocene, **48**) has been prepared in 14% yield, together with polyferrocenylene, by reaction of a mixture of bromo- and 1,1′-dibromo-ferrocene with copper powder.[291] It has also been prepared by the direct route from ferrocenylcyclopentadiene.[292]

Br Br
Fe + Fe $\xrightarrow{Cu}$ **47** + Fe Fe Fe + polymer
Br
48

Acetylation of biferrocenyl has recently been studied and the three possible monoacetyl biferrocenyls have been isolated.[293–295] Ullmann coupling was used as a route to both mono- and di-acetyl biferrocenyls of known structure.[296]

[284] S. I. Goldberg and D. W. Mayo, *Chem. Ind.* (*London*), 671 (1959).

[285] S. I. Goldberg, D. W. Mayo and J. A. Alford, *J. Org. Chem.*, **28**, 1708 (1963).

[286] E. G. Perevalova and O. A. Nesmeyanova, *Dokl. Akad. Nauk SSSR*, **132**, 1093 (1960) [*C.A.*, **54**, 21027 (1960)].

[287] M. D. Rausch, U.S. pat. 3,010,981 [*C.A.*, **56**, 8750 (1962)].

[288] M. D. Rausch, *J. Org. Chem.*, **26**, 1802 (1961).

[289] K. Hata, I. Motoyama, and H. Watanabe, *Bull. Chem. Soc. Japan*, **37**, 1719 (1964).

[290] I. J. Spilners and J. P. Pellegrini, Jr., *J. Org. Chem.*, **30**, 3800 (1965).

[291] A. N. Nesmeyanov, V. N. Drozd V. A. Sazonova, V. I. Romanenko, A. K. Prokofev, and L. A. Nikonova, *Izv. Akad. Nauk, SSSR, Otd. Khim. Nauk*, 667 (1963) [*C.A.*, **59**, 7556 (1963)].

[292] K. L. Rinehart, Jr., D. G. Ries, C. H. Park, and P. A. Kittle, National Am. Chem. Soc. Meeting, Denver, January, 1964, Abstracts, p. 23C.

[293] K. Yamakawa, N. Ishibashi, and K. Arakawa, *Chem. Pharm. Bull* (*Japan*), **12**, 119 (1964).

[294] S. I. Goldberg and J. S. Crowell, *J. Org. Chem.*, **29**, 996 (1964).

[295] M. D. Rausch, *J. Org. Chem.*, **29**, 1257 (1964).

[296] S. I. Goidberg and R. L. Matteson, *J. Org. Chem.*, **29**, 323 (1964).

Cyano Ferrocenes

Cyanoferrocene (ferrocenonitrile, **49**) was reported by three groups of workers[139, 141, 155] who utilized various reagents to dehydrate ferrocenecarboxaldoxime. Subsequent methods of synthesis have involved reaction of halo ferrocenes with cuprous cyanide[109, 297] or direct cyanation of ferricenium chloride by hydrogen cyanide in the presence of ferric chloride.[169, 170] Syntheses of cyanoferrocene and substituted cyano ferrocenes are presented in Table XXI.

$$\text{Ferrocene} \xrightarrow{\text{HCN, FeCl}_3} \text{Cyanoferrocene } (\mathbf{49}) \xrightarrow{\text{HCN, FeCl}_3} \text{1,1'-Dicyanoferrocene } (\mathbf{50})$$

The direct cyanation of substituted ferrocenes has been little investigated.[169, 170] A mechanism has been suggested[171] and the simplicity of the reaction makes it appealing. Electronic effects greatly influence the position of substitution; thus alkyl groups give homoannular and heteroannular substitution products while halo and cyano groups only give heteroannular products.[169, 170, 298, 299]

Acetylation of cyanoferrocene gives exclusively 1′-acetylcyanoferrocene[189] because of the ring-deactivating effect of the cyano group. When cyanoferrocene is allowed to react with Grignard reagents or organolithium compounds, the corresponding ketones are obtained.[168]

1,1′-Dicyanoferrocene (**50**) has been obtained by dehydration of 1,1′-dicarboxamidoferrocene,[300] and better by dehydration of 1,1′-ferrocenedicarboxaldoxime.[91] Cyanation of cyanoferrocene or ferrocene in tetrahydrofuran-liquid hydrogen cyanide containing ferric chloride has been shown to produce 1,1′-dicyanoferrocene in 23–27% or 68% yield, respectively.[169, 170, 301]

[297] A. N. Nesmeyanov, V. A. Sazonova, and V. N. Drozd, *Dokl. Akad, Nauk SSSR*, **130,** 1030 (1960) [*C.A.*, **54,** 12089 (1960)].

[298] A. N. Nesmeyanov, E. G. Perevalova, and K. I. Grandberg, *Izv. Akad. Nauk SSSR, Ser. Khim.*, 1903 (1964) [*C.A.*, **62,** 2793 (1965)].

[299] A. N. Nesmeyanov, E. G. Perevalova, L. P. Yur'eva, and L. N. Kakurina, *Izv. Akad. Nauk SSSR, Ser. Khim.*, 1897 (1964) [*C A.*, **62,** 2793 (1965)].

[300] N. A. Nesmeyanov and O. A. Reutov, *Dokl. Akad. Nauk SSSR*, **120,** 1267 (1958) [*C.A.*, **53,** 1292 (1959)].

[301] N. A. Nesmeyanov, E. G. Perevalova, L. P. Yuryeva, and K. I. Grandberg, *Izv. Akad. Nauk SSSR, Otd. Khim. Nauk*, 1772 (1962) [*C.A.*, **58,** 7971 (1963)].

Ferrocenecarboxylic Acids and Their Derivatives

Ferrocenoic acid (ferrocenecarboxylic acid, **34**) has been prepared in numerous ways (see Table XXII). The majority of the methods involve one of the key intermediate ferrocenes mentioned earlier, and the most common preparations of ferrocenoic acid or substituted ferrocenoic acids involve carbonation of lithio ferrocenes[53, 93–95, 117, 302] or oxidation of acetyl ferrocenes with a variety of reagents.[142, 164, 182, 192, 303–305] Commercial production of ferrocenoic acid is accomplished by oxidation of acetylferrocene.[147]

As noted on p. 14, lithiation of ferrocene gives a mixture of the corresponding acids upon carbonation. Separation of the mixed acids is generally accomplished by extraction with ether, in which only the monobasic acid is soluble.[94] Methods of effecting monolithiation have also been discussed earlier.

Three derivatives of ferrocenoic acid have been prepared by substitution on ferrocene—ferrocenecarboxamides,[306, 307] alkyl and aryl thiolferrocenoates,[308] and ferrocenonitrile (see preceding section). One method for the preparation of ferrocenecarboxamides involves the reaction of ferrocene or certain of its alkyl derivatives with either N,N-diphenylcarbamyl chloride or carbamyl chloride itself in the presence of aluminum chloride.[55, 307] The second method involves the reaction of ferrocene with an isocyanate in the presence of aluminum chloride.[306] Amides prepared by these reactions are found in Table XXIV. Substitution reactions on ferrocene have not led to ferrocene diamides, in accord with the view that the carbamyl chloride-aluminum chloride complex is a weak acylating reagent.[307] The amides formed in the reaction involving isocyanates were found to be resistant to hydrolysis by acid or base, with the exception of the ethyl compound, which was hydrolyzed by acid.[306] In contrast the N,N-diphenylamide **51** formed in the diphenylcarbamyl chloride route is hydrolyzed by base, the overall yield of acid being 46%.[307]

Treatment of ferrocene with an alkyl or aryl chlorothiolformate and aluminum chloride produces the corresponding alkyl or aryl thiolferrocenoate **52**.[308] Hydrolysis of methyl thiolferrocenoate provides

[302] J. Tirouflet, E. Laviron, R. Dabard, and J. Komenda, *Bull. Soc. Chim. France*, 857 (1963).

[303] V. Weinmayr, U.S. pat. 2,683,157 [*C.A.*, **49**, 10364 (1955)].

[304] L. Wolf and M. Beer, *Naturwiss.*, **44**, 442 (1957).

[305] A. N. Nesmeyanov, V. A. Sazonova, and V. N. Drozd, *Izv. Akad. Nauk SSSR, Otd. Khim. Nauk*, 45 (1962) [*C.A.*, **57**, 865 (1962)].

[306] M. Rausch, P. Shaw, D. Mayo, and A. M. Lovelace, *J. Org. Chem.*, **23**, 505 (1958).

[307] W. F. Little and R. Eisenthal, *J. Am. Chem. Soc.*, **82**, 1577 (1960).

[308] D. E. Bublitz and G. H. Harris, *J. Organometal. Chem.*, **4**, 404 (1965).

ferrocenoic acid in yields up to 80%, and replacement of the methylthio group by hydrazine and amines has also been reported.[308]

$$C_5H_5FeC_5H_5 \xrightarrow[AlCl_3]{ClCON(C_6H_5)_2} \underset{\mathbf{51}}{C_5H_5FeC_5H_4CON(C_6H_5)_2}$$

$$C_5H_5FeC_5H_5 \xrightarrow[AlCl_3]{ClCOSR} \underset{\mathbf{52}}{C_5H_5FeC_5H_4COSR} \xrightarrow{KOH} \underset{\mathbf{34}}{C_5H_5FeC_5H_4CO_2H}$$

$$\mathbf{51} \xrightarrow{NaOH} \mathbf{34}$$

Two methods for the preparation of ferrocenoic acid via acid derivatives give especially good yields. They are by way of cyanoferrocene[169, 170] followed by hydrolysis[309] (69% overall) and via methyl thiolferrocenoate[308] followed by hydrolysis (70% overall). Methods of preparation of ferrocene mono- and di-carboxylic acid are presented in Table XXII, and substituted ferrocenoic acids are listed in Table XXIII.

Ferrocenoic acid has been converted to the usual acid derivatives (refs. 91, 93, 117, 142, 164, 189, 204, 304–307, 313–318). When heated to 250–350° in an atmosphere of nitrogen, it gives the ferrous salt of ferrocenoic acid in 75% yield.[319]

Of the three possible isomeric ferrocene dicarboxylic acids only the 1,1′ and 1,2 diacids are known. The 1,2 isomer was obtained by hypochlorite oxidation of 1,2-diacetylferrocene, and its conversion to a cyclic anhydride was utilized in the proof of structure of 1,2-diacetylferrocene.[173]

The 1,1′ isomer has been obtained in various ways, most of which utilize the key intermediates (see Table XXIII). The preparation of pure ferrocene-1,1′-dicarboxylic acid by hypohalite oxidation of 1,1′-diacetylferrocene[63, 228, 320] is the commercial method.[147] The most common method, carbonation of a dimetalated ferrocene, simultaneously produces monocarboxylated material which must be removed by solvent extraction. Even molar ratios of 25:1 of *n*-butyllithium to ferrocene still

[309] A. N. Nesmeyanov, E. G. Perevalova, L. P. Yur'eva, and K. I. Grandberg, *Izv. Akad. Nauk SSSR, Ser. Khim.*, 1377 (1963) [*C.A.*, **59**, 15310 (1963)].

[310] G. R. Knox and P. L. Pauson, *J. Chem. Soc.*, 692 (1958).

[311] A. N. Nesmeyanov and B. N. Strunin, *Dokl. Akad. Nauk SSSR*, **137**, 106 (1961) [*C.A.*, **55**, 19885 (1961)].

[312] D. E. Bublitz and K. L. Rinehart, Jr., *Tetrahedron Letters*, 827 (1964).

[313] H. L. Lau and H. Hart, *J. Org. Chem.*, **24**, 280 (1959).

[314] N. Baggett, A. B. Foster, A. H. Haines, and M. Stacey, *J. Chem. Soc.*, 3528 (1960).

[315] E. M. Acton and R. M. Silverstein, *J. Org. Chem.*, **24**, 1487 (1959).

[316] A. C. Haven, Jr., U.S. pat. 2,816,904 [*C.A.*, **52**, 5479 (1958)].

[317] A. C. Haven, Jr., U.S. pat. 3,035,074 [*C.A.*, **57**, 13806 (1962)].

[318] K. Schlögl and H. Seiler, *Naturwiss.*, **45**, 337 (1958).

[319] R. L. Schaaf, *J. Org. Chem.*, **27**, 107 (1962).

[320] F. W. Knobloch and W. H. Rauscher, *J. Polymer Sci.*, **54**, 651 (1961).

produce 23% of monobasic acid.[94] Hydrolysis of the dinitrile may offer an attractive, still little investigated, alternative. Another route to the dibasic acid is from 1,1′-dicarbomethoxyferrocene **(53)** which is prepared directly in 28% yield from cyclopentadienylsodium and methyl chloroformate in the presence of ferrous chloride.[91]

$$C_5H_5Na \xrightarrow[\text{2. } FeCl_2]{\text{1. } ClCO_2CH_3} \underset{\mathbf{53}}{CH_3O_2CC_5H_4FeC_5H_4CO_2CH_3}$$

Ferrocene-1,1′-dicarboxylic acid forms the usual derivatives of a dibasic acid (refs. 63, 93, 105, 117, 192, 228, 277, 300, 320). However, reports of the preparation of the 1,1′-dialdehyde from the corresponding acid chloride seem dubious,[277] and attempts to form the intramolecular, bridged anhydride with dicyclohexylcarbodiimide have been unsuccessful.[228] The dimethyl ester **53** can be reduced by lithium aluminum hydride to 1,1′-bis(hydroxymethyl)ferrocene[104] or to 1,1′-dimethylferrocene by lithium aluminum hydride-aluminum chloride.[167, 234]

Ferrocenesulfonic Acids and Their Derivatives

Ferrocenesulfonic acid **(54)** is obtained by two equally useful methods involving treatment of ferrocene either with an equimolar amount of chlorosulfonic acid in acetic anhydride (66% yield)[310] or with an equimolar amount of the dioxane-sulfur trioxide complex in ethylene dichloride (62% yield).[321] In both methods the sulfonic acid is isolated as the

$$C_5H_5FeC_5H_5 \xrightarrow[\text{or 1 mole } C_5H_5N\cdot SO_3]{\text{1 mole } ClSO_3H,\ (CH_3CO)_2O} \underset{\mathbf{54}}{C_5H_5FeC_5H_4SO_3H}$$

dihydrate. A less important route, from the standpoint of yield, is the reaction of ferrocene with 0.25 molar equivalent of 100% sulfuric acid in acetic anhydride. The sulfonic acid was isolated as the ammonium salt in 25% yield.[164]

Ferrocene-1,1′-disulfonic acid has been obtained as the free acid from the reaction of ferrocene, either with 2.0 molar equivalents of chlorosulfonic acid in acetic anhydride,[310] or with 3.0 molar equivalents of the dioxane-sulfur trioxide complex,[321] or with 1.5 molar equivalents of 100% sulfuric acid in acetic anhydride.[164] The disulfonic acid has also been reported as a product from treatment of acetyl- or 1,1′-diacetyl-ferrocene with the sulfur trioxide complex, with apparent replacement of the acetyl groups.[311] It is noteworthy that in the sulfonation reactions no mixtures of mono- and di-substitution products have been reported under the conditions employed.

[321] A. N. Nesmeyanov, E. G. Perevalova, and S. S. Churanov, *Dokl. Akad. Nauk SSSR*, **114**, 335 (1957) [*C.A.* **52**, 368 (1958)].

Sulfonation reactions have been carried out on acetylferrocene,[311] phenylferrocene,[128] 1,1′-dicarbomethoxyferrocene,[311] ferrocenoic acid,[322] and methyl ferrocenoate.[322] From the monosubstituted ferrocenes named, the product formed has the 1,1′ orientation, as would be anticipated from the deactivating influence of the substituent on the ring in which it is present.

Ferrocenesulfonic acids undergo the functional group modifications characteristic of aryl sulfonic acids.[311, 321–324] The lead salt of the disulfonic acid on treatment with phosphorus oxychloride gives the disulfonyl chloride.[323]

Hydrogenation of ferrocenesulfonyl chloride leads to ferrocenesulfinic acid (**55**);[310, 321, 324] reduction by lithium aluminum hydride gives ferrocenethiol (**56**).[310, 323] Both are convertible to a large number of derivatives. Ferrocenyl disulfide, diferrocenyl sulfone, and ferrocenyl phenyl sulfone, as mentioned earlier (see Table VIII-B), have been prepared from diferrocenylmercury.[325, 326]

$$C_5H_5FeC_5H_4SO_2Cl \xrightarrow{H_2} \underset{\mathbf{55}}{C_5H_5FeC_5H_4SO_2H}$$

$$C_5H_5FeC_5H_4SO_2Cl \xrightarrow{LiAlH_4} \underset{\mathbf{56}}{C_5H_5FeC_5H_4SH}$$

Bridged Ferrocenes

Ferrocene is theoretically capable of conversion to compounds having three possible types of cyclized structures, **57**, **58**, and **59**. Of the three types shown, only **57** and **59** have been synthesized. Compounds of the type **59** are called bridged ferrocenes and are discussed here. They are also included in a recent book on bridged aromatic compounds.[327] Structures of type **57** are discussed in the following section on ferroçocarbocyclic compounds.

[322] A. N. Nesmeyanov and O. A. Reutov *Izv. Akad. Nauk SSSR, Otd. Khim. Nauk*, 926 (1959) [*C.A.*, **54**, 469 (1960)].

[323] A. N. Nesmeyanov E. G. Perevalova, S. S. Churanov, and O. A. Nesmeyanova, *Dokl. Akad. Nauk SSSR*, **119**, 949 (1958) [*C.A.*, **52**, 17225 (1958)].

[324] E. G. Perevalova, O. A. Nesmeyanova, and I. G. Luk'yanova, *Dokl. Akad. Nauk SSSR* **132**, 853 (1960) [*C.A.*, **54**, 21025 (1960)].

[325] A. N. Nesmeyanov, E. G. Perevalova, and O. A. Nesmeyanova, *Izv. Akad. Nauk SSSR, Otd. Khim. Nauk*, 47 (1962) [*C.A.*, **57**, 12532 (1962)].

[326] A. N. Nesmeyanov, E. G. Perevalova, and O. A. Nesmeyanova, *Dokl. Akad. Nauk SSSR*, **119**, 288 (1958) [*C.A.*, **52**, 14579 (1958)].

[327] B. H. Smith, *Bridged Aromatic Compounds*, Academic, New York, 1964.

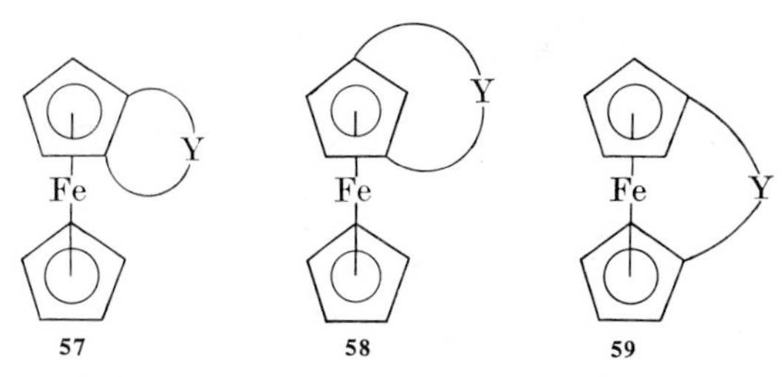

β-Ferrocenylpropionic acid **(60)**, an important precursor of bridged ferrocenes, is available by any of four synthetic routes. Two of these are of equal utility and have been noted earlier, that via ethyl ferrocenoylacetate (68% overall yield of the propionic acid from acetylferrocene)[163, 205, 211] in Fig. 5 and that via the Doebner condensation of ferrocenecarboxaldehyde with malonic acid [142] in Fig. 4.

A third method, of lesser importance, involves the hydrolysis and decarboxylation of ferrocenylmethylmalonic ester[328] prepared from N,N-dimethylaminomethylferrocene methiodide. β-Ferrocenylpropionic acid is obtained in 56% yield from the methiodide. The fourth method of preparation is the Willgerodt reaction of propionylferrocene (4% yield).[211]

The first published synthesis of a bridged ferrocene, 1,1′-(α-ketotrimethylene)ferrocene **(61)**, utilized the cyclization of β-ferrocenylpropionic acid with polyphosphoric acid.[329] The synthesis has been improved by the use of trifluoroacetic anhydride in carbon tetrachloride.[140, 176, 205]

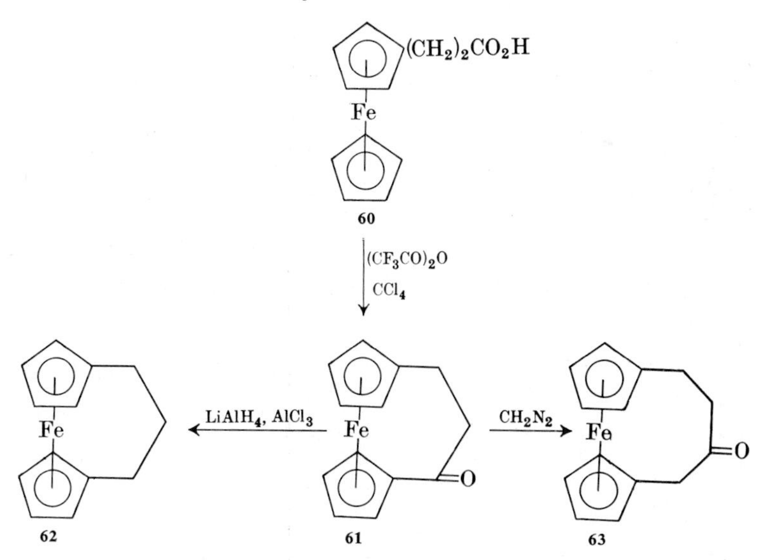

[328] C. R. Hauser and J. K. Lindsay, *J. Org. Chem.*, **22**, 1246 (1957).
[329] K. L. Rinehart, Jr., and R. J. Curby, Jr., *J. Am. Chem. Soc.*, **79**, 3290 (1957).

1,1′-(α-Ketotrimethylene)ferrocene **(61)** has been shown to undergo a number of the reactions of other acyl ferrocenes—conversion to the alcohol,[140, 205] the enol acetate,[140] and the olefin.[330] Reduction to 1,1′-trimethyleneferrocene **(62)** is accomplished by lithium aluminum hydride-aluminum chloride,[150, 205, 331] by catalytic hydrogenation,[176, 332] by the Wolff-Kishner[140] and by the Clemmensen[140] procedures in yields of 99%, 70–90%, 70%, and 90%, respectively. The trimethyleneferrocene **62** has also been prepared in a direct fashion from 1,3-dicyclopentadienylpropane (see Table II).[333, 334] Structural and conformational studies of certain monobridged ferrocenes have been reported.[335–337]

Treatment of **61** with diazomethane gives a mixture of two insertion products, mostly 1,1′-(β-ketotetramethylene)ferrocene **(63)**.[140] Wolff-Kishner reduction of the latter gives a 75% yield of 1,1′-tetramethyleneferrocene.[140]

1,1′-Trimethyleneferrocene **(62)** undergoes acetylation to a mixture of the 2- and 3-acetyl derivatives,[140, 166, 176] whereas formylation apparently gives only the 3-formyl derivative.[148, 150] Both 3-formyl- and 3-acetyl-1,1′-trimethyleneferrocene have been converted via the corresponding propionic acid to 1,1′,3,3′-bis(trimethylene)ferrocene **(64)**.[148, 150, 166] Similarly, 2-acetyl-1,1′-trimethyleneferrocene gives 1,1′,2,2′-bis(trimethylene)ferrocene **(65)**.[166] Synthesis of the bis bridged compound **64** has also been accomplished by stepwise closure of ferrocene-1,1′-dipropionic acid followed by reduction.[148, 166]

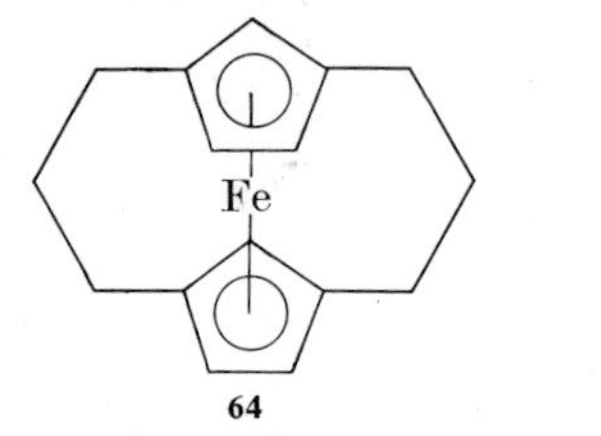

64

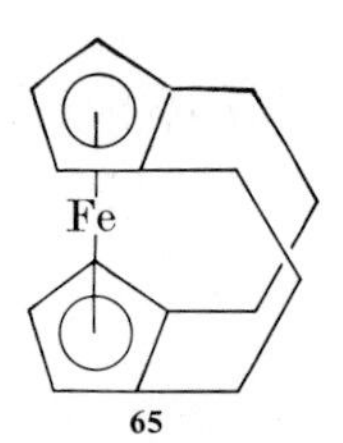

65

Repetition of these sequences gives the tris(trimethylene)ferrocenes **66** and **67**.[149, 161, 166] The reported synthesis of a tetrabridged ferrocene by a similar series of reactions[149] has been disproved.[312] Considerable amounts of homoannular cyclized products (of the general type **57**) are obtained in

[330] M. N. Applebaum, R. W. Fish, and M. Rosenblum, *J. Org. Chem.*, **29,** 2452 (1964).
[331] K. Schlögl, A. Mohar, and M. Peterlik, *Monatsh. Chem.*, **92,** 921 (1961).
[332] K. Schlögl and H. Seiler, *Monatsh. Chem.*, **91,** 79 (1960).
[333] A. Lüttringhaus and W. Kullick, *Makromol. Chem.*, **44–46,** 669 (1961).
[334] A. Lüttringhaus and W. Kullick, *Angew. Chem.*, **70,** 438 (1958).
[335] M. Rosenblum, A. K. Banerjee, N. Danieli, and L. K. Herrick, *Tetrahedron Letters*, 423 (1962).
[336] N. D. Jones, R. E. Marsh, and J. H. Richards, *Acta Cryst.*, **19,** 330 (1965).
[337] M. B. Lange and K. N. Trueblood, *Acta Cryst.*, **19,** 373 (1965).

these sequences; their formation, attributed to steric hindrance in the cyclization step,[140, 166] is discussed below.

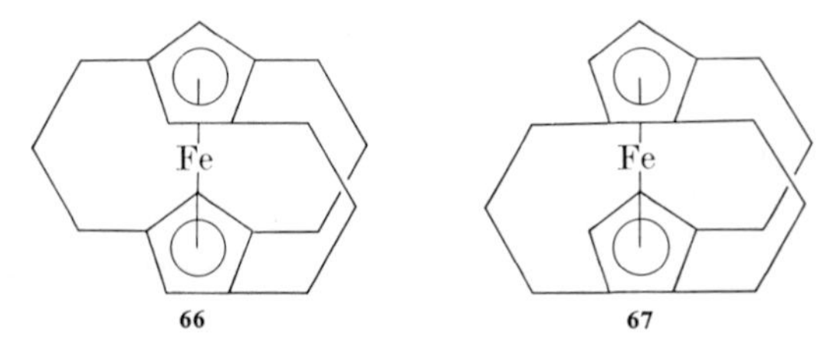

66 **67**

Treatment of α,α-dimethyl-β-ferrocenylpropionic acid with trifluoroacetic anhydride gave a mixture of 1,1′- and 1,2-(α-keto-β,β-dimethyltrimethylene)ferrocene.[140] Reduction of the former gave 1,1′-(β,β-dimethyltrimethylene)ferrocene.

Acyloin condensations of ferrocene-1,1′-dialiphatic acid esters have given a series of bridged ferrocenes in which the bridges contain from four to nine carbon atoms,[332] and a Dieckmann condensation of dimethyl ferrocene-1,1′-diacetate gave 1,1′-(β-ketotrimethylene)ferrocene.[338]

The first bridged ferrocene reported containing an oxygen atom in the bridge was the intramolecular ether **(68)** of 1,1′-bis(hydroxymethyl)-ferrocene, formed when the diol was treated with *p*-toluenesulfonic acid;[89] other chemists later demonstrated that 1,1′-bis(α-hydroxyethyl)ferrocene forms a similar compound.[226–228]

68 **69** **70**

Bridged ferrocenes containing both silicon and oxygen in the bridge have been prepared principally from 1,1′-bis(methyldiethoxysilyl)ferrocene **(69)**.[339] Treatment of **69** with acid in aqueous dioxane gave 1,3-(1,1′-ferrocenylene)-1,3-dimethyl-1,3-diethoxydisiloxane in 77% yield, while treatment with acid in aqueous ethanol gave 1,3-(1,1′-ferrocenylene)-1,3-dimethyl-1,3-dihydroxydisiloxane in 21% yield.[339] Treatment of **69** with acidic aqueous dioxane at 80° gave the cyclic tetrasiloxane **70** in yields of 11–24%. Treatment of 1,1′-bis(dimethylethoxysilyl)ferrocene

[338] W. Mock and J. H. Richards, *J. Org. Chem.*, **27**, 4050 (1962).
[339] R. L. Schaaf, P. T. Kan, and C. J. Lenk, *J. Org. Chem.*, **26**, 1790 (1961).

with acid in aqueous dioxane gave a 90% yield of 1,3-(1,1′-ferrocenylene)-tetramethyldisiloxane, for which a direct synthesis has also been reported (see Table II).[339]

Monobridged ferrocenes prepared by the same routes are presented in Table XXV. Those prepared in a direct manner from a substituted cyclopentadiene and an iron source are found in the latter portions of Tables II and III and are discussed on pp. 11, 13.

Numerous bridged ferrocenes of the type **71** prepared by the reaction of 1,1′-diacylferrocenes with various aldehydes are summarized in Table XXVI.[214, 340–343]

$$\text{Fe}(C_5H_4COCH_2R)_2 \xrightarrow[\text{Base}]{R'CHO} \text{Fe}(C_5H_4COCHR)_2CHR'$$

71

Ferroçocarbocyclic Compounds

Treatment of γ-ferrocenylbutyric acid with polyphosphoric acid[196, 205, 329, 344] or, better, trifluoroacetic anhydride[176, 205, 329] leads to 2,3-ferroçocyclohexenone, [1,2-(α-ketotetramethylene)ferrocene (**72a**)], in yields of 15–86%. The synthesis of γ-ferrocenylbutyric acid is readily accomplished by reaction of ferrocene with succinic anhydride[196, 211] in the presence of aluminum chloride to give, in 87–91% yield, β-ferrocenoylpropionic acid which is then reduced. When the methyl ester acid chloride was used in place of succinic anhydride the corresponding ester was obtained in 32% yield.[344] The keto acid can be reduced by various methods to the desired γ-ferrocenylbutyric acid in 71–98% yield.[196, 205, 211, 344] 1,2-(α-Ketotetramethylene)ferrocene (**72a**) has been reduced to **72b** by the Clemmensen reduction[345] or lithium aluminum hydride-aluminum chloride,[167] and to the corresponding alcohol by lithium aluminum hydride.[176, 208] Dehydration of the alcohol with alumina gave

[340] M. Furdik, S. Toma, and J. Suchy, *Chem. Zvesti*, **15**, 789 (1961).

[341] M. Furdik, S. Toma, J. Suchy, and P. Elecko, *Chem. Zvesti*, **15**, 45 (1961) [*C.A.*, **55**, 18692 (1961)].

[342] M. Furdik and S. Toma, *Acta Fac. Rerum Nat. Univ. Comenianae, Chimia*, **7**, 545 (1963) [*C.A.*, **61**, 10706 (1964)].

[343] M. Furdik, M. Dzurilla, S. Toma, and J. Suchy, *Acta Fac. Rerum Nat. Univ. Comenianae, Chimia*, **8**, 569 (1964) [*C.A.*, **61**, 12033 (1964)].

[344] A. N. Nesmeyanov, N. A. Vol'kenau, and V. D. Vil'chevskaya, *Dokl. Akad. Nauk SSSR*, **118**, 512 (1958) [*C.A.*, **52**, 11019 (1958)].

[345] A. N. Nesmeyanov, L. A. Kazitsyna, B. V. Lokshin, V. D. Vil'chevskaya, *Dokl. Akad. Nauk SSSR*, **125**, 1037 (1959) [*C.A.*, **53**, 21857 (1959)].

1,2-ferroço-1,3-cyclohexadiene.[57] Addition of Grignard reagents or hydrazoic acid to **72a** proceeds normally,[56, 57, 224] and the ketone, a racemic mixture, has been resolved via the (−) menthydrazide.[54] The absolute configuration of **72a** has also been established.[58, 346] Oxidation of the ketone by manganese dioxide gives the novel ferroçobenzoquinone **(73)**.[242]

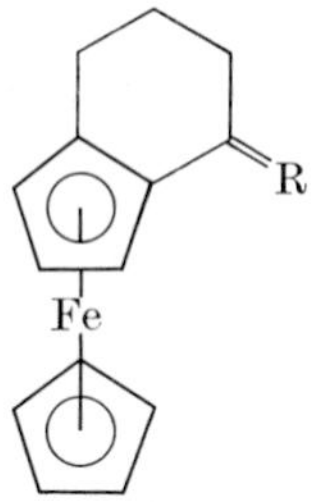

72a, b
a: R = O
b: R = H_2

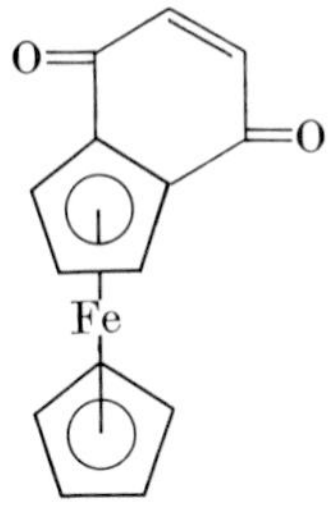

73

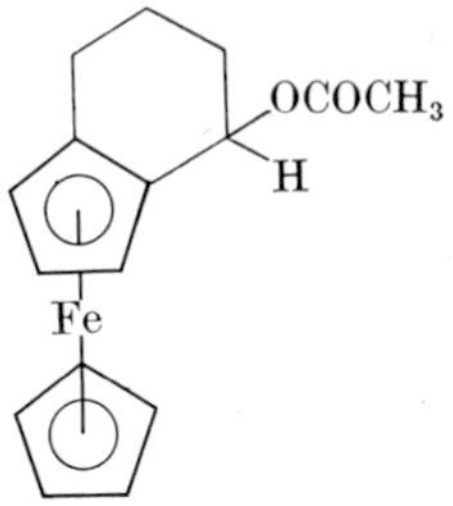

74

The epimeric acetates, **74**, have been extremely useful in studies of the stereochemical course of the solvolysis of groups on carbon atoms adjacent to the ferrocene nucleus.[208, 347]

Cyclization of α-phenyl- and β-phenyl-γ-ferrocenylbutyric acid in the presence of polyphosphoric acid leads to mixtures of *exo*- and *endo*-1,2-(α-keto-β-phenyltetramethylene)ferrocene and *exo*- and *endo*-1,2-(α-keto-γ-phenyltetramethylene)ferrocene, respectively.[348, 349] The absolute configurations of these products have recently been reported.[60] Cyclization reactions have also been studied for γ-(2,1′-diethylferrocenyl)- and γ-(3,1′-diethylferrocenyl)-butyric acid.[184, 350] Similar treatment of δ-ferrocenylvaleric acid with either polyphosphoric acid or trifluoroacetic anhydride gives 2,3-ferroçocycloheptenone in 14–28% yield.[205, 329]

While cyclization of β-ferrocenylpropionic acid reportedly gives only 1,1′-(α-ketotrimethylene)ferrocene **(61)**, cyclization of other ferrocenylpropionic acids with trifluoroacetic anhydride leads to homocyclic ketones, **75**, **76**, **77**, **78**, in addition to the corresponding bridged ketones.[140, 166] It is thus evident that steric factors begin to play an important part in determining the course of the reaction when substituents are present either in the propionic side chain, as in the precursor to **78**, or adjacent to the position of potential electrophilic attack, as in the precursors to **75**, **76**, and **77**. Similarly, when the two carbon atoms adjacent to the

[346] H. Falk and K. Schlögl, *Monatsh. Chem.*, **96**, 266 (1965).
[347] J. H. Richards, *Proc. Paint Res. Inst.*, 1433 (1964).
[348] R. Dabard and B. Gautherson, *Bull. Soc. Chim. France*, 667 (1963).
[349] J. Tirouflet, R. Dabard, and B. Gautheron, *Compt. Rend.*, **256**, 1315 (1959).
[350] R. Dabard, G. Tainturier, and J. Tirouflet, *Bull. Soc. Chim. France*, 2009 (1963).

ferrocene nucleus are constrained in a plane as in *o*-ferrocenylbenzoic acid, only the homoannular product of cyclization **(80)** appears to be formed (see below).[186, 187]

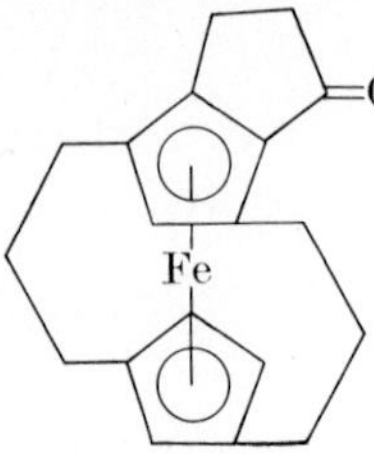

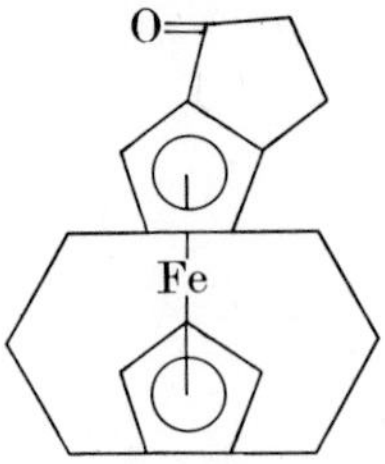

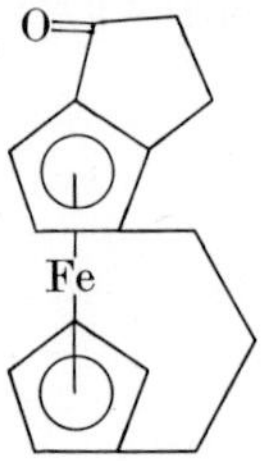

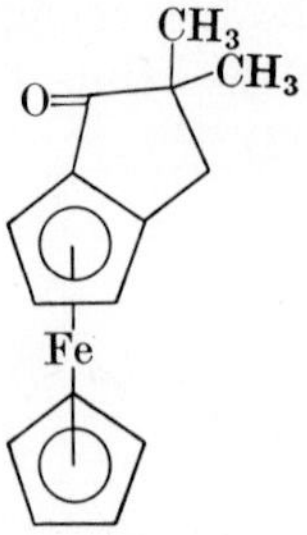

O CH_3 CH_3 Fe

75 **76** **77** **78**

Treatment of ferrocene-1,1′-dibutyric acid with polyphosphoric acid gave a 46% yield of 1,2,1′,2′-bis(α-ketotetramethylene)ferrocene as a mixture of diastereoisomers.[351] The separation of these isomers into a racemic pair **(79a)** and a meso compound **(79b)** has been accomplished by two groups.[177, 205, 241] The absolute configurations of these compounds have been investigated.[346] Reduction of the mixed isomers leads to bis(tetrahydroindenyl)iron.[351]

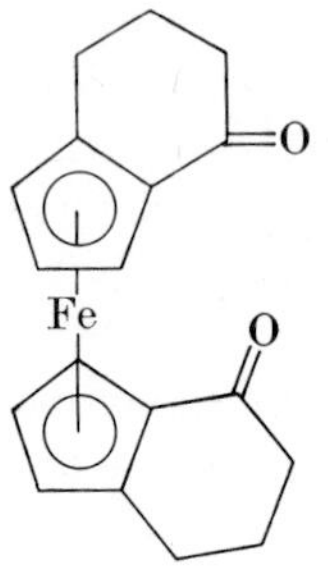

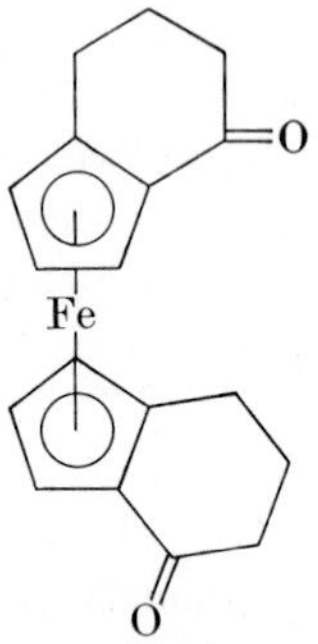

Fe O

79a **79b**

The acid chlorides of *o*-ferrocenylbenzoic acid[187] and 1′,2-diphenylferrocenoic acid[186] react with aluminum chloride to give 2,3-ferroçoindenones **80a** and **80b**, respectively. The former compound, **80a**, has served as a starting material for studies of benzopentalenylcyclopentadienyliron.[352–354]

[351] A. N. Nesmeyanov, N. A. Vol'kenau, and V. D. Vil'chevskaya, *Dokl. Akad. Nauk SSSR*, **111**, 362 (1956) [*C.A.*, **51**, 9600 (1957)].

[352] M. Cais, R. Raveh, and A. Modiano, *Israel J. Chem.*, **1**, 228 (1963).

[353] M. Cais, A. Modiano, N. Tirosh, and A. Eisenstadt, *Proc. Internat. Conf. Coord. Chem. 8th Conf.*, Vienna September 7, 1964, p. 229.

[354] M. Cais, A. Modiano, and A. Raveh, *J. Am. Chem. Soc.*, **87**, 5607 (1965).

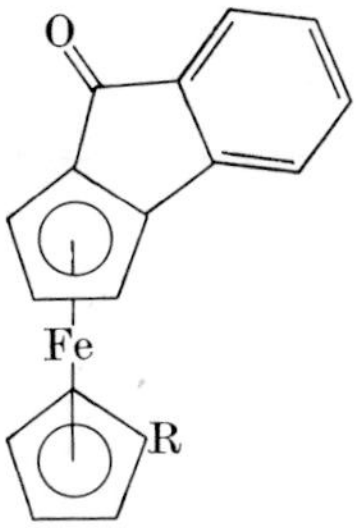

80a, b
a: R = H
b: R = C_6H_5

Cyclization of *o*-carboxyphenylacetylferrocene in polyphosphoric acid has been reported to yield the cyclic ketone **81**;[224, 355] however, later investigation revealed it to be the ferrocenylisocoumarin **81a.**[355]

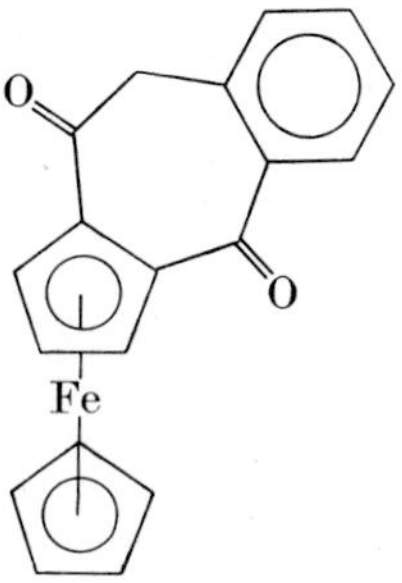

81 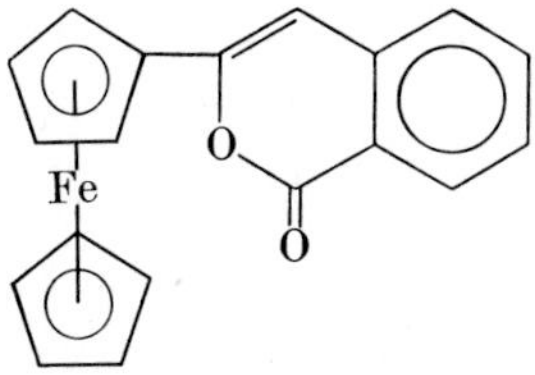 **81a**

When *o*-ferrocenoylbenzoic acid[344, 356] or its methyl ester[344] was treated with sulfuric acid, the product was reported to be 2,3-ferroçonaphthoquinone; however, further investigation has shown the product to have the structure **82**.[357]

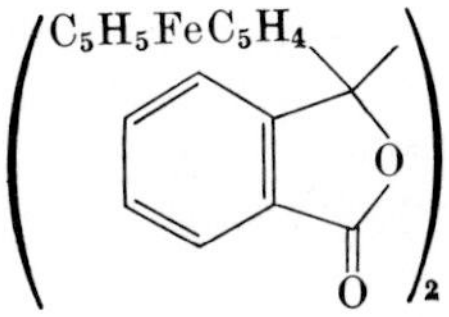

$C_5H_5FeC_5H_4$

82

Some mono- and di-ferroçocarbocylic compounds have been prepared by routes not involving substitution reactions on ferrocene. Thus

[355] J. Boichard, *Compt. Rend.* **253,** 2702 (1961).
[356] W. B. Hardy and E. Klingsberg, U.S. Pat. 2,900,401 [*C.A.*, **54,** 569 (1960)].
[357] A. N. Nesmeyanov, V. D. Vil'chevskaya, and N. S. Kochetkova, *Dokl. Akad. Nauk SSSR*, **138,** 390 (1961) [*C.A.*, **55,** 21080 (1961)].

benzoferrocene was prepared directly from the sodium salts of cyclopentadiene and indene by treatment with ferrous chloride (see Table IV).[358, 359] When bis(indenyl)iron is treated with sodium cyclopentadienide, ligand exchange occurs, giving benzoferrocene (4%) and ferrocene (34%).[359] This reaction is said to account for the lack of formation of bis(indenyl)iron during the preparation of benzoferrocene as mentioned above. Catalytic reduction of benzoferrocene leads to 1,2-tetramethyleneferrocene.[358, 359]

The reaction of azulene with lithium aluminum hydride followed by addition of ferrous chloride gave the fused-ring compound **15** in 17% yield.[88] Bis(indenyl)iron and substituted bis(indenyl)iron compounds prepared from indenes are included in Table II. Catalytic reduction of bis(indenyl)iron produces bis(tetrahydroindenyl)iron [1,2;1′,2′-bis(tetramethylene)ferrocene];[360] similar reduction of **15** gives 1,2;1′,2′-bis-(pentamethylene)ferrocene.[88]

Ferroçocarbocyclic compounds are listed in Table XXVII.

Ferrocenyl Heterocycles and Ferroço Heterocyclic Compounds

The first reported ferroço heterocyclic compound, in which the cyclopentadienyl and heterocyclic rings are fused, was N-methyltetrahydropyridoferrocene **(83)**, prepared from β-ferrocenylethylamine, formaldehyde, and formic acid.[260, 361] Similarly, Bischler-Napieralski ring closure of N-acetyl- or N-formyl-β-ferrocenylethylamine gives the corresponding dihydropyridoferrocenes (e.g., **84**),[362, 363] which are reduced by lithium aluminum hydride to the corresponding tetrahydropyridoferrocenes.[362, 363]

Numerous heterocyclic compounds containing the ferrocene nucleus have been reported; this discussion is limited to those which contain a ferrocenyl substituent bonded directly to the heterocyclic ring. Reaction of ferrocenyllithium with pyridine or quinoline produces the 2-ferrocenyl substituted compound (e.g., **85**) in yields of 24–32% and 3%, respectively.[364, 365] The three isomeric (methylferrocenyl)pyridines formed by reaction of methylferrocenyllithium and pyridine are also known.[365] Catalytic reduction of 2-ferrocenylpyridine leads to 2-ferrocenylpiperidine

[358] R. B. King and M. B. Bisnette, *Angew. Chem.*, **75**, 642 (1963).

[359] R. B. King and M. B. Bisnette, *Inorg. Chem.*, **3**, 796 (1964).

[360] E. O. Fischer and D. Seus, *Z. Naturforsch.*, **9b**, 386 (1954).

[361] J. M. Osgerby and P. L. Pauson, *Chem. Ind.* (*London*), 196 (1958).

[362] J. M. Osgerby and P. L. Pauson, *J. Chem. Soc.*, 4600 (1961).

[363] J. M. Osgerby and P. L. Pauson, *Chem. Ind.* (*London*), 1144 (1958).

[364] A. N. Nesmeyanov, V. A. Sazonova, and A. V. Gerasimenko, *Dokl. Akad. Nauk SSSR* **147**, 634 (1962) [*C.A.*, **58**, 9133 (1963)].

[365] K. Schlögl and M. Fried, *Monatsh. Chem.*, **94**, 537 (1963).

CH₂CH₂NH₂ ferrocene $\xrightarrow[HCO_2H]{CH_2O}$ **83**

CH₂CH₂NHCHO ferrocene — Bischler-Napieralski → **84** — 1. Alkylation 2. Reduction → **83**

(86) in 59% yield.[365] Diazotization of 3-aminopyridine followed by reaction with ferrocene gives 3-ferrocenylpyridine in 27% yield.[365]

$$C_5H_5FeC_5H_4Li \xrightarrow{C_5H_5N} C_5H_5FeC_5H_4(C_5H_4N)\ \mathbf{85} \xrightarrow{H_2} C_5H_5FeC_5H_4(C_5H_9NH)\ \mathbf{86}$$

Reaction of 1,2-diferrocenoylethane with hydrazine produces 3,6-diferrocenyl-4,5-dihydropyridazine.[366] The formation of 2-ferrocenylchromanone **(87)** proceeds as shown. 3-Ferrocenylisocoumarin **(81a)** is also known.[355] The resolution of certain optically active ferroço heterocyclic compounds has recently been reported.[56]

$$C_5H_5FeC_5H_4CHO + CH_3CO(C_6H_4)OH \rightarrow C_5H_5FeC_5H_4CH{=}CHCO(C_6H_4)OH \xrightarrow[Heat]{CH_3CO_2Na} \mathbf{87}$$

[366] N. Sugiyama and T. Teitei, *Bull. Chem. Soc. Japan*, **35**, 1423 (1962).

A larger number of five-membered heterocyclic compounds bearing a ferrocenyl substituent has been reported. Representative syntheses of this family are shown in the accompanying equations.

$$C_5H_5FeC_5H_4COC{\equiv}CH \xrightarrow{H_2NOH} C_5H_5FeC_5H_4\text{(N, O ring)}$$

$$C_5H_5FeC_5H_4COCH_2COR \xrightarrow{R'NHNH_2} C_5H_5FeC_5H_4\text{(R; N, NR' ring)}$$

Ferrocenyl and ferroço heterocyclic compounds and their methods of synthesis are listed in Table XXVIII.

No compounds are known which contain heterocyclic rings fused to both ferrocene rings. The known disubstituted compounds (listed at the end of Table XXVIII) are all ferrocenyl substituted heterocyclic structures, prepared either by reaction with dilithioferrocene or by addition of hydrazine to bis-(β-diketone) derivatives of ferrocene.

Nitro and Azo Ferrocenes

Nitroferrocene (**88**) has never been obtained by electrophilic nitration of ferrocene. The ease with which ferrocene undergoes oxidation to ferricenium salts precludes the use of the nitrating agents common in benzene chemistry, since these are also strongly oxidizing. The synthesis of nitroferrocene has been achieved, however, by two routes: the reaction of lithioferrocene in diethyl ether at $-70°$ with dinitrogen tetroxide[367] or *n*-propyl nitrate.[368] Reduction of nitroferrocene by chemical[367, 368] or with electrolytic[369] processes gives aminoferrocene.

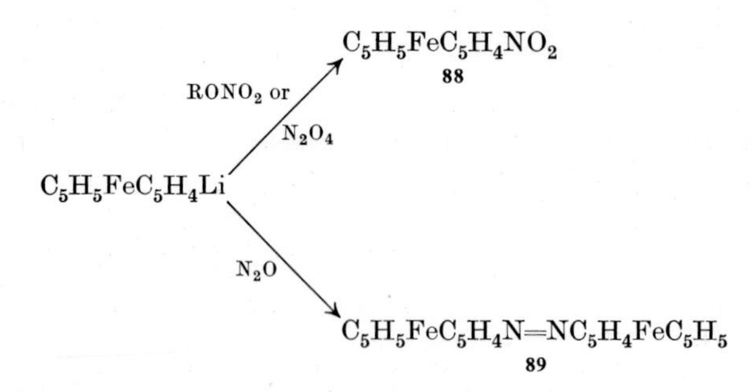

[367] J. F. Helling and H. Shechter, *Chem. Ind.* (*London*), 1157 (1959).
[368] H. Grubert and K. L. Rinehart, Jr., *Tetrahedron Letters*, No. 12, 16 (1959).
[369] A. M. Hartley and R. E. Visco, *Anal. Chem.*, **35**, 1871 (1963).

When lithioferrocene is treated with nitrous oxide, azoferrocene **(89)** can be isolated in 25–30% yield.[370, 371] Air oxidation of aminoferrocene in the presence of cuprous bromide provides azoferrocene in yields up to 99%.[116] Treatment of azoferrocene with sulfuric or hydrochloric acid gave aminoferrocene in yields of 10–44%,[371] and its reduction by catalytic hydrogenation[370] or zinc in sodium hydroxide[371] gave aminoferrocene in yields of 23–76%. Attempts to reduce azoferrocene to hydrazoferrocene have been unsuccessful. The synthesis of phenylazoferrocene has been achieved directly from phenylazocyclopentadiene and cyclopentadiene in the presence of an iron salt and a base, as mentioned in Table IV. It is also formed, in addition to phenylazoxyferrocene, from aminoferrocene and azobenzene.[372] Catalytic reduction of phenylazoferrocene by hydrogen in methanol gives phenylhydrazoferrocene, while reduction in acetic acid gives aminoferrocene.[373]

Dinitroferrocene has not been reported, but syntheses of two bis-(arylazo) ferrocenes have been reported utilizing the appropriately substituted azocyclopentadienes and an iron source (see Table II). 1,1′-Bis(phenylazo)ferrocene is reduced catalytically to either bis(phenylhydrazo)ferrocene[373] or 1,1′-diaminoferrocene.[373, 374]

Halo Ferrocenes

Although the ease with which the iron atom in ferrocene is oxidized from Fe(II) to Fe(III) precludes the formation of halo ferrocenes by halogenation of ferrocene, these compounds have been synthesized by displacement reactions on either ferrocenylmercuric chloride or ferroceneboronic acid. The reaction of halogens with diferrocenylmercury (Fig. 2) or ferrocenylmercuric chloride (Fig. 3)[110, 114, 286, 375, 376] is of value only for the preparation of iodoferrocene; bromoferrocene has been made by this route, but the yields are low.[375] The reaction of N-halosuccinimides with ferrocenylmercuric chloride (Fig. 2) is useful for synthesis of both bromo- and iodo-ferrocene.[120] The second method, the reaction of ferroceneboronic acid with cuprous halides (Fig. 1),[108, 109, 114] is useful for the preparation of chloro- and bromo-ferrocene.

[370] A. N. Nesmeyanov, E. G. Perevalova, and T. V. Nikitina, *Tetrahedron Letters*, No. 1, 1 (1960).

[371] A. N. Nesmeyanov, E. G. Perevalova, and T. V. Nikitina, *Dokl. Akad. Nauk SSSR*, **138,** 1118 (1961) [*C.A.*, **55,** 24707 (1961)].

[372] A. N. Nesmeyanov, T. V. Nikitina, and E. G. Perevalova, *Izv. Akad. Nauk SSSR, Ser. Khim.*, 197 (1964) [*C.A.*, **60,** 9310 (1964)].

[373] G. R. Knox, *Proc. Chem. Soc.*, 56 (1959).

[374] G. R. Knox and P. L. Pauson, *J. Chem. Soc.*, 4615 (1961).

[375] A. N. Nesmeyanov, E. G. Perevalova, and O. A. Nesmeyanova, *Dokl. Akad. Nauk SSSR*, **100,** 1099 (1955) [*C.A.*, **50,** 2558 (1956)].

[376] M. D. Rausch, *J. Org. Chem.*, **26,** 3579 (1961).

Halo ferrocenes undergo the usual reactions characteristic of an aryl halide. Cyanation of chloroferrocene has been shown to occur exclusively in the unsubstituted ring.[169, 170] Chloroferrocene and bromoferrocene have been acetylated with acetyl chloride and aluminum chloride.[188, 189] Both chloro- and bromo-ferrocene have been converted to acetoxyferrocene with cupric acetate,[109, 112, 188, 297] to the nitrile with cuprous cyanide,[297] and to diferrocenyl by the Ullmann reaction.[296] Bromoferrocene reacts with a number of sodium and potassium tetraarylboron compounds to form the corresponding aryl ferrocenes.[268] It has also been converted to N-ferrocenylphthalimide,[188, 297] to ferrocenyl azide,[377] and to a variety of sulfones[379] and amines.[116] Substituted bromoferrocenes have been shown to react similarly.[115, 305] Iodoferrocene has been converted to diferrocenyl[122, 286–288] and to various ferrocenyl aryl ethers and sulfides.[288, 380] Both bromo- and iodo-ferrocene form Grignard reagents.[110] Recent studies of the reaction between chloroferrocene and certain organolithium compounds have demonstrated the possible existence of a ferrocyne intermediate.[381]

The dihalo ferrocenes in which both halogens are identical are prepared from bis(chloromercuri)ferrocene or ferrocenylene-1,1′-diboronic acid, while mixed halides are prepared from bis(haloferrocenyl)mercury or haloferrocenylmercuric chlorides. 1′-Iodochloroferrocene under Ullmann conditions gave an 88% yield of bis-(1′-chloroferrocenyl).[291] Similar treatment of a mixture of bromo- and dibromo-ferrocene gives 1,1′-diferrocenylferrocene (1,1′-terferrocenyl, **48**).[291]

Syntheses of mono- and di-halo ferrocenes are listed in Table XXIX.

Aminoferrocene

The most common synthesis of aminoferrocene **(90)** is the hydrogenolysis of an N-ferrocenylurethane.[204, 315, 317, 318] It may also be prepared by reaction of ferrocenyllithium with either O-benzylhydroxylamine[278] or methoxyamine (see equations on p. 57).[315]

Aminoferrocene is also obtained by the reduction of phenylazoferrocene (35–100%),[372–374] nitroferrocene,[367–369] azoferrocene (23–76%),[370, 371] ferrocenylazide (72%),[377] and phenylazoxyferrocene (65%).[372] Reaction of

[377] A. N. Nesmeyanov, V. N. Drozd, and V. A. Sazonova, *Dokl. Akad. Nauk SSSR*, **150,** 321 (1963) [*C.A.*, **59,** 5196 (1963)].

[378] A. N. Nesmeyanov, V. N. Drozd, and V. A. Sazonova, *Dokl. Akad. Nauk SSSR*, **150,** 102 (1963) [*C.A.*, **59,** 7558 (1963)].

[379] V. N. Drozd, V. A. Sazonova, and A. N. Nesmeyanov, *Dokl. Akad. Nauk SSSR*, **159,** 591 (1964) [*C.A.*, **62,** 7794 (1965)].

[380] M. D. Rausch, U.S. pat. 3,604,026 [*C.A.*, **58,** 10241 (1963)].

[381] J. W. Huffman, L. H. Keith, and R. L. Asbury, *J. Org. Chem.*, **30,** 1600 (1965).

[382] S. I. Goldberg and L. H. Keith, *J. Chem. Eng. Data*, **9,** 250 (1964).

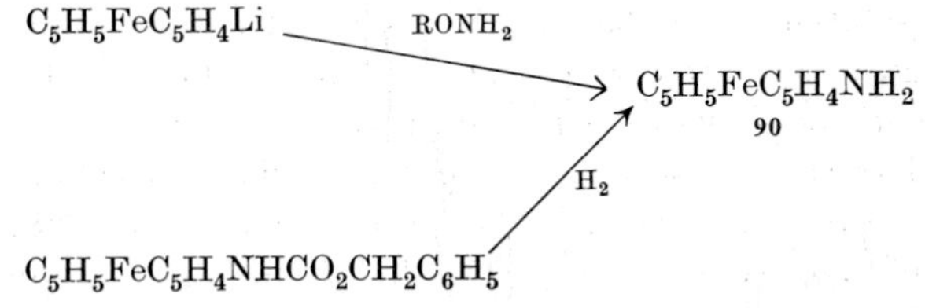

bromoferrocene with the copper salt of phthalimide leads to N-ferrocenylphthalimide which, upon cleavage by hydrazine, affords aminoferrocene in 82% yield.[109, 297]

N-Acylferrocenylamines are, in low yield, the products of the Schmidt reaction of acyl ferrocenes.[383]

Aminoferrocene is reported to decompose on standing but is stable as its N-acetyl derivative.[204, 278] N-Ethylation of **90** to the mono- and di-ethylated derivatives has been accomplished using ethyl fluoride in the presence of boron trifluoride etherate.[384] Reactions of **90** with isocyanates,[318] chloroformates,[189, 374] phosgene,[315, 318] acid chlorides,[204, 220, 278, 315] and aldehydes[158, 220] have been reported. In addition Diels-Alder adducts of N-ferrocenylmaleimide have been investigated.[385] The lack of a report of a diazonium salt is noteworthy.

The preparation of 1,1′-diaminoferrocene in 62% yield has been reported from reduction of ferrocenylene-1,1′-diazide[377] and from catalytic hydrogenation of 1,1′-bis(phenylazo)ferrocene.[373, 374] The material was reported to be unstable on standing and was not characterized except as its reaction products with methyl-, ethyl-, and phenyl-chlorocarbonate which produced stable carbamate derivatives.

Hydroxyferrocene and Its Derivatives

Hydroxyferrocene was first obtained from treatment of ferroceneboronic acid with cupric acetate, followed by alkaline hydrolysis of acetoxyferrocene (Fig. 1) to give hydroxyferrocene in 88% yield.[112, 113] The yield was only 60% when acetoxyferrocene was treated with phenylmagnesium bromide.[112] Cupric propionate has been used in place of cupric acetate, but with no apparent advantage.[112] Hydroxyferrocene has also been obtained in 30% yield by treatment of benzenediazaminoferrocene with sulfuric acid.[378] When benzenediazaminoferrocene was treated with hydrochloric acid followed by sodium 2-naphtholate, a 28% yield of ferrocenyl 2-naphthyl ether resulted.[378] Hydroxyferrocene, liberated

[383] M. Cais and N. Narkis, *J. Organometal. Chem.*, **3**, 188 (1965).

[384] A. N. Nesmeyanov, V. A. Sazonova, and V. I. Romanenko, *Dokl. Akad. Nauk SSSR*, **152**, 1358 (1963) [*C.A.*, **60**, 1793 (1964)].

[385] M. Furdik, S. Toma, and J. Suchy, *Chem. Zvesti*, **17**, 21 (1963) [*C.A.*, **59**, 10116 (1963)].

[386] M. R. Barusch and E. G. Lindstrom, U.S. pat. 2,834,796 [*C.A.*, **52**, 16366 (1958)].

from basic solutions by carbon dioxide, is unstable in air and decomposes on standing over periods of several days but is stable under nitrogen. Esters and ethers of hydroxyferrocene appear to be stable.

Acetoxyferrocene in alkali with dimethyl sulfate or chloroacetic acid gave the methyl ether (90%) or ferrocenyloxyacetic acid (82%).[109, 115] Reaction of hydroxyferrocene with allyl bromide in acetone in the presence of potassium carbonate gave an 84% yield of the allyl ether.[115] The allyl ether does not rearrange under the usual conditions for the Claisen rearrangement.

Substituted halo ferrocenes react with cupric acetate to give the corresponding substituted acetoxy ferrocenes. Thus 1′-acetoxyferrocenecarboxylic acid and its methyl ester[305] and also 1′-acetoxy-1-ethyl- and 1′-acetoxy-1-acetyl-ferrocene have been obtained from their respective bromo-substituted precursors.[188] In a double reaction, 1′-bromoferroceneboronic acid gives 1,1′-diacetoxyferrocene.[109] Other derivatives of 1,1′-dihydroxyferrocene reported are the dimethyl derivatives **9a** and **9b**, obtained by the reaction sequence on p. 10.[61, 82] The dibenzoate was hydrolyzed to the extremely air-sensitive dihydroxydimethylferrocene.

1,1′-Dihydroxyferrocene is also very sensitive to air oxidation.[109, 115] The compound was prepared by treatment of ferrocene-1,1′-diboronic acid or 1′-bromoferroceneboronic acid with cupric acetate to give 1,1′-diacetoxyferrocene (in yields of 40–83%) which was hydrolyzed to the free dihydroxy compound. Although the latter was unstable, it could be converted to the more stable dibenzoate, bis(benzene sulfonate), or bis(oxyacetic acid).

Ferrocenes Substituted with Phosphorus-, Sulfur-, Selenium-, Silicon-, Germanium-, or Arsenic-Containing Groups

The preparation and chemical properties of many types of substituted ferrocenes have been treated in earlier sections. While most of these ferrocenes contain only carbon atoms bonded directly to the ferrocene nucleus, a significant number of substituents containing heteroatoms bonded to ferrocene have been noted. They include the oxygen-, nitrogen- and halogen-containing substituents of the immediately preceding sections, the lithio, sodio, mercuri, and boronic acid derivatives useful as synthetic intermediates, and the ferrocenyl Grignard reagents (pp. 14–20, 56). This section describes substituted ferrocenes in which still other heteroatoms are bonded directly to a ferrocene ring.

Ferrocenylphosphorus Compounds. Of ferrocenes containing heteroatoms bonded to the nucleus, only compounds having a ferrocene-phosphorus bond or ferrocene-arsenic bond (see p. 60) have been prepared by substitution on ferrocene itself. The reaction of ferrocene with phenyldichlorophosphine or diphenylchlorophosphine in the presence of aluminum chloride leads to the formation of diferrocenylphenylphosphine and ferrocenyldiphenylphosphine, respectively.[387] Treatment of ferrocene with phosphorus trichloride in the presence of aluminum chloride gives triferrocenylphosphine oxide and diferrocenylphosphinic acid as the major products.[388] When ferrocene is treated with N,N-diethylphosphoramidous dichloride [$(C_2H_5)_2NPCl_2$] and aluminum chloride, triferrocenylphosphine and its oxide are obtained. Variation of the amounts of reactants leads to other phosphorus derivatives of ferrocene.[389] No derivatives have been reported in which two phosphorus atoms are bonded to the ferrocene molecule.

Ferrocenylsulfur Compounds. Compounds containing ferrocene bonded directly to a sulfur atom in lower oxidation states are obtained by three methods. The first is the reduction of ferrocenesulfonyl chloride (p. 44) to mercaptoferrocene and its subsequent reaction to give other products. The second is the reaction between a halo ferrocene (p. 56) and the sodium salt of an aryl thiol to produce aryl ferrocenyl sulfides. The third method (Fig. 3, p. 20) involves the reaction between diferrocenylmercury and various sulfur-containing reagents. References to these preparative methods are found in the appropriate sections above.

Ferrocenylselenium Compounds. Diferrocenyl selenide, the only selenium-containing ferrocene, is prepared from diferrocenylmercury (Fig. 3) and selenium tetrabromide in chloroform.[326]

Ferrocenylsilicon Compounds. Many derivatives of ferrocene have been prepared in which a silicon atom is bonded to the ferrocene ring. Some of them have already been mentioned (pp. 11, 47).

The preparation of triphenylsilylferrocene from ferrocenyllithium and triphenylchlorosilane[93] has been described. The reaction can be extended to include trimethylchlorosilane[118] and tri-(*n*-hexyl)chlorosilane,[103, 285] and to the use of sodioferrocene.[98]

The ferrocene-silicon bond is rather weak, and silicon substituents may often be removed by acidic hydrolysis.[390] Dimethylphenylsilylferrocene has been prepared by acidic hydrolysis of lithium (1′-phenyldimethylsilyl-1-ferrocenyl)dimethylsilanolate **(91)**.[391]

Treatment of 1,3-(1,1′-ferrocenylene)tetramethyldisiloxane with sodium

[387] G. P. Sollott, H. E. Mertwoy, S. Portnoy, and J. L. Snead, *J. Org. Chem.*, **28**, 1090 (1963).

[388] G. P. Sollott and E. Howard, Jr., *J. Org. Chem.*, **27**, 4034 (1962).

[389] G. P. Sollott and W. R. Peterson, Jr., *J. Organometal. Chem.*, **4**, 491 (1965).

[390] G. Marr and D. E. Webster, *J. Organometal. Chem.*, **2**, 99 (1964).

[391] P. T. Kan, C. T. Lenk, and R. L. Schaaf, *J. Org. Chem.*, **26**, 4038 (1961).

amide gives the intermediate **92** which, upon reaction with dimethylphenylchlorosilane followed by hydrolysis, gives 1-ferrocenyl-3-phenyltetramethyldisiloxane **(93)**.[391] When trimethylsilylferrocene is treated with alcoholic acid, ferrocene results.[390]

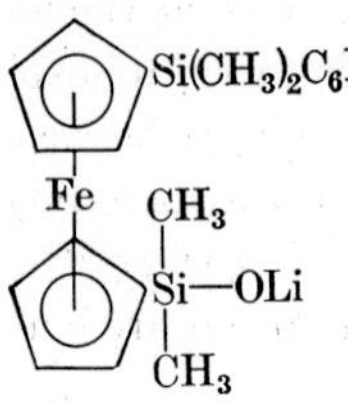

$Si(CH_3)_2C_6H_5$ / CH_3 / Fe / Si—OLi / CH_3

91

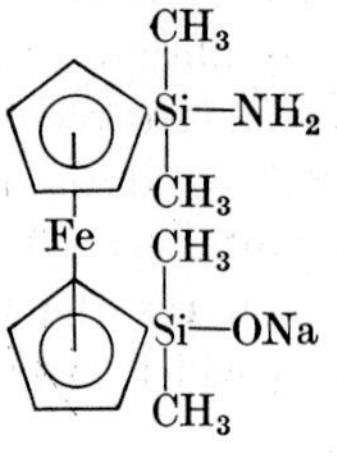

CH_3 / Si—NH_2 / CH_3 / Fe / CH_3 / Si—ONa / CH_3

92

CH_3 CH_3 / Si—O—Si—C_6H_5 / CH_3 CH_3 / Fe

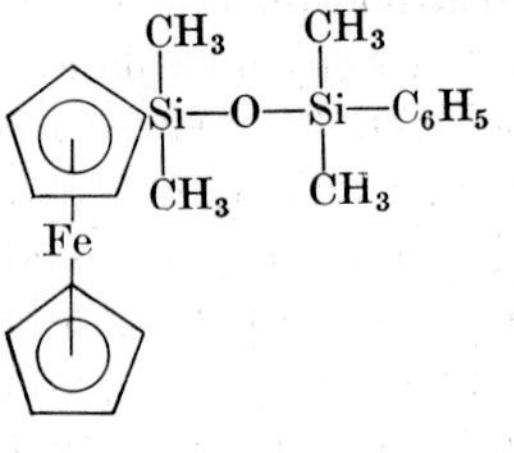

93

Numerous disubstituted ferrocenes have been prepared in which the substituent is silicon. However, the majority of these compounds were prepared directly from the appropriately substituted cyclopentadiene and ferrous chloride and are found in Table II-A. The other derivatives that have been reported were made by the reaction of the appropriately substituted silyl chloride with either 1,1′-disodio- or 1,1′-dilithio-ferrocene.

Ferrocenylgermanium Compounds. Triphenylgermylferrocene, the only monosubstituted ferrocenylgermanium compound reported, was prepared from sodioferrocene or 1,1′-ferrocenylenedimagnesium bromide by reaction with triphenylbromogermane.[98] 1,1′-Bis(triphenylgermyl)-ferrocene has been prepared by three routes. One of these, utilizing triphenylgermylcyclopentadiene and ferrous chloride,[98] is listed in Table II. Another method, treatment of a mixture of mono- and di-sodioferrocene with triphenylgermanium bromide, gives the material in 13% yield.[98] The last method, the reaction between 1,1′-ferrocenylenedimagnesium bromide and triphenylbromogermane, produces the material in unstated yield.[98]

Ferrocenylarsenic Compounds. Treatment of ferrocene with arsenic(III) chloride in the presence of aluminum chloride gives arsenosoferrocene **(94)** in 22% yield.[392] Treatment of this material with hydrochloric acid converts it to ferrocenyldichloroarsine (95%) in a reversible

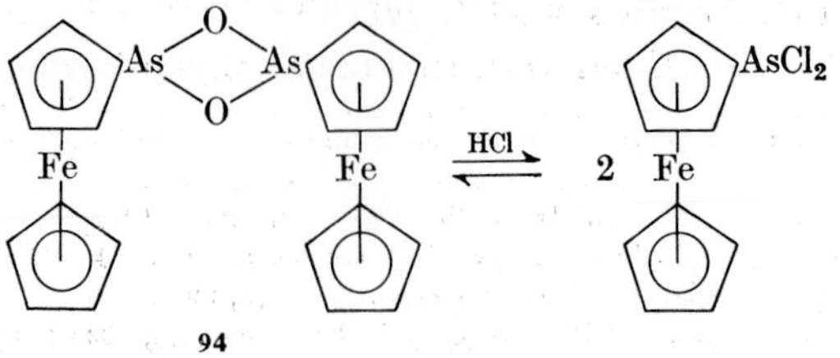

O / As As / O / Fe Fe $\underset{}{\overset{HCl}{\rightleftharpoons}}$ 2 Fe / $AsCl_2$

94

[392] G. P. Sollott and W. R. Peterson, Jr., *J. Org. Chem.*, **30**, 389 (1965).

reaction. The conversion of these products to other derivatives is the only route reported for the preparation of other ferrocenylarsenic compounds.

The metalated ferrocenes discussed in this section are shown in Table XXX.

Azaferrocene (π-Cyclopentadienyl-π-pyrrolyliron)

Azaferrocene (**95**) has been prepared by two methods. The more useful of the two involves the reaction of potassiumpyrrole with cyclopentadienyliron dicarbonyl iodide[393] in benzene solution to provide the desired compound in 22% yield.[394] This method has also been applied successfully to 2,4- and 2,5-dimethylpyrrole to give the dimethyl derivatives in 20% and 32% yield, respectively.

The second method of preparation involves the reaction of a mixture of sodiumpyrrole and cyclopentadienylsodium with ferrous chloride.[359] The yield is only 0.85% and no substituted pyrroles have been shown to react.

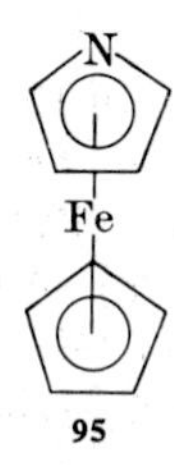

95

Ferrocene-Containing Polymers

Numerous polymers have been prepared in which a ferrocene is the repeating unit. They can be grouped into three general structural categories: (a) polymethylene- and polyalkyl-ferrocenylenes; (b) polyferrocenyleneamides; (c) ferrocenylene-containing polyurethanes and polyureas.

When ferrocene is allowed to react with an aldehyde in the presence of a catalyst such as sulfuric acid or zinc chloride, polymers of the type **96** are formed.[396–399] The same products are obtained when the correspondingly

[393] R. B. King and F. G. A. Stone, *Inorg. Syn.*, **8,** 110 (1963).
[394] K. K. Joshi, P. L. Pauson, A. R. Qazi, and W. H. Stubbs, *J. Organometal. Chem.*, **1,** 471 (1964).
[395] Imperial Chemical Industries Ltd., Fr. pat. 79,755 (Addn. to 1,305,312) [*C.A.*, **59,** 11567 (1963)].
[396] H. Valot, *Compt. Rend.*, **258,** 5870 (1964).
[397] E. W. Neuse, *Nature*, **204,** 179 (1964).
[398] E. W. Neuse, K. Koda, and E. Carter, *Macromol. Chem.*, **84,** 213 (1965).
[399] E. W. Neuse and K. Koda, *J. Organometal. Chem.*, **4,** 475 (1965).

substituted ferrocenylcarbinols are polymerized.[400–402] Infrared analysis shows that the polymer contains all three possible substitution arrangements (1,2; 1,3; and 1,1′).[400] Polyferrocenylenemethylenes of type **97** have been obtained in unstated yields by treatment of the corresponding α,ω-bis(cyclopentadienyl)alkanes with a base and ferrous chloride.[334] Treatment of N,N-dimethylaminomethylferrocene with zinc chloride has produced polymers of type **96** (R = H).[403–405]

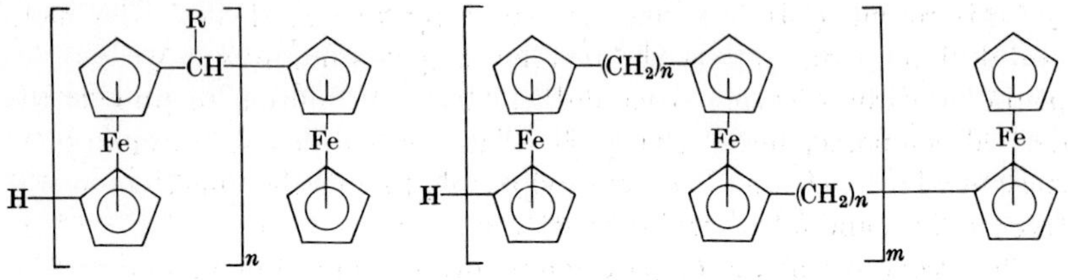

96

(R = H, CH_3, C_6H_5, 2-C_4H_3O)

97

Polyferrocenylalkylenes have been obtained by treating alkyl-substituted ferrocenes with di-*t*-butyl peroxide.[406–409] When ferrocene itself is heated with di-*t*-butyl peroxide in the presence of haloaromatic compounds such as *p*-dichlorobenzene or 1-bromonaphthalene, polymers are formed which incorporate the haloaryl moiety as well.[407, 408] When the polymerization of ferrocene is carried out in acetone, four isopropylidene groups are incorporated per ferrocene molecule.[408]

Homopolymerization of acetylferrocene with zinc chloride at 200° gives polyferrocenylacetylene,[408] and when 1,1′-diacetylferrocene is treated with zinc chloride at 180° the resulting polymer is crosslinked by 1,1′-ferrocenylene groups.[408] *trans*-Cinnamoylferrocene does not form a homopolymer in the presence of azoisobutyronitrile or benzoyl peroxide, but it can be copolymerized with a variety of monomers.[410] In contrast

[400] E. W. Neuse and D. S. Trifan, *J. Am. Chem. Soc.*, **85**, 1952 (1963).

[401] A. Wende and H. J. Lorkowski, *Plaste Kautschuk*, **18**, 32 (1963) [*C.A.*, **59**, 4050 (1963)].

[402] E. W. Neuse and E. Quo, *Bull. Chem. Soc. Japan*, **38**, 931 (1965).

[403] E. W. Neuse and E. Quo, *Nature*, **205**, 494 (1965).

[404] E. W. Neuse and E. Quo, *J. Polymer Sci.*, **3A**, 1499 (1965).

[405] E. W. Neuse, E. Quo, and W. G. Howells, *J. Am. Chem. Soc.*, **30**, 4071 (1965).

[406] A. N. Nesmeyanov, V. V. Korshak, V. V. Voevodski, N. S. Kochetkova, S. L. Sosin, R. B. Materikova, T. N. Bolotnikova, V. M. Chibrikin, and N. M. Bazhin, *Dokl. Akad. Nauk SSSR*, **137**, 1370 (1961) [*C.A.*, **55**, 21081 (1961)].

[407] Ya. M. Paushkin, T. P. Vishnyakova, I. I. Patalakh, T. A. Sokolinskaya, and F. F. Machus, *Dokl. Akad. Nauk SSSR*, **149**, 856 (1963) [*C.A.*, **59**, 4049 (1963)].

[408] I. M. Paushkin, L. S. Polak, T. P. Vishnyakova, I. I. Patalakh, F. F. Machus, and T. A. Sokolinskaya, *J. Polymer Sci.*, **1C**, 1481 (1963)

[409] S. L. Sosin, V. V. Korshak, and V. P. Alekseeva, *Dokl. Akad. Nauk SSSR*, **149**, 327 (1963) [*C.A.*, **59**, 4049 (1963).]

[410] L. E. Coleman, Jr., and M. D. Rausch, *J. Polymer Sci.*, **28**, 207 (1958).

to cinnamoylferrocene, vinylferrocene has been shown to form a homopolymer as well as copolymers with other monomers.[204] Crosslinking of these polymers has been accomplished by carrying out the reaction in the presence of formaldehyde.[204]

The product obtained from diazotized poly-(*p*-aminostyrene) and ferricenium ion is poly-(*p*-ferricenyliumstyrene), in which the iron atom is in the +3 oxidation state.[411] It has been proposed for use as an oxidation-reduction resin.

The polyferrocenylene amides have been prepared by standard methods from diphenylferrocene-1,1′-dicarboxylate,[412] or 1,1′-bis(chlorocarbonyl)-ferrocene[105, 320] and a variety of amines.

Polyurethanes containing ferrocenylene groups have been prepared from various 1,1′-di-(α-hydroxyalkyl)ferrocenes and diisocyanates or ferrocene-1,1′-diisocyanate and a variety of diols.[105] Reaction of ferrocene diisocyanate with diamines leads to the formation of polyureas.[105] Silicon-containing ferrocene polymers have also been reported.[413, 414]

Ligand Exchange Reactions

An early investigation of the reaction of ferrocene, 1,2-dichloroethane, and aluminum chloride yielded a product (originally described as "pentaethanodiferrocene")[415] which, later investigations demonstrated, could be obtained from ferrocene and aluminum chloride in the absence of 1,2-dichloroethane.[416–418] Several products are formed; one was recently assigned the structure **100**[418, 419] and it was suggested that the material was formed from **99**. The precursor **99** has been observed previously

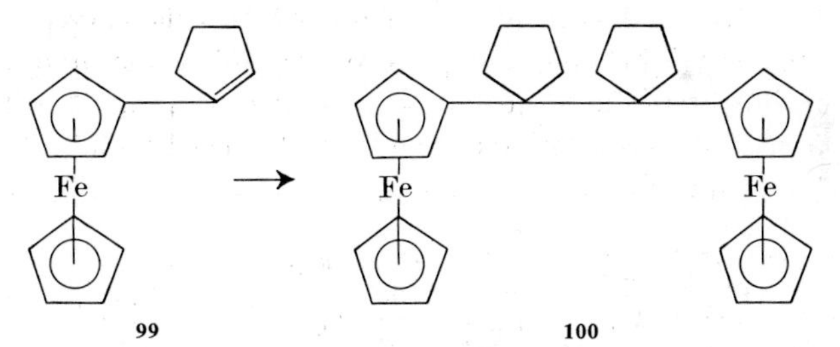

[411] B. Sansoni and O. Sigmund, *Angew. Chem.*, **73,** 299 (1961).

[412] L. Plummer and C. S. Marvel, *J. Polymer Sci.*, **2A.,** 2559 (1964).

[413] E. V. Wilkus and W. H. Rauscher, *J. Org. Chem.*, **30,** 2889 (1965).

[414] G. Greber and M. L. Hallensleben, *Makromol. Chem.*, **83,** 148 (1965).

[415] A. N. Nesmeyanov and N. S. Kochetkova, *Dokl. Akad. Nauk SSSR*, **126,** 307 (1959) [*C.A.* **53,** 21856 (1959)].

[416] S. I. Goldberg *J. Am. Chem. Soc.*, **84,** 3022 (1962).

[417] S. G. Cottis and H. Rosenberg, *Chem. Ind.* (*London*), 860 (1963).

[418] S. G. Cottis and H. Rosenberg, *J. Polymer Sci.*, **2B,** 295 (1964).

[419] A. N. Nesmeyanov, N. S. Kochetkova, P. V. Petrovskii, and E. I. Fedin, *Dokl. Akad. Nauk SSSR*, **152,** 875 (1963) [*C.A.*, **60,** 6367 (1964)].

in the reaction between ferrocene and hydrogen fluoride at 100°.[164] Although these reactions are not true ligand exchanges, they demonstrate the ability of Lewis acids to transform ferrocene to substituted derivatives.

True ligand exchange reactions of ferrocene have been observed more recently in the presence of basic and Lewis acid catalysts.[239, 359, 420–422] In addition, thermal exchange has been reported.[423] When ferrocene is heated in benzene solution in the presence of aluminum chloride, the cyclopentadienyliron(II)benzene cation **(101)** is formed. It may be isolated as the tetraphenylborate salt.[420–422] When acetyl- and ethylferrocene are used, the major products are acetylcyclopentadienyliron(II)benzene and cyclopentadienyliron(II)benzene cations, respectively. Other benzenoid compounds such as tetralin and mesitylene give similar compounds.

Fe $\xrightarrow[C_6H_6]{AlCl_3}$ [Fe]$^{\oplus}$

101

When the reaction is carried out in a solvent incapable of forming a stable complex like **101**, the reaction follows a different course. Thus, when ethylferrocene is treated with aluminum chloride in methylene chloride, analysis of the reaction mixture indicates the presence of ferrocene, ethylferrocene, and 1,1′-diethylferrocene.[239] This reaction has been shown to be an equilibrium, and proof of ring migration as opposed to alkyl group migration has been presented since no product with a 1,3 orientation was observed to have been formed when 1,2-diethylferrocene **(102)** was rearranged.

2 Fe (C_2H_5, C_2H_5) $\underset{CH_2Cl_2}{\overset{AlCl_3}{\rightleftharpoons}}$ Fe (C_2H_5, C_2H_5; C_2H_5, C_2H_5) + Fe

102

[420] A. N. Nesmeyanov, N. A. Vol'kenau, and I. N. Bolesova, *Dokl. Akad. Nauk SSSR*, **149,** 615 (1963) [*C.A.*, **59,** 6438 (1963)].

[421] A. N. Nesmeyanov, N. A. Vol'kenau, and I. N. Bolesova, *Tetrahedron Letters*, 1725 (1963).

[422] A. N. Nesmeyanov, N. A. Vol'kenau, and L. S. Shilovtseva, *Dokl. Akad. Nauk SSSR*, **160,** 1327 (1965) [*C.A.*, **62,** 14722 (1965)].

[423] H. P. Fritz and L. Schaefer, *Z. Naturforsch.*, **19b,** 169 (1964).

Base-catalyzed ring migration occurs when *bis*(indenyl)iron is treated with cyclopentadienylsodium in tetrahydrofuran; the products formed are ferrocene and benzoferrocene.[359] The equilibrium has been shown to be to the right in boiling tetrahydrofuran; as noted above (p. 52), this accounts for the lack of formation of *bis*(indenyl)iron in the preparation of benzoferrocene.

When ferrocene and hexachlorocyclopentadiene are heated together at 120°, a product of the empirical formula $C_5H_5Cl_2Fe$ is formed.[423] Its structure has not been conclusively proved, but it is known to be dimeric, ligand exchange of chloride for cyclopentadiene having occurred. Thus in this case the order of stability is the opposite of that observed for ferrocene versus dicyclopentadienyldiiron tetracarbonyl. The latter compound gives ferrocene and iron carbonyl on heating.

RUTHENOCENE

Ruthenocene (dicyclopentadienylruthenium, **103**) was first prepared, in 20% yield, by the reaction of cyclopentadienylmagnesium bromide with ruthenium(III) acetylacetonate.[424] Later its synthesis from anhydrous ruthenium(III) chloride and cyclopentadienylsodium in 1,2-dimethoxyethane was reported.[425, 426] When a mixture of anhydrous ruthenium(III) chloride and metallic ruthenium was employed in the reaction with

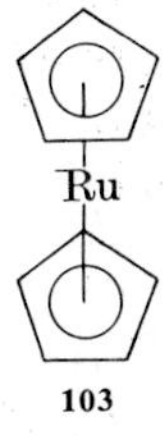

103

[424] G. Wilkinson, *J. Am. Chem. Soc.*, **74,** 6146 (1952).
[425] E. O. Fischer, M. D. Rausch, and H. Grubert, *Chem. Ind.* (*London*), 765 (1958).
[426] E. O. Fischer and H. Grubert, *Chem. Ber.*, **92,** 2302 (1959).

cyclopentadienylsodium,[427] yields of 56–69% based on the total ruthenium available were realized. Labeled ruthenocene (^{106}Ru) has been prepared in a carrier-free state in unspecified yields.[428]

Ruthenocene is somewhat more stable thermally than ferrocene and undergoes electrophilic substitution less readily. Acylation gives both mono- and di-substitution products under conditions where ferrocene gives disubstitution.[425, 429] A homoannular dibenzoylruthenocene, obtained in unspecified amounts, has been assigned the 1,2 structure on the basis of nuclear magnetic resonance studies.[187, 430, 431]

A number of substituents other than acetyl have also been introduced. Ruthenocenecarboxaldehyde has been obtained both by formylation using N-methylformanilide and phosphorus oxychloride (9%)[187] and by acid-catalyzed decomposition of phenylruthenocenylcarbinyl azide (7%). The acid-catalyzed decomposition of the azide gave, in addition to the aldehyde, 5% of 2,3-ruthenoçoindenone **(104)** along with the other usual products of the reaction.[187]

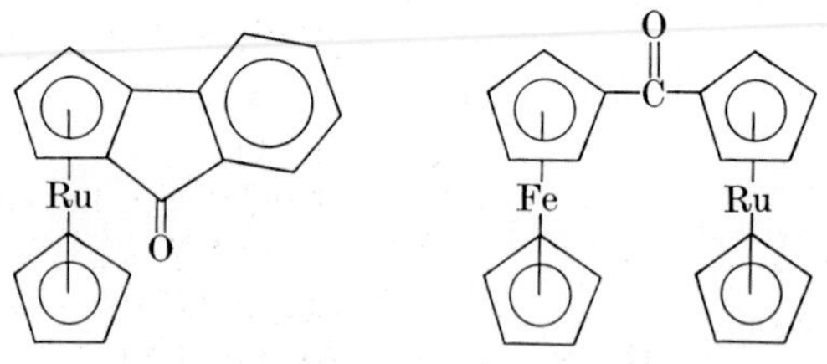

Reaction of ruthenocene with phenyl isocyanate in the presence of aluminum chloride gave N-phenylruthenocenecarboxamide in 20% yield,[429] metalation with *n*-butyllithium followed by carbonation gave a mixture of the mono- and di-carboxylic acids,[429] mercuric acetate in acetic acid gave the mono- and di-mercurated products.[429] Acylation of ruthenocene with ferrocenoyl chloride in the presence of aluminum chloride yielded ferrocenyl ruthenocenyl ketone **(105)**.

Like ferrocenes, substituted ruthenocenes undergo the usual reactions characteristic of their organic substituents. For example, vinylruthenocene can be prepared in 47% over-all yield by the reduction of acetylruthenocene to the corresponding alcohol followed by dehydration with alumina.[255] Similarly, acetyl- and benzoyl-ruthenocene give ethyl- and

[427] D. E. Bublitz, W. E. McEwen, and J. Kleinberg, *Org. Syntheses*, **41**, 96 (1961).

[428] H. Götte and M. Wenzel, W. German pat. 1,049,860 and 1,059,452 [*C.A.*, **55**, 2685, 6495 (1961)].

[429] M. D. Rausch, E. O. Fischer, and H. Grubert, *J. Am. Chem. Soc.*, **82**, 76 (1960).

[430] D. E. Bublitz, J. Kleinberg, and W. E. McEwen, *Chem. Ind.* (*London*), 936 (1960).

[431] M. D. Rausch and V. Mark, *J. Org. Chem.*, **28**, 3225 (1963).

benzyl-ruthenocene upon reduction with lithium aluminum hydride-aluminum chloride.[432] Ruthenocene compounds which have been prepared by substitution reactions on ruthenocene are presented in Table XXXI.

Bis(indenyl)ruthenium has been prepared in 12% yield from indenylsodium and a mixture of ruthenium(III) chloride and ruthenium metal,[434] and in unstated yield from indenylmagnesium bromide and ruthenium(III) acetylacetonate.[433] Studies of its magnetic and physical properties have also been reported.[434–437] Catalytic reduction to bis(tetrahydroindenyl)-ruthenium proceeds in nearly quantitative yield.[434]

OSMOCENE

The preferred synthesis of osmocene (dicyclopentadienylosmium) involves treatment of cyclopentadienylsodium with osmium(IV) chloride in 1,2-dimethoxyethane to obtain the desired material in 22–23% yield.[425, 426] The same reaction in tetrahydrofuran gave the product in 10% yield.[438] The reaction of cyclopentadienyl Grignard reagent with osmium chloride, apparently in a mixture of oxidation states, gave osmocene in poor yield.[14, 439]

Osmocene has been shown to undergo Friedel-Crafts acylation to give acetyl- and benzoyl-osmocene.[425, 429, 438, 440] Whereas ferrocene and ruthenocene give disubstitution or mixtures of mono- and di-substitution products, osmocene gives only a monoacylated product, even under the forcing conditions of high temperature and use of a large excess of acylating reagent.[429] A recently reported exception is 1,1′-diacetylosmocene, which has been obtained under vigorous conditions; however, the yield was only 0.1%.[431] Other comparisons in this metal series have also been reported.[187, 255, 390] Metalation with *n*-butyllithium followed by carbonation led to a mixture of the mono- and di-carboxylic acids of osmocene.[429] Reaction of osmocenyllithium with trimethylchlorosilane gave trimethylsilylosmocene in unstated yield.[390]

Substituted osmocenes, like the other metallocenes, undergo reactions characteristic of the substituents. Thus acetylosmocene can be converted

[432] V. Mark and M. D. Rausch, *Inorg. Chem.*, **3,** 1067 (1964).

[433] M. Dub, *Organometallic Compounds Literature Survey 1937–1958,* Vol. **1,** 193, Springer-Verlag (1961).

[434] J. H. Osiecki, C. J. Hoffman, and D. P. Hollis, *J. Organometal. Chem.*, **3,** 107 (1965).

[435] E. O. Fischer, *Rec. Trav. Chim.*, **75,** 629 (1956).

[436] E. O. Fischer and H. Leipfinger, *Z. Naturforsch.*, **10b,** 353 (1955).

[437] E. O. Fischer and U. Piesbergen, *Z. Naturforsch.*, **11b,** 758 (1956).

[438] E. A. Hill and J. H. Richards, *J. Am. Chem. Soc.*, **83,** 3840 (1961).

[439] Ascertained at an earlier date by D. E. Bublitz in a personal communication from E. R. Lippincott.

[440] R. Riemschneider, *Monatsh. Chem.*, **90,** 658 (1959).

to vinylosmocene by reduction followed by dehydration of the resulting carbinol with alumina[255] or to ethylosmocene by reduction with lithium aluminum hydride-aluminum chloride.[432] Although benzoylosmocene can be converted to osmocenylphenylcarbinylazide, treatment of this azide with concentrated sulfuric acid gives benzoylosmocene as the only isolable organometallic compound.[187] Osmocene compounds prepared by substitution on osmocene are presented in Table XXXI.

CYCLOPENTADIENYLMANGANESE TRICARBONYL

Synthesis of the Parent Compound

Cyclopentadienylmanganese tricarbonyl (**106**) is, in some respects, simpler than ferrocene since it contains only one cyclopentadienyl ring. It can be prepared by the reaction of bis(cyclopentadienyl)manganese with carbon monoxide under various conditions of elevated temperature and pressure[441–446] in yields up to 80%.[443] The bis(cyclopentadienyl)-manganese may be previously prepared or generated *in situ*, the latter approach being preferred. Cyclopentadienylmanganese tricarbonyl has also been prepared by reaction of dicyclopentadiene with manganese

$C_5H_5^{\ominus}\,Na^{\oplus} \xrightarrow{MnCl_2} (C_5H_5)_2Mn \xrightarrow[\text{Heat}]{CO} C_5H_5Mn(CO)_3$

106

pentacarbonyl,[447] by reaction of cyclopentadienylsodium with manganese pentacarbonyl bromide,[448] and by reaction of bis(cyclopentadienyl)manganese with iron pentacarbonyl.[449]

Direct Synthesis of Cyclopentadienylmanganese Tricarbonyl Derivatives

Several substituted cyclopentadienylmanganese tricarbonyl derivatives have been prepared directly from the appropriately substituted

[441] T. S. Piper, F. A. Cotton, and G. Wilkinson, *J. Inorg. Nucl. Chem.*, **1**, 165 (1955).
[442] J. E. Brown and H. Shapiro, U.S. pat. 2,818,417 [*C.A.*, **52**, 8535 (1958)].
[443] E. O. Fischer and K. Plesske, *Chem. Ber.*, **91**, 2719 (1958).
[444] D. C. Freeman, Jr., Brit. pat. 858,442 [*C.A.*, **56**, 3514 (1962)].
[445] E. O. Fischer and R. Jira, *Z. Naturforsch.*, **9b**, 618 (1954).
[446] H. Shapiro, U.S. pat. 2,916,504 [*C.A.*, **54**, 15402 (1960)].
[447] Ethyl Corp., Brit. pat. 782,738 [*C.A.*, **52**, 3851 (1958)].
[448] J. Kozikowski, U.S. pat. 3,015,668 [*C.A.*, **57**, 866 (1962)].
[449] J. Kozikowski and M. L. Larson, U.S. pat. 2,870,180 [*C.A.*, **53**, 11407 (1959)].

bis(cyclopentadienyl)manganese compounds. Others have been obtained from the substituted cyclopentadiene and a mixture of manganese(II) chloride and carbon monoxide or a complex manganese carbonyl.[446, 448]

A unique synthesis of π-dihydropentalenylmanganese tricarbonyl **(107)** resulted from the tetramerization of acetylene in the presence of dimanganese decacarbonyl at elevated temperature and pressure; the yield was 40%.[450, 451] Methylcyclopentadienylmanganese tricarbonyl was suggested as a possible intermediate in the formation of **107** for, when

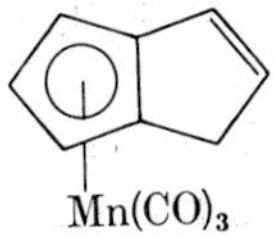

$Mn(CO)_3$

107

the former was heated at elevated temperatures with acetylene at 600 p.s.i., the dihydropentalene resulted in 27% yield. The dihydropentalene derivative was also formed from cyclooctatetraene and dimanganese decacarbonyl in unspecified yields.

The compounds which have been prepared via the direct synthetic route are listed in Table XXXII.

Routes to Substituted Cyclopentadienylmanganese Tricarbonyl Derivatives via Substitution Reactions

Metalation

Cyclopentadienylmanganese tricarbonyl and its methyl homolog both reacted with mercuric acetate in the presence of alcoholic calcium chloride[452] to give the monochloromercurated derivative in 37% yield. In absolute ethanol, in addition to the monosubstituted compounds, bischloromercurated derivatives of undesignated structures were obtained in 14% and 18% yields, respectively.[452] The chloromercurated derivative of cyclopentadienylmanganese tricarbonyl has also been obtained by treatment of the sulfinic acid with mercuric chloride in ethanol; the yield was 80%.[453] Chloromercuricyclopentadienylmanganese tricarbonyl reacted with sodium thiosulfate in aqueous acetone to give bis(cyclopentadienylmanganese tricarbonyl)mercury in 78% yield[452] and gave the same compound with n-butyllithium.[453] Chloromercuricyclopentadienylmanganese tricarbonyl[452, 453] has also been used for the preparation of

[450] T. H. Coffield, K. J. Ihrman, and W. Burns, *J. Am. Chem. Soc.*, **82**, 4209 (1960).

[451] T. H. Coffield, K. J. Ihrman, and W. Burns, *J. Am. Chem. Soc.*, **82**, 1251 (1960).

[452] A. N. Nesmeyanov, K. N. Anisimov, and Z. P. Valueva, *Izv. Akad. Nauk SSSR, Otd. Khim. Nauk*, 1683 (1962) [*C.A.*, **58**, 4590 (1963)].

[453] M. Cais and J. Kozikowski, *J. Am. Chem. Soc.*, **82**, 5667 (1960).

iodocyclopentadienylmanganese tricarbonyl (96% yield) by treatment with iodine in refluxing carbon tetrachloride.[452]

Acylation

Cyclopentadienylmanganese tricarbonyl undergoes the usual Friedel-Crafts type reactions with acid chlorides and anhydrides.[443, 454] Acetylcyclopentadienylmanganese tricarbonyl[443, 454–456] and benzoylcyclopentadienylmanganese tricarbonyl[443, 455–457] have been obtained and undergo the usual reactions characteristic of the functional groups present. Their ability to undergo pinacol formation is noteworthy.[458, 459] Acylation by methyl chlorothiolformate gives the corresponding methyl thiolester of the metallocene.[308]

Treatment of cyclopentadienylmanganese tricarbonyl with succinic or glutaric anhydride in the presence of aluminum chloride gave the β-metallocenoylpropionic acid[443, 460] and the γ-metallocenoylbutyric acid.[460] Clemmensen reduction of these acids provided the metallocenylbutyric and -valeric acids which were cyclized with polyphosphoric acid to the cyclic ketones **108** and **109** in 70% and 85% yields.[460] The absolute configuration of the cyclic ketone **108** has recently been determined.[461]

The aldehyde of cyclopentadienylmanganese tricarbonyl, prepared in 81% yield by lithium aluminum tri-(*t*-butoxy)hydride reduction of the corresponding acid chloride, has been converted to the β-metallocenylpropionic acid via the corresponding acrylic acid. When the propionic acid was treated with polyphosphoric acid, the homocyclic ketone **110** was obtained in 65% yield.[450, 462] Reduction followed by dehydration led to the dihydropentalenyl derivative **107**.[462]

The carboxylic acid chloride of cyclopentadienylmanganese tricarbonyl reacted with cyclopentadienylmanganese tricarbonyl in the presence of aluminum chloride to give the dimetallocenyl ketone **111** in 28% yield,[463] and with ferrocene to give the mixed dimetallocenyl ketone **112** in unstated yield.[463] The latter material was further acylated by the carboxylic

[454] F. A. Cotton and J. B. Leto, *Chem. Ind. (London)*, 1368 (1958).

[455] A. N. Nesmeyanov, K. N. Anisimov, and Z. P. Valueva, *Izv. Akad. Nauk SSSR, Otd. Khim. Nauk*, 1780 (1961) [*C.A.*, **56**, 8733 (1962)].

[456] E. O. Fischer and K. Plesske, Brit. pat. 864,834 [*C.A.*, **55**, 22338 (1961)].

[457] J. Kozikowski, R. E. Maginn, and M. S. Klove, *J. Am. Chem. Soc.*, **81**, 2995 (1959).

[458] E. Cuingnet and M. Adalberon, *Compt. Rend.*, **257**, 461 (1963).

[459] E. Cuingnet and M. Tarterat-Adalberon, *Bull. Soc. Chim. France*, 3721 (1965).

[460] M. Cais and A. Modiano, *Chem. Ind. (London)*, 202 (1960).

[461] S. G. Cottis, H. Falk, and K. Schlögl, *Tetrahedron Letters*, 2857 (1965).

[462] M. Cais, A. Modiano, N. Tirosh, A. Eisenstadt, and A. Rubinstein, *Intern. Union Pure Appl. Chem., Abstr. A*, 166 (1963) (London).

[463] M. Cais and M. Feldkimel, *Tetrahedron Letters*, 440 (1961).

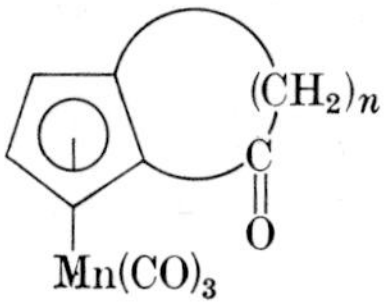

acid chloride of cyclopentadienylmanganese tricarbonyl to give the dimetallocenoyl ferrocene **113** in 82% yield.[463]

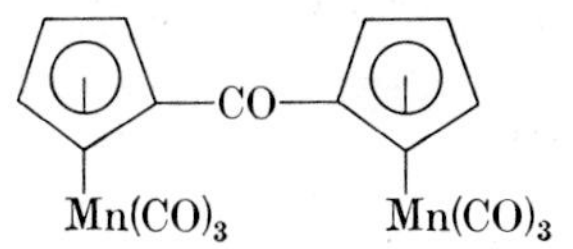

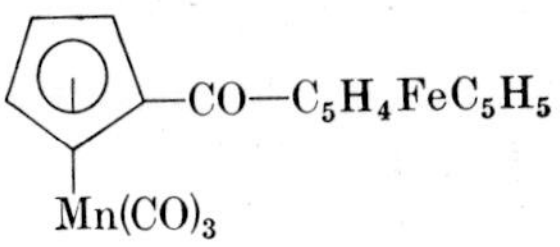

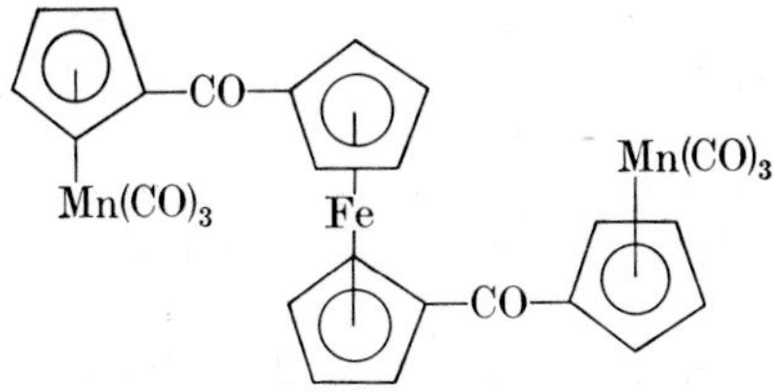

Methylcyclopentadienylmanganese tricarbonyl reacts similarly with acetyl chloride,[464, 465] benzoyl chloride,[457, 459, 464, 465] *p*-chlorobenzoylchloride,[466] succinyl chloride,[466] and phthalyl chloride,[466] except that a mixture of isomers is obtained in each case. Some of these mixture have been separated and positional assignments for certain isomers have been made.[467–470]

Alkylation

Alkylation of cyclopentadienylmanganese tricarbonyl has been studied, but not so thoroughly as alkylation of ferrocene. Ethyl bromide and aluminum chloride gave a 63% yield of ethylcyclopentadienylmanganese tricarbonyl,[455] while cyclohexyl chloride gave a mixture of mono- and disubstituted products[457] and isobutylene a mixture of mono-, di, and tri-*t*-butylated products. When *t*-butyl chloride was used as the alkylating agent, a 69% yield of *t*-butylcyclopentadienylmanganese tricarbonyl was obtained.[471] Repeated acetylation and reduction to the corresponding ethyl compound have led to pentaethylcyclopentadienylmanganese

[464] R. Riemschneider and H. G. Kassahn *Z. Naturforsch.* **14b**, 348 (1959).
[465] J. Kozikowski, U.S. pat. 2,916,503 [*C.A.*, **54**, 5693 (1960)].
[466] R. Riemschneider and H. G. Kassahn, *Chem. Ber.*, **92**, 3208 (1959).
[467] R. Riemschneider, K. Petzoldt, and W. Herrmann, *Z. Naturforsch.*, **16b**, 279 (1961).
[468] R. Riemschneider and W. Herrmann, *Ann. Chem.*, **648**, 68 (1961).
[469] R. Riemschneider and K. Petzoldt, *Z. Naturforsch.*, **15b**, 627 (1960).
[470] E. Cuingnet and M. Adalberon, *Compt. Rend.*, **258**, 3053 (1964).
[471] K. N. Anisimov and N. E. Kolobova, *Izv. Akad. Nauk SSSR, Otd. Khim.* 721 (1962) [*C.A.*, **57**, 15135 (1962)].

tricarbonyl.[472] Chloromethylation of the parent compound **106** has been effected by means of hydrogen chloride, zinc chloride, and formaldehyde[473] or bis(chloromethyl) ether.[474] The chloromethylcyclopentadienylmanganese tricarbonyl thus obtained is a versatile intermediate to many other derivatives.[475]

Alkylation reactions of methylcyclopentadienylmanganese tricarbonyl have also been reported.[465]

Miscellaneous Reactions

Although formylcyclopentadienylmanganese tricarbonyl has been prepared, as mentioned above, via the acid chloride, its synthesis has not been effected by substitution. However, formylation of the methyl homolog was accomplished by means of hydrogen cyanide, hydrogen chloride, and aluminum chloride.[476]

Sulfonation with sulfuric acid (100%) in the presence of acetic anhydride and *p*-toluidine at 100° gave a 93% yield of the corresponding cyclopentadienylmanganese tricarbonyl sulfotoluidide **114**, from which a host of other sulfonic esters and amides have been prepared.[453]

$$C_5H_5Mn(CO)_3 + H_2SO_4 + H_2NC_6H_4CH_3\text{-}p \xrightarrow{(CH_3CO)_2O} p\text{-}CH_3C_6H_4NHO_2S\text{-}C_5H_4Mn(CO)_3$$

114

Aminocyclopentadienylmanganese tricarbonyl has been synthesized by two routes, Curtius degradation of the acylazide[383, 477] and the Schmidt reaction applied to the acetyl or benzoyl derivative of the parent metallocene.[383, 478, 479] The best yield (64%) is obtained by the Curtius method.

A series of halo derivatives has recently been synthesized from aminocyclopentadienylmanganese tricarbonyl via the diazonium salt intermediate[480] which, in contrast to that of ferrocene, is stable.

[472] A. N. Nesmeyanov, K. N. Anisimov, N. E. Kolobova, and I. B. Zlotina, *Izv. Akad. Nauk SSSR, Ser. Khim.*, 1326 (1964) [*C.A.*, **61**, 12024 (1964)].

[473] A. N. Nesmeyanov, K. N. Anisimov, and Z. P. Valueva, *Izv. Akad. Nauk SSSR, Ser. Khim.*, 2233 (1963) [*C.A.*, **60**, 9303 (1964)].

[474] A. N. Nesmeyanov, K. N. Anisimov, and Z. P. Valueva, *Dokl. Akad. Nauk SSSR*, **157**, 622 (1964) [*C.A.*, **61**, 9519 (1964)].

[475] A. N. Nesmeyanov, K. N. Anisimov, and Z. P. Valueva, *Dokl. Akad. Nauk SSSR*, **162**, 112 (1965) [*C.A.*, **63**, 5666 (1965)].

[476] R. Riemschneider and K. Petzoldt, *Z. Naturforsch.*, **17b**, 785 (1962).

[477] E. Cuingnet and M. Adalberon, *Compt. Rend.*, **257**, 181 (1963).

[478] E. Cuingnet and M. Adalberon, *Compt. Rend.*, **258**, 5884 (1964).

[479] E. Cuingnet and M. Tarterat-Adalberon, *Bull. Soc. Chim. France*, 3734 (1964).

[480] M. Cais and N. Narkis, *J. Organometal. Chem.*, **3**, 269 (1965).

Ethynylcyclopentadienylmanganese tricarbonyl has been prepared by the dehydrohalogenation of α-chlorovinylcyclopentadienylmanganese tricarbonyl.[262] Presumably other acylcyclopentadienylmanganese tricarbonyl derivatives may serve as precursors for the latter compounds and subsequently be converted to other ethynyl derivatives.

Derivatives of cyclopentadienylmanganese tricarbonyl prepared via substitution reactions are listed in Table XXXIII.

The ability of the cyclopentadienylmanganese tricarbonyl nucleus to stabilize an α-carbonium ion was found to be less than that of ferrocene.[481] Deuterium exchange reactions for the protons of cyclopentadienylmanganese tricarbonyl have been investigated, the rate being comparable to that of benzene.[482]

Azacyclopentadienylmanganese Tricarbonyl (π-Pyrrolylmanganese Tricarbonyl)

Azacyclopentadienylmanganese tricarbonyl has been prepared in unstated yield by reaction of pyrrole with dimanganese decacarbonyl in 1,2-dimethoxyethane at 130°,[494] and in 66% yield by the reaction between the potassium salt of pyrrole and dimanganese decacarbonyl in ligroin.[394] The latter reaction has also been successfully applied to mono-, di-, and tetra-methyl pyrroles.

Substitution reactions of azacyclopentadienylmanganese tricarbonyl have not been reported.

Ligand Exchange Reactions

Numerous photochemical substitution reactions have been reported for cyclopentadienylmanganese tricarbonyl and its homologs that involve not the cyclopentadienyl ring but the carbonyl ligands.[483–492] Under appropriate conditions the reaction of cyclopentadienylmanganese

[481] W. E. McEwen, J. A. Manning, and J. Kleinberg, *Tetrahedron Letters*, 2195 (1964).

[482] A. N. Nesmeyanov, D. N. Kursanov, V. N. Setkina, N. V. Kislyakova, N. S. Kochetkova, and R. B. Materikova, *Dokl. Akad. Nauk SSSR*, **143**, 351 (1962) [*C.A.*, **57**, 5938 (1963)].

[483] W. Strohmeier and K. Gerlach, *Z. Naturforsch.*, **15b**, 675 (1960).

[484] W. Strohmeier and D. Von Hobe, *Z. Naturforsch.*, **16b**, 402 (1961).

[485] W. Strohmeier and C. Barbeau, *Z. Naturforsch.*, **17b**, 848 (1962).

[486] W. Strohmeier, H. Laporte, and D. von Hobe, *Chem. Ber.*, **95**, 455 (1962).

[487] G. E. Schroll, U.S. pat. 3,054,740 [*C.A.*, **58**, 5723 (1963)].

[488] R. S. Nyholm, S. S. Sandhu, and M. H. B. Stiddard, *J. Chem. Soc.*, 5916 (1963).

[489] W. Strohmeier, J. F. Guttenberger, and H. Hellman, *Z. Naturforsch.*, **19b**, 353 (1964).

[490] W. Strohmeier and J. F. Guttenberger, *Chem. Ber.*, **96**, 2112 (1963).

[491] W. Strohmeier and J. F. Guttenberger, *Chem. Ber.*, **97**, 1256 (1964).

[492] W. Strohmeier and J. F. Guttenberger, *Chem. Ber.*, **97**, 1871 (1964).

tricarbonyl with 1,2-bis(diphenylphosphino)ethane can give either **115** or **116**.[488] Compounds which have been prepared by a photochemically

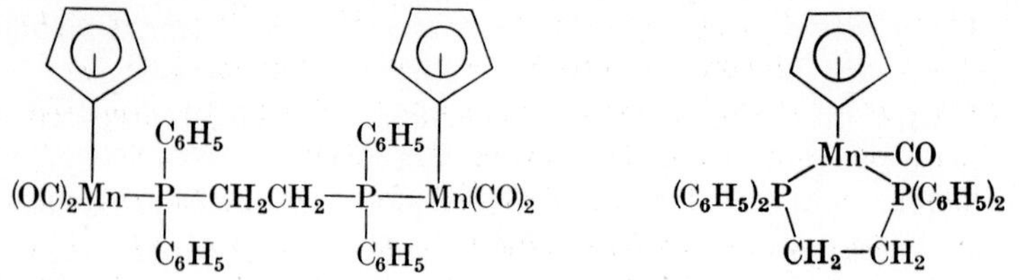

$(OC)_2Mn-P(C_6H_5)_2-CH_2CH_2-P(C_6H_5)_2-Mn(CO)_2$ (each Mn bearing a cyclopentadienyl ring)

115

$(C_6H_5)_2P$, $P(C_6H_5)_2$, $Mn-CO$, CH_2-CH_2

116

induced ligand exchange reaction of cyclopentadienylmanganese tricarbonyl are presented in Table XXXIV.

When cyclopentadienylmanganese tricarbonyl is allowed to react with phosphorus trichloride in the presence of aluminum chloride, the binuclear reaction product **117** is formed.[493]

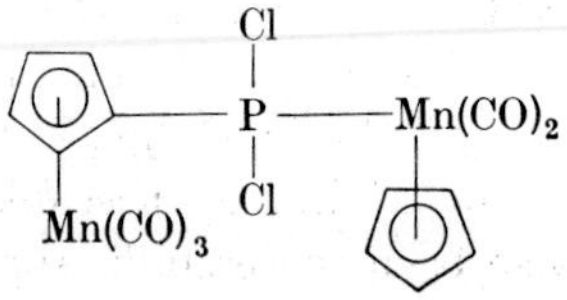

Cl, P, Cl, $Mn(CO)_2$, $Mn(CO)_3$

117

CYCLOPENTADIENYLRHENIUM TRICARBONYL

Recent studies have shown that cyclopentadienylrhenium tricarbonyl, like the related manganese compound, is aromatic. The parent compound has been prepared from both dicyclopentadiene and rhenium pentacarbonyl (50% yield)[495] and from sodium cyclopentadienide and rheniumpentacarbonyl chloride (60–89% yield).[496, 497] Cyclopentadienylthallium has also been employed in the latter reaction.[496] Use of the simple rhenium(III) chloride with cyclopentadienylsodium gave only 16% of the desired compound.[499] When indenylsodium and rheniumpentacarbonyl chloride were utilized, an 80% yield of indenylrhenium tricarbonyl was obtained.[497]

[493] A. N. Nesmeyanov, K. N. Anisimov, and Z. P. Valueva, *Izv. Akad. Nauk SSSR, Ser. Khim.*, 763 (1964) [*C.A.*, **61**, 3132 (1964)].

[494] K. K. Joshi and P. L. Pauson, *Proc. Chem. Soc.*, 326 (1962).

[495] M. L. H. Green and G. Wilkinson, *J. Chem. Soc.*, 4314 (1958).

[496] A. N. Nesmeyanov, K. N. Anisimov, N. E. Kolobova, and L. I. Baryshnikov, *Izv. Akad. Nauk SSSR, Otd. Khim. Nauk*, 193 (1963) [*C.A.*, **58**, 12588 (1963)].

[497] E. O. Fischer and W. Fellman, *J. Organometal. Chem.*, **1**, 191 (1964).

[498] A. N. Nesmeyanov, K. N. Anisimova, N. E. Kolobova, and L. I. Baryshnikov, *Dokl. Akad. Nauk SSSR*, **154**, 646 (1964) [*C.A.*, **60**, 10713 (1964)].

[499] R. L. Pruett and E. L. Morehouse, *Chem. Ind.* (*London*), 980 (1958).

Substitution reactions carried out on cyclopentadienylrhenium tricarbonyl include acylation, sulfonation, and metalation. The usual Friedel-Crafts conditions gave the acetyl and benzoyl derivatives in 43–67% and 41–61% yields, respectively.[497, 498] Sulfonation in acetic anhydride-sulfuric acid provided the corresponding cyclopentadienylrhenium tricarbonyl sulfonic acid, isolated as the *p*-toluidine salt,[500] while both the mono- and di-chloromercuri derivatives were obtained by treatment with alcoholic mercuric acetate followed by a source of chloride ion.[500] The yields of mercuration and sulfonation reaction products were unstated. Deuterium exchange reactions for the protons of cyclopentadienylrhenium tricarbonyl have been investigated, the rate being comparable to that of the manganese analog.[501]

CYCLOPENTADIENYLTECHNETIUM TRICARBONYL

Cyclopentadienyltechnetium tricarbonyl has been prepared in 86% yield from technetium(IV) chloride, carbon monoxide, and cyclopentadienylsodium in the presence of copper powder[502] and has been shown to be capable of undergoing substitution reactions similar to those of the related manganese and rhenium compounds. Thus benzoylcyclopentadienyltechnetium tricarbonyl is obtained in 50% yield by the usual acylation procedure.[497]

Further studies should reveal a parallel between the degree of reactivity of the group VIIB metal cyclopentadienyl tricarbonyls and that of the similarly related group VIII metallocenes of iron, ruthenium, and osmium.

CYCLOPENTADIENYLVANADIUM TETRACARBONYL

Cyclopentadienylvanadium tetracarbonyl **(118)** has been prepared, in yields up to 94%, by treatment of bis(cyclopentadienyl)vanadium with carbon monoxide under pressure.[503, 504] A standard preparation reports over-all yields of 15–25% from cyclopentadienylsodium and vanadium(III) chloride.[505] Acetylation and propionylation have been reported.[506, 507]

[500] A. N. Nesmeyanov, N. E. Kolobova, K. N. Anisimov, and L. I. Baryshnikov, *Izv. Akad. Nauk SSSR, Ser. Khim.*, 1135 (1964) [*C.A.*, **61**, 7034 (1964)].

[501] A. N. Nesmeyanov, D. N. Kursanov, V. N. Setkina, N. V. Kislyakova, N. E. Kolobova, K. N. Anisimov, *Izv. Akad. Nauk SSSR, Ser. Khim.*, 762 (1965) [*C.A.*, **63**, 3669 (1965)].

[502] C. Palm, E. O. Fischer, F. Baumgärtner, *Naturwiss.*, **49**, 279 (1962).

[503] E. O. Fischer and W. Hafner, *Z. Naturforsch.*, **9b**, 503 (1954).

[504] R. L. Pruett in W. I. Jolly, *Preparative Inorganic Reactions*, Vol. 2, p. 187, Interscience, New York, 1965.

[505] R. B. King and F. G. A. Stone, *Inorg. Syn.*, **7**, 100 (1963).

[506] F. Calderazzo, G. Rebaudo, and R. Ercoli, *Chim. Ind.* (*Milan*), **42**, 52 (1960).

[507] R. Riemschneider, O. Goehring, and M. Krüger, *Monatsh. Chem.*, **91**, 305 (1960).

The best yields (65%) of acetylcyclopentadienylvanadium tetracarbonyl were obtained using acetic anhydride-aluminum chloride in methylene chloride.[508] With the exception of methylcyclopentadienyl-, 1-acetyl-*x*-methylcyclopentadienyl-, and 1-propionyl-*x*-methylcyclopentadienylvanadium tetracarbonyl,[507] all prepared in unspecified yields, the derivatives mentioned above are the only ones reported for this compound.

$V(CO)_4$

118

CYCLOPENTADIENYLCHROMIUM NITROSYL DICARBONYL

Cyclopentadienylchromium nitrosyl dicarbonyl was first prepared, in 80% yield, from bis(cyclopentadienylchromium tricarbonyl) and nitric oxide at room temperature.[509] Its ability to undergo electrophilic substitution was demonstrated by its acetylation under the usual conditions to give a 75% yield of acetylcyclopentadienylchromium nitrosyl dicarbonyl **(119)**.

$COCH_3$

$ONCr(CO)_2$

119

EXPERIMENTAL PROCEDURES

Ferrocene.[74, 75] The detailed procedure in *Organic Syntheses* is also useful for the preparation of many 1,1′-disubstituted compounds from the correspondingly substituted cyclopentadienes. The use of sodium hydride dispersed in mineral oil obviates the necessity of dispersing the sodium or cutting it into small chips.[167] Sodium hydride dispersion is available from Metal Hydrides, Inc., Beverly, Massachusetts.

Mixed Lithiated Ferrocenes.[93] A solution of 5.0 g. (0.027 mole) of ferrocene in 75 ml. of anhydrous diethyl ether is placed in an oven-dried three-necked flask fitted with a stirrer, a dropping funnel, and a water-cooled condenser, and 0.08 mole of commercial *n*-butyllithium is added dropwise with stirring. The stirring is stopped after 1 hour, and the mixture is allowed to stand for 24 hours under an atmosphere of nitrogen.

Analysis by carbonation to the corresponding acids indicates a 35% conversion to metalated products, consisting of 70% mono- and 30%

[508] E. O. Fischer and K. Plesske, *Chem. Ber.*, **93**, 1006 (1960).

[509] E. O. Fischer and K. Plesske, *Chem. Ber.*, **94**, 93 (1961).

di-lithioferrocene. The mixture of mono- and di-lithioferrocene can be converted to any of a number of useful compounds.

Ferrocenyllithium.[98] A Schlenk tube containing 1.77 g. (4.2 mmoles) of chloromercuriferrocene is evacuated and filled with argon. Diethyl ether (9 ml.) and ethyllithium (9.3 mmoles, 9.2 ml. of 1.01 *M* solution) in ether are added in succession at 27°. The mixture is stirred for 45 minutes, then cooled to 0°. Analysis by carbonation to ferrocenecarboxylic acid indicates a 64% yield of monolithioferrocene.

1,1′-Dilithioferrocene.[98] To 2.76 g. (4.2 mmoles) of 1,1′-bis-(chloromercuri)ferrocene in 9 ml. of diethyl ether is added 18.1 mmoles of 0.99 *M* ethyllithium solution. The reaction mixture is stirred at room temperature for 1 hour. Analysis by carbonation indicates a 44% yield of 1,1′-dilithioferrocene, while characterization of the lithium reagent solution with trimethylchlorosilane gives 1,1′-bis(trimethylsilyl)ferrocene in 36% yield and trimethylsilylferrocene in 8% yield.

Ferroceneboronic Acid and 1,1′-Ferrocenediboronic Acid.[110] A mixture of lithiated ferrocenes is prepared in 220 ml. of tetrahydrofuran-diethyl ether (1:1 by volume) from *n*-butyllithium (~0.27 mole) and 16.7 g. (0.09 mole) of ferrocene. The mixture is filtered through glass wool and then added dropwise in 2 hours to 72.5 g. (0.315 mole) of tri-*n*-butyl borate in 50 ml. of diethyl ether at −70°. A solid forms. The mixture is warmed during 1.5 hours to room temperature, decomposed with 100 ml. of 10% aqueous sodium hydroxide, and filtered. The ether solution is extracted nine times with 10% aqueous sodium hydroxide (total volume 400 ml.). Acidification of the basic solution with 10% sulfuric acid at 0° gives a yellow precipitate which is washed with water. Soxhlet extraction of the precipitate with diethyl ether for 4 days removes 9.02 g. (44%) of ferroceneboronic acid, which is obtained as a yellow powder after evaporation of solvent; m.p. 136–140° dec. Remaining as an insoluble yellow powder from the Soxhlet extraction is 4.42 g. (18%) of 1,1′-ferrocenediboronic acid, dec. ~200°.

The ether solution which has been washed with alkali is concentrated; filtration and vacuum sublimation of the precipitate gives 4.93 g. (29%) of crude ferrocene.

Chloromercuriferrocene.[119] This well-documented procedure gives both mono- and di-mercurated derivatives, which are separable.

N,N-Dimethylaminomethylferrocene.[127] This derivative is one of the most useful. The procedure described in *Organic Syntheses* has also been applied to aminomethylation of alkyl ferrocenes.

Ferrocenecarboxaldehyde.[140] Ferrocene (11.16 g., 0.06 mole) is added in small portions over a period of 15 minutes to a vigorously stirred solution prepared from 21.6 g. (0.16 mole) of N-methylformanilide and

15.3 g. (0.10 mole) of phosphorus oxychloride. The purple viscous mixture is stirred for 1 hour at room temperature and then at 65–70° for 2 hours in a nitrogen atmosphere. The mixture is then cooled to 0°, and 50 g. of sodium acetate dissolved in 400 ml. of water is added and the solution stirred overnight. The mixture is extracted twice with 400 ml. of diethyl ether, and the ether extracts are combined and washed successively with 1 *N* HCl, water, saturated sodium bicarbonate solution, and finally water, all saturated with sodium chloride. The ether extract is concentrated to 50 ml. and shaken with a cold solution of 40 g. of sodium bisulfite in 100 ml. of water. The bisulfite addition compound is collected, washed with cold saturated aqueous sodium bisulfite then with ether, and is finally dried (yield 15.40 g.).

The bisulfite addition compound is taken up in 200 ml. of 2 *N* sodium hydroxide, and the liberated aldehyde is extracted into ether. The combined ether extract is washed with saturated sodium chloride solution, then dried over magnesium sulfate. Removal of solvent leaves 10.2 g. of dark purple crystals of the aldehyde (80%).

Acetylferrocene. *Phosphoric Acid Method.*[139] A mixture of 93 g. (0.5 mole) of ferrocene, 250 ml. of acetic anhydride, and 20 ml. of 85% phosphoric acid is heated at 100° for 10 minutes.* The reaction mixture is cooled slightly and poured onto ice. After standing overnight, the mixture is neutralized with 200 g. of sodium carbonate monohydrate in 200 ml. of water. The resulting brown pasty mass is cooled in an ice bath and filtered. The tan product is washed four times with 100-ml. portions of water and filtered. The granular product is dried in a vacuum desiccator over phosphorus pentoxide. Sublimation of the crude product at 100°/1 mm. gives 81.5 g. (71%) of an orange crystalline product, m.p. 85–86° after recrystallization from *n*-heptane.

Hydrogen Fluoride Method.[164] A mixture of 84 g. of ferrocene, 270 ml. of acetic anhydride, and 400 g. of hydrogen fluoride is stirred at 40–45° for 12 hours. Upon dilution with 3 l. of cold water to ensure complete precipitation of the product, 89 g. (86%) of acetylferrocene is obtained as an orange solid melting at 80–84°. Crystallization from isooctane gives orange needles, m.p. 85–86°.

Boron Trifluoride Method.[163] A stirred solution of 93.0 g. (0.5 mole) of ferrocene and 51 g.† (0.5 mole) of acetic anhydride in 800 ml. of methylene chloride, cooled in an ice bath, is saturated with gaseous boron trifluoride. (There is a copious evolution of white fumes.) The amber

* Slightly better results are obtained if the reaction mixture is carefully heated to 100° (exothermic) then quenched.[167]

† The amount of acetic anhydride has been changed by the present authors from that in the reference. Use of the original amount gives a mixture of mono- and di-acetylferrocene.

solution soon becomes deep purple. After the reaction mixture has been stirred for 0.5 hour, it is allowed to come to room temperature during 4 hours. Excess aqueous sodium acetate is then added with stirring and cooling, and the two layers are separated. The methylene chloride layer is combined with a methylene chloride extract of the aqueous layer and is washed with water, saturated sodium bicarbonate solution, and dried over magnesium sulfate. After filtration, the solvent is removed and the residue is recrystallized from hexane to give 102 g. (90%) of orange needles of acetylferrocene, m.p. 85–86°.

1,1′-Diacetylferrocene.[172] Ferrocene (30 g., 0.16 mole) dissolved in 100 ml. of dry methylene chloride is added over a period of 15 minutes to a stirred mixture of 53 g. (0.04 mole) of aluminum chloride and 32 ml. (0.45 mole) of acetyl chloride in 200 ml. of dry methylene chloride. The reaction, carried out in an atmosphere of dry nitrogen to avoid oxidation of ferrocene, proceeds readily at room temperature with immediate evolution of hydrogen chloride and the formation of an intensely violet solution.

The mixture is stirred at room temperature for 2 hours, then cooled, decomposed with ice, and filtered from aluminum hydroxide. The aluminum hydroxide is washed with chloroform until colorless. The separated aqueous phase is extracted several times with chloroform, and these extracts, combined with the main body of organic solution, are washed to neutrality. The pale blue color of the aqueous solution suggests little oxidation of ferrocene. The deep red organic solution is concentrated to about 200 ml. and dried by distilling the solvent on the steam bath. At this point 100 ml. of cyclohexane is added, and the solution is again concentrated to a volume of 200 ml. When cooled, large ruby-red rectangular plates of 1,1′-diacetylferrocene (24.4 g.), m.p. 127.5–128.5°, deposit. The second crop, after one recrystallization, weighs 7.0 g., m.p. 127.5–128.5°. The total yield is thus 31.4 g. (72%). An analytical sample of this substance melts at 130–131°.

1,1′-Di-*t*-butylferrocene.[88] Dimethylfulvene (2.12 g., 0.02 mole) in 10 ml. of diethyl ether is added to methyllithium prepared in a nitrogen atmosphere from 0.33 g. (0.05 g. atom) of lithium and 3.15 g. (0.0248 mole) of methyl iodide. The mixture is stirred for 0.5 hour, 1.625 g. (0.01 mole) of ferric chloride in 20 ml. of ether is added to the pale yellow precipitate to produce a blue mixture, and stirring is maintained for a further 2.5 hours. The ferricenium salt mixture is poured on 100 g. of ice and 100 ml. of 12.5% titanous chloride solution, and the red-yellow ethereal solution is removed and dried over anhydrous sodium sulfate and zinc dust. Removal of the solvent in a nitrogen atmosphere affords 3.65 g. of an oil, which is chromatographed on alumina with ligroin as solvent; the eluate is evaporated, yielding 1,1′-di-*t*-butylferrocene as a red-brown oil (2.8 g., 94%).

TABULAR SURVEY

The tables which follow are arranged in the order in which the compounds with which they deal are treated in the text, and the titles of the tables correspond rather closely to those of the paragraphs. Thus tables dealing with the preparation of ferrocenes from cyclopentadienes or fulvenes appear first. Next come tables dealing with the preparation or utilization of the key intermediates like chloromercuriferrocene, then tables summarizing ferrocenes bearing particular substituents—the acyl ferrocenes, alkenyl ferrocenes, etc. Finally are listed derivatives of other organometallic compounds like cyclopentadienylmanganese tricarbonyl.

Within individual tables there is, inevitably, some variation in order, reflecting the differing contents of the tables. For example, entries in Table I, dealing with the direct preparation of ferrocene from cyclopentadiene, are arranged in the general order of decreasing strength of the base employed, while entries in Table XI, dealing with various routes to ferrocenecarboxaldehyde, are arranged in order of the complexity of the starting material, and the entries of Table XXVIII, containing ferrocene heterocycles, are arranged primarily in order of heterocycle ring size and secondarily in order of number of heteroatoms in the ring.

For the bulk of the tables, in which a number of substituted ferrocenes are recorded together, the compounds are arranged in the order of substituents as follows: hydrogen, alkyl, cycloalkyl, alkenyl, alkynyl, aralkyl, and aryl groups, followed by substituents containing other functional groups but with a carbon atom bonded to the ring, followed by substituents with a heteroatom bonded to the ring. The latter substituents are arranged primarily according to group in the periodic table—alkali metals first, halogens last—and secondarily according to period in the periodic table—lithium before sodium, etc. Among alkyl groups the order of increasing total number of carbon atoms is followed primarily and the order of increasing branching secondarily. Many tables are subdivided into parts A, B, C, etc., for example according to whether the ferrocenes are mono-, di, or multi-substituted.

Most of the tables are regarded as reasonably complete, but Tables XII and XV give only representative examples of derivatives prepared from ferrocenecarboxaldehyde and acyl ferrocenes, respectively.

In this tabular survey, as well as in the text, the literature through December 1965 has been covered.

TABLE I. SYNTHESES OF FERROCENE FROM CYCLOPENTADIENE

Iron Source	Base	Solvent/Conditions	Yield (%)	Refs.
$FeCl_2$	Na	THF	67–73	75
$FeCl_3$ or $FeCl_2$	Na	$(C_2H_5)_2O$	44	510
$FeCl_2$	Na	THF or $(CH_3OCH_2)_2$	85–90	511
$FeCl_2$	Na	CH_3OH or C_2H_5OH	ca. 50–90	386, 512
$FeCl_2$	Na	THF, C_2H_5OH	90	513
$FeCl_2 \cdot 4\ H_2O$	Na	C_2H_5OH	30	513
$FeCl_2$	K	$(C_2H_5)_2O$, C_6H_6	50	514
$FeCl_2 \cdot 4\ H_2O$	Tl	$(C_2H_5)_2O$	—	515
$FeCl_3$	Tl	THF	98	516
$FeCl_2$	Tl	$(C_2H_5)_2NH$	68	516
$FeCl_2$	Tl	THF	49	516
$FeCl_2$	$MAlH_4$ (M = Li, Na, or K)	—	—	517
$FeCl_3$	$NaC{\equiv}CH$	NH_3, $(C_2H_5)_2O$	44	518
$FeCl_3$	CaC_2	NH_3, $(CH_3OCH_2)_2$	12	519
$FeCl_3$	C_5H_5MgBr	THF	51	520
$FeCl_2$	—*	THF	100	521
$FeCl_3$	C_2H_5MgBr	$(C_2H_5)_2O$	60	76
$FeCl_3$	C_2H_5MgBr	$(n\text{-}C_4H_9)_2O$	61	76
$FeCl_3$	C_2H_5MgBr	$(i\text{-}C_5H_{11})_2O$	46	76
$FeCl_3$	C_2H_5MgBr	$C_6H_5OCH_3$, $(C_2H_5)_2O$	48	76
$FeCl_3$	C_2H_5MgBr	$C_6H_5OCH_3$, dioxane	38	76
$FeCl_3$	C_2H_5MgBr	$C_6H_5OC_2H_5$, dioxane	40	76

Note: References 510–659 are on pp. 151–154.

* Prepared from $(C_5H_5)_3M$ where M = Sc, Y, La, Ce, Pr, Nd, Sm, Gd.

TABLE I. SYNTHESES OF FERROCENE FROM CYCLOPENTADIENE (*Continued*)

Iron Source	Base	Solvent/Conditions	Yield (%)	Refs.
$FeCl_2$	C_2H_5MgBr	$(C_2H_5)_2O$	71	76
$FeCl_3$	C_2H_5MgBr	$(C_2H_5)_2O$	ca. 20	522
$FeCl_2$	C_2H_5MgBr	$(C_2H_5)_2O$	70	523
$FeCl_2$	C_2H_5MgBr	THF	71	76
$FeCl_2$	C_2H_5MgBr	$C_6H_5OCH_3$, dioxane	37	76
$FeCl_2$	$NaNH_2$	NH_3	52	524
$FeCl_3$	$(C_2H_5)_2NLi$	$(C_2H_5)_2O$	72	78
$FeCl_2$	NH_3	Excess NH_3	3	77
$FeCl_2$	$(C_2H_5)_2NH$	Excess amine	73–88	74, 77, 511, 525, 529
$FeBr_2$	$(C_2H_5)_2NH$	Excess amine	85	77
$FeCl_3$	$(C_2H_5)_2NH$, zinc metal added	Excess amine	70	526
$FeBr_3$	$(C_2H_5)_2NH$†	CH_3OH	ca. 50	527
Fe, Br_2	$(C_2H_5)_2NH$	$(CH_3OCH_2)_2$	ca. 90	528
$FeCl_2$	$(C_2H_5)_2NH$	THF	61	529
$FeCl_2$	$(C_2H_5)_2NH$	$CH_3CO_2C_4H_9$-*n*	65	529
$FeCl_2$	$(C_2H_5)_2NH$	*n*-$C_3H_7CO_2C_2H_5$	54	529
$FeCl_2$	$(C_2H_5)_2NH$	$(CH_3OCH_2)_2$	ca. 50–85	528
$FeCl_2$	$(C_2H_5)_2NH$	(*n*-$C_4H_9)_2O$	42	529
$FeCl_2$	$(C_2H_5)_2NH$	Petroleum ether	ca. 80–90	528
$FeCl_2$	$(C_2H_5)_2NH$	$C_6H_5OCH_3$	40	529
$FeCl_2$	$(C_2H_5)_2NH$	$C_6H_5OC_2H_5$	38	529
$FeCl_2$	$(C_2H_5)_2NH$	$CH_3CO_2C_4H_9$-*i*	27	529
$FeCl_2$	$(C_2H_5)_2NH$	Dioxane	15	529
$FeCl_2$	$(C_2H_5)_2NH$	(*i*-$C_5H_{11})_2O$	11	529
$FeCl_2$	$(C_2H_5)_3N$	Excess amine	14–20	77, 529
$FeCl_2$	$(C_2H_5)_3N$	THF	6	529

$FeCl_2$	$(C_2H_5)_3N$	$CH_3CO_2C_4H_9$-*n*	5	529
$FeCl_3$	$(C_2H_5)_3N$	Dioxane	33	76
$FeCl_2$ or $Fe(OCOCH_3)_2$	$(C_2H_5)_3N$, $(C_2H_5)_2NH$, C_5H_5N, or $NaOCH_3$	C_6H_6	4	530
$FeCl_2$	Piperidine	Excess amine	57–69	77, 531
$FeCl_2$	*n*-$C_3H_7NH_2$	Excess amine	38	77
$FeCl_2$	(*n*-$C_3H_7)_2NH$	Excess amine	29	77
$FeCl_2$	*n*-$C_4H_9NH_2$	Excess amine	7	77
Ferrous acetylacetonate-pyridine complex	—	C_6H_6	Quant.	532
$Fe(CO)_5$	—	250–350°, 30–250 atm.	60–70	80
$Fe(CO)_5$	—	C_6H_6, 275°	15–22	533
$Fe(CO)_5$	—	Mineral oil, 360°	50	534
Fe_2O_3	—	275–400°	ca. 40	535
Fe_2O_3	—	300°	ca. 50	79
FeO (from $FeCO_3$)	‡	350–400°	15	536
FeC_2O_4	—	360°	—	537, 538
Fe	Al_2O_3, K_2O, Mo_2O_3	300°	40–50	2
Fe	CH_3CO_2Na (with added $HgCl_2$)	THF	24–30	539

Note: References 510–659 are on pp. 151–154.

† Benzyltrimethylammonium hydroxide was also present.

‡ The yield was increased by the addition of $Cd(C_2O_4)$ or $Fe(NH_4C_2O_4)_2$ to 34% or 28%, respectively.

TABLE II. SYMMETRICALLY SUBSTITUTED FERROCENES PREPARED FROM SUBSTITUTED CYCLOPENTADIENES

The iron source was ferrous chloride unless otherwise noted.

A. 1,1′-Disubstituted Ferrocenes

Substituents in Ferrocene	Base	Solvent/Conditions	Yield (%)	Refs.
CH_3	Li	$(C_2H_5)_2O$*	37	175
	Na	THF	63	540
	Na	THF*	—	541
	Na	CH_3OH†	81	512
	NaK	THF	22	234
	Na	$(CH_3OCH_2)_2$‡	—	542
	$(C_2H_5)_2NH$	$(CH_3OCH_2)_2$, excess amine	—	525
	$(C_2H_5)_2NH$	$(CH_3OCH_2)_2$	—	528
	—	300°§	—	79
	—	375–390°‖	—	543
$C(CH_3)_3$	Na	Ammonia	61	238, 544
	K	Ammonia	—	545
	Piperidine	THF	—	238
C_5H_{11}-n	Na	Ammonia	ca. 30	510
$CH_2C(CH_3)_3$	Na	THF	63	544, 546
	RMgCl	THF	—	546
$CH(CH_3)C_3H_7$-n	Na	THF	—	395
$C(CH_3)_2C_2H_5$	Na	Ammonia	—	238
$C(CH_3)_2C_4H_9$-t	Na	THF	57	544
$C(CH_3)_2CH_2C(CH_3)_3$	Na	Ammonia	—	238
$C(CH_3)(C_2H_5)_2$	Na	THF	—	395, 547
$C(CH_3)_2(CH_2)_2CH_3$	Na	THF*	—	547
$C(CH_3)(C_2H_5)(CH_2)_3CH_3$	Na	THF*	—	395, 547
1-Methylcyclohexyl	Na	THF*	—	395, 547
$CH_2CH{=}CH_2$	Na	$(CH_3OCH_2)_2$‡	—	542

$CH_2C_6H_5$	Na	$(CH_3OCH_2)_2$‡	—	542
	CH_3Li	THF	—	240
	—	180°¶	—	240
$CH(C_6H_5)_2$	C_6H_5Li	$(C_2H_5)_2O$*	—	270
$C(C_6H_5)_3$	C_6H_5Li	THF, $(C_2H_5)_2O$	18**	86, 548
$CH_2C_5H_4FeC_5H_5$	Na(Li)	THF	12** (53**)	131
C_6H_5	C_6H_5Li	$(C_2H_5)_2O$*	—	270
$COCH_3$	Na	THF	2††	91
	Na	$(CH_3OCH_2)_2$‡	—	542
CO_2CH_3	Na	THF	30††	91
CH_2CH_2OH	n-C_4H_9Li	THF	19	549
$CH_2CH(OC_2H_5)_2$	i-C_3H_7MgCl	THF	37	549
$Si(CH_3)_3$	n-C_4H_9Li	THF, C_6H_6	50	103, 550
$Si(C_2H_5)_2H$	i-C_3H_7MgBr	THF	—	551
$Si(C_6H_{13}$-$n)_3$	n-C_4H_9Li	THF, $(C_2H_5)_2O$	4	103
$Si(C_{12}H_{25}$-$n)_3$	Na	THF	53**	552, 553
CH_2CH_2 $Si(CH_3)_2N$ CH_2 CH_2CH_2	i-C_3H_7MgCl	THF	69	339, 391

Note: References 510–659 are on pp. 151–154.

* The iron source was ferric chloride.

† The iron source was ferric chloride plus iron.

‡ The iron source was FeX_n.

§ The iron source was Fe_2O_3.

‖ The iron source was ferrous oxalate.

¶ The iron source was $Fe(CO)_5$.

** The cyclopentadiene starting material is formed *in situ*; the yield given is that of the ferrocene based on the immediate precursor of the cyclopentadiene.

†† The yield is based on 3-methyl-2-cyclopentenone. The hydroxy product is unstable and was isolated as the benzoate ester (shown) or, alternatively, as the oxyacetic acid ether.

TABLE II. SYMMETRICALLY SUBSTITUTED FERROCENES PREPARED FROM SUBSTITUTED CYCLOPENTADIENES (*Continued*)
The iron source was ferrous chloride unless otherwise noted.

Substituents in Ferrocene	Base	Solvent/Conditions	Yield (%)	Refs.
A. 1,1′-Disubstituted Ferrocenes (Continued)				
$Si(CH_3)_2OC_2H_5$	i-C_3H_7MgCl	THF	62	339, 554
$Si(OC_2H_5)_2CH_3$	i-C_3H_7MgCl	THF	57	339
$Si(CH_3)_2CH_2Cl$	n-C_4H_9Li	$(C_2H_5)_2O$	72	555
$Si(CH_3)_2Si(CH_3)_3$	n-C_4H_9Li	—	65	556
$Si(CH_3)_2OSi(CH_3)_3$	n-C_4H_9Li	$(C_2H_5)_2O$	55	555, 557
$Si(CH_3)_2OSi(C_6H_5)_3$	n-C_4H_9Li	$(C_2H_5)_2O$	37	555, 557
$Si(CH_3)_2OSi(CH_3)_2C_6H_5$	n-C_4H_9Li	$(C_2H_5)_2O$	59	555, 557
$[Si(CH_3)_2O]_3C_2H_5$	n-C_4H_9Li	THF	23	339
$[Si(CH_3)_2O]_2Si(CH_3)_2C_6H_5$	n-C_4H_9Li	$(C_2H_5)_2O$	50	555, 557
$[Si(CH_3)_2O]_3Si(CH_3)_2C_6H_5$	n-C_4H_9Li	$(C_2H_5)_2O$	36	555, 557
$Ge(C_6H_5)_3$	Na	THF	—	98
$N{=}N{-}CH_3$	Li	THF, $(C_2H_5)_2O$	26	373, 374
$N{=}N{-}C_6H_5$	Li	THF, $(C_2H_5)_2O$	80	373, 374
B. Multisubstituted Ferrocenes				
1,2,4-$(C_6H_5)_3$	—	180°‡‡	37	558
1,1′,2,2′-$(CH_2C_6H_5)_4$	$LiAlH_4$	THF, $(C_2H_5)_2O$	13**	81
1,1′,2,2′-$[CH_2(C_6H_4CH_3$-$p)]_4$	$LiAlH_4$	THF, $(C_2H_5)_2O$	26**	81
1,1′,2,2′-$[CH_2(C_6H_4Cl$-$o)]_4$	$LiAlH_4$	THF, $(C_2H_5)_2O$	11**	81
1,1′,2,2′-$[CH_2(C_6H_4Cl$-$p)]_4$	$LiAlH_4$	THF, $(C_2H_5)_2O$	26**	81
1,1′,2,2′-$[CH_2(C_6H_4F$-$p)]_4$	$LiAlH_4$	THF, $(C_2H_5)_2O$	30**	81
1,1′,2,2′-$[CH_2(C_6H_4Br$-$m)]_4$	$LiAlH_4$	THF, $(C_2H_5)_2O$	24**	81
1,1′,2,2′-$[CH_2(C_6H_4Br$-$p)]_4$	$LiAlH_4$	THF, $(C_2H_5)_2O$	10**	81
1,2,3,4-$(C_6H_5)_4$	—	180°‡‡	43	558

1,1′,3,3′-$(C_6H_5)_4$	RMgBr	$(C_2H_5)_2O$, C_6H_6	—	270
	—	150°¶	ca. 10	240
	—	—§§	ca. 20	240
	$NaNH_2$	NH_3, $C_6H_4(CH_3)_2$	40	558
	$NaCH_2SOCH_3$	THF	68	558
1,1′,2,2′,3,3′-$(C_6H_5)_6$	$NaCH_2SOCH_3$	THF	45	558
	—	180°¶	21	558
	—	180°‡‡	10	558
1,1′,2,2′,4,4′-$(C_6H_5)_6$	—	180°‡‡	7	558
	n-C_4H_9Li	$(C_2H_5)_2O$, C_6H_6*	—	270
sym-$(C_6H_5)_8$	Na	Ammonia	—	510
	$NaNH_2$	Xylene	40	271
	$NaCH_2SOCH_3$	THF	20	558
	—	180°¶	32	558
	—	180°‡‡	14	558
1,1′,4,4′-$(CH_3)_4$-2,2′,3,3′-$(C_6H_5)_4$	C_6H_5Li	$(C_2H_5)_2O$	46	558
1,2,3-$(CH_3)_3$-4,5-$(C_6H_5)_2$	—	180°‡‡	40	558
1,1′,2,2′,3,3′-$(CH_3)_6$-4,4′,5,5′-$(C_6H_5)_4$	—	180°‡‡	6	558
1,1′-$(C_5H_5FeC_5H_4)_2$, 3,3′$(C_6H_5)_2$	$NaNH_2$	Ammonia	16–22	291
1,1′-$(OCOC_6H_5)_2$, 3,3′-$(CH_3)_2$	Na	Ammonia	34**,††	61, 82

Note: References 510–659 are on pp. 151–154.

* The iron source was ferric chloride.

¶ The iron source was $Fe(CO)_5$.

** The cyclopentadiene starting material is formed *in situ*; the yield given is that of the ferrocene based on the immediate precursor of the cyclopentadiene.

†† The yield is based on 3-methyl-2-cyclopentenone. The hydroxy product is unstable and was isolated as the benzoate ester (shown) or, alternatively, as the oxyacetic acid ether.

‡‡ The iron source was $(C_5H_5Fe)_2(CO)_4$.

§§ The iron source was $(R_2C_5H_3Fe)_2(CO)_4$.

TABLE II. SYMMETRICALLY SUBSTITUTED FERROCENES PREPARED FROM SUBSTITUTED CYCLOPENTADIENES (*Continued*)

The iron source was ferrous chloride unless otherwise noted.

C. Fused-Ring Ferrocenes

Substituents in Ferrocene	Base	Solvent/Conditions	Yield (%)	Refs.
1,2;1′,2′-(—$CH_2CH{=}CH$—)$_2$	n-C_4H_9Li	Hexane	54	84
1,2;1′,2′-dibenzo-	Na	THF	27–95	359, 434
	Na	$(CH_3OCH_2)_2$‡	—	542
	CH_3Li	$(C_2H_5)_2O$, C_6H_6	—	559
	RMgBr	$(C_2H_5)_2O$, C_6H_6*	—	560
1,2;1′,2′-dibenzo-3,3′-di-CH_3	C_2H_5MgBr	$(n$-$C_4H_9)_2O$	Low	561
	n-C_4H_9Li	$(C_2H_5)_2O$	23	561
	$(C_2H_5)_2NH$	Excess amine	Low	561
1,2;1′,2′-dibenzo-3,3′-di-C_2H_5	n-C_4H_9Li	$(C_2H_5)_2O$	—	562
1,2;1′,2′-dibenzo-3,3′-di-C_4H_9-n	n-C_4H_9Li	$(C_2H_5)_2O$	20	562
1,2;1′,2′-dibenzo-3,3′-di-$CH_2CH{=}CH_2$	n-C_4H_9Li	$(C_2H_5)_2O$	—	562
1,2;1′,2′-dibenzo-3,3′-di-$CH_2C_6H_5$	n-C_4H_9Li	$(C_2H_5)_2O$	10	562
1,2;1′,2′-dibenzo-3,3′-di-C_6H_5	n-C_4H_9Li	$(C_2H_5)_2O$	18	562
1,2:1′,2′-(—$CH_2CH{=}CHCH{=}CH$—)$_2$	Li	THF, $(C_2H_5)_2O$	17	88
1,2:1′;2′-(—$CH_2CHCH_2CHCH_2$—)$_2$ (bridged CH–CH)	—	$(C_2H_5)_2O$, xylene‖ ‖	20	563
[Structure: bis-fused diferrocene, two Fe] (Complete structure)	Li	THF	10	83

D. Bridged Ferrocenes				
1,1′-$(CH_2)_3$-	Na	THF	2.5**	333, 334
1,1′-$(CH_2)_4$-	Na	THF	0.05**	333, 334
1,1′-$(CH_2)_5$-	Na	THF	0.02**	333, 334
1,1′-$Si(CH_3)_2OSi(CH_3)_2$-	n-C_4H_9Li	THF, $(C_2H_5)_2O$	22**	339
1,1′-$[Si(CH_3)_2O]_2Si(CH_3)_2$-	n-C_4H_9Li	THF	6**	339

Note: References 510–659 are on pp. 151–154.

* The iron source was ferric chloride.

‡ The iron source was ferrous FeX_n.

** The cyclopentadiene starting material is formed *in situ*; the yield given is that of the ferrocene based on the immediate precursor of the cyclopentadiene

‖ ‖ The iron source was acetonylacetonate.

TABLE III. FERROCENES PREPARED FROM FULVENES

The iron source was ferrous chloride unless otherwise noted.

Substituents in Ferrocene*			Nucleophile, Base or Metal	Solvent	Yield (%)	Refs.
R_1	R_2	R_3				
A. 1,1′-Dialkylferrocenes ($R_1R_2R_3CC_5H_4FeC_5H_4CR_1R_2R_3$)						
CH_3	CH_3	H	$LiAlH_4$	$(C_2H_5)_2O$	24	86, 548
CH_3	CH_3	H	$LiAlH_4$	$(C_2H_5)_2O$†	74	87, 88
CH_3	CH_3	H	C_2H_5MgX	$(C_2H_5)_2O$†	—	564
CH_3	CH_3	CH_3	CH_3Li	$(C_2H_5)_2O$†	94	87, 88
CH_3	CH_3	C_6H_5	C_6H_5Li	$(C_2H_5)_2O$, THF	73	86, 548
CH_3	CH_3	C_6H_5	C_6H_5Li	$(C_2H_5)_2O$†	90	87, 88
C_2H_5	C_2H_5	H	$LiAlH_4$	$(C_2H_5)_2O$	23	565
C_6H_5	C_6H_5	H	$LiAlH_4$	$(C_2H_5)_2O$, THF	45	86, 548
C_6H_5	C_6H_5	C_6H_5	C_6H_5Li	$(C_2H_5)_2O$, THF	18	548
CH_3	C_2H_5	H	$LiAlH_4$	$(C_2H_5)_2O$	23	565
CH_3	C_6H_5	H	$LiAlH_4$	$(C_2H_5)_2O$	26	565
$C_5H_4FeC_5H_5$	H	H	$LiAlH_4$	$(C_2H_5)_2O$, THF	53	131
$C_5H_4FeC_5H_5$	H	C_6H_5	C_6H_5Li	$(C_2H_5)_2O$, THF	21	90
—$(CH_2)_5$—		H	$LiAlH_4$	$(C_2H_5)_2O$†	77	88
—$(CH_2)_5$—		C_6H_5	C_6H_5Li	$(C_2H_5)_2O$	96	88
$N(CH_3)_2$	H	H	$LiAlH_4$	$(C_2H_5)_2O$, THF	58	90
$N(CH_3)_2$	H	CH_3	CH_3Li	$(C_2H_5)_2O$, THF	71	90
B. 1,1′-Dialkenylferrocenes						
—$C(CH_3)$=CH_2			$NaNH_2$	Ammonia	65	87–88
			Na	THF	—	89
			$NaNH_2$	Ammonia	71	87–88

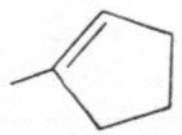

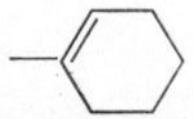

C. 1,1'-Ethyleneferrocenes

R_1	R_2				
CH_3	CH_3	Na	$(CH_3OCH_2)_2$	70	566
CH_3	CH_3	Na	THF	Low	89
CH_3	CH_3	Na	$(C_2H_5)_2O$†	—	567
$(CH_3)_2CHCH_2$	H	Na	$(CH_3OCH_2)_2$	69	566, 568
3-Heptyl	H	Na	$(CH_3OCH_2)_2$	65	566, 568
CH_3	i-C_3H_7	Na	$(CH_3OCH_2)_2$	13	566, 568
CH_3	C_6H_5	Na	$(CH_3OCH_2)_2$	73	566, 568
CH_3	$C_6H_4NH_2$-m	Na	$(CH_3OCH_2)_2$	21	566, 568
CH_3	$CH_2N(C_2H_5)_2$	Na	$(C_2H_5)_2O$†	—	567

Note: References 510–569 are on pp. 151–154.

* R_1 and R_2 were substituents in the 6-position of the fulvene; R_3 is introduced by the nucleophile.
† The iron source was ferric chloride.

TABLE IV. UNSYMMETRICALLY SUBSTITUTED FERROCENES PREPARED FROM SUBSTITUTED CYCLOPENTADIENES
The substituent(s) in the ferrocene were present in the substituted cyclopentadiene.

Ferrocene	Base	Iron Source	Solvent	Yield (%)	Refs.
Benzyl	—	$[C_5H_5Fe(CO)_2]_2$	—	—	240
Phenyl*	CH_3Li	$FeCl_3$	$(C_2H_5)_2O$	—	270
1,3-Diphenyl	—	$[C_5H_5Fe(CO)_2]_2$	—	ca. 20	240
1,2-Di-*p*-anisyl	CH_3MgI	$FeCl_3$	$(C_2H_5)_2O$, C_6H_6	—	272
1,3-Di-*p*-anisyl	CH_3MgI	$FeCl_3$	THF	—	272
1-Ferrocenyl-3-phenyl	$NaNH_2$	$FeCl_2$	—	15	291
Benzo†	Na	$FeCl_2$	THF	9	358, 359
Phenylazo‡	C_6H_5Li or CH_3Li	$FeCl_2$	$(C_2H_5)_2O$, THF	6	373, 374

Note: References 510–659 are on pp. 151–154

* This compound was also prepared in 20% yield from cyclopentadienylsodium and ferrous chloride in *p*-dibromobenzene.
† The starting material was indene.
‡ Phenylazocyclopentadienyllithium was formed *in situ* from phenyllithium and azocyclopentadiene.

TABLE V. FERROCENE DERIVATIVES PREPARED FROM LITHIATED FERROCENES

*A. Monosubstituted Ferrocenes**

Substituent(s) in Ferrocene	Reagent	Yield (%)	Refs.
CO_2H	CO_2	25†	96
	CO_2	18–64	93, 94, 98,‡ 569, 570
$CONHC_6H_5$	C_6H_5NCO	10	306
$C(C_6H_5)_2OH$	$(C_6H_5)_2CO$	65	98
$CH_2CHOHCH_2Cl$	$ClCH_2CH—CH_2$ (epoxide, O bridging CH—CH₂)	—	571
$CH_2CH_2N(CH_3)_2$	$ClCH_2CH_2N(CH_3)_2$	ca. 20	132
2-Pyridyl	C_5H_5N	24–32	364, 365
2-Quinolyl	C_9H_7N	22	572
$C_5H_4FeC_5H_5$	$(n\text{-}C_6H_{13})_3SiBr$	0.11	285
	$n\text{-}C_4H_9Br$, $CoBr_2$	11	289
$B(OH)_2$	$(n\text{-}C_4H_9O)_3B$	11–44	108, 109,‡ 296
$SiH(CH_3)C_2H_5$	$ClSiH(CH_3)C_2H_5$	—	551
$SiH(C_2H_5)_2$	$ClSiH(C_2H_5)_2$	—	551
$Si(CH_3)_3$	$ClSi(CH_3)_3$	12–19	118, 234
$Si(C_2H_5)_3$	$HSi(C_2H_5)_3$	—	551
$Si(C_3H_7\text{-}n)_3$	$ClSi(C_3H_7\text{-}n)_3$	—	573
$Si(CH_2CH_2CF_3)_3$	$ClSi(CH_2CH_2CF_3)_3$	ca. 30	573
$Si(C_6H_{13}\text{-}n)_3$	$BrSi(C_6H_{13}\text{-}n)_3$	32	103,‡ 285
$Si(CH_3)_2CH_2CH{=}CH_2$	$ClSi(CH_3)_2CH_2CH{=}CH_2$	—	574
$Si(CH_3)_2C_6H_4CF_3\text{-}3$	$ClSi(CH_3)_2C_6H_4CF_3\text{-}3$	ca. 25	573
$Si(C_6H_5)_3$	$ClSi(C_6H_5)_3$	20–65	93, 98,‡ 575
NO_2	$n\text{-}C_3H_7ONO_2$	—	368
	N_2O_4	—	367
N_3	$p\text{-}CH_3C_6H_4SO_2N_3$	28	377
$N{=}NC_5H_4FeC_5H_5$	N_2O	25–30	370,‡ 371
$N{=}NNHC_6H_5$	$C_6H_5N_3$	57	377
$N{=}NNHC_5H_4FeC_5H_5$	$C_5H_5FeC_5H_4N_3$	—	377

*B. 1,1′-Disubstituted Ferrocenes**

Substituent(s) in Ferrocene	Reagent	Yield (%)	Refs.
CO_2H	CO_2	19–40	93, 94,‡ 302, 569, 570
$C(C_6H_5)_2OH$	$(C_6H_5)_2CO$	—	104
$CH_2CHOHCH_2Cl$	$ClCH_2CH—CH_2$ (epoxide, O bridging CH—CH₂)	—	571
2-Pyridyl	C_5H_5N	3	364, 365
2-Quinolyl	C_9H_7N	1–7	365, 572
$B(OH)_2$	$(n\text{-}C_4H_9O)_3B$	13–18	108–110
$SiH(CH_3)C_2H_5$	$ClSiH(CH_3)C_2H_5$	—	551
$SiH(C_2H_5)_2$	$ClSiH(C_2H_5)_2$	—	551
$Si(CH_3)_3$	$ClSi(CH_3)_3$	10–27	118, 234
$Si(C_2H_5)_3$	$HSi(C_2H_5)_3$	—	551
$Si(C_3H_7\text{-}n)_3$	$ClSi(C_3H_7\text{-}n)_3$	ca. 40	573
$Si(CH_2CH_2CF_3)_3$	$ClSi(CH_2CH_2CF_3)_3$	—	573
$Si(C_6H_{13}\text{-}n)_3$	$BrSi(C_6H_{13}\text{-}n)_3$	35	103
$Si(CH_3)_2C_6H_4CF_3\text{-}3$	$ClSi(CH_3)_2C_6H_4CF_3\text{-}3$	—	573
$Si(C_6H_5)_3$	$ClSi(C_6H_5)_3$	7–50	93, 575
N_3	$p\text{-}CH_3C_6H_4SO_2N_3$	6	377
$N{=}NNHC_6H_5$	$C_6H_5N_3$	24	377

Note: References 510–659 are on pp. 151–154.

* The products from most reactions were isolated from a mixture of mono- and di-substituted products.

† No dicarboxylic acid was obtained under the conditions of this experiment.

‡ The highest yield is reported in this reference.

TABLE VI. SUBSTITUTED FERROCENES PREPARED FROM FERROCENEBORONIC ACIDS

Substituent(s) in Ferrocene 1	1′	Substituent in Boronic Acid	Reagent	Yield (%)	Refs.
A. From Ferroceneboronic Acid					
H	—	—	$Zn(H_2O)$	82	108
HgCl	—	—	$HgCl_2$	76	108, 109
$C_5H_4FeC_5H_5$	—	—	$Ag_2O(NH_3)$	52	108, 109
$C_5H_4FeC_5H_5$	—	—	$(CH_3CO_2)_2Cu(H_2O)$	21*	109
$N(CO)_2C_6H_4$-*o*	—	—	$Cu[N(CO)_2C_6H_4\text{-}o]_2$	47	111
$OCOCH_3$	—	—	$(CH_3CO_2)_2Cu$	59†	112, 113
$OCOC_2H_5$	—	—	$(C_2H_5CO_2)_2Cu$	53†	109, 112
Cl	—	—	$CuCl_2$	84	108, 109
Br	—	—	$CuBr_2$	74–80	108, 109, 296
B. From 1′-Substituted Ferroceneboronic Acids					
H	Cl	Cl	$ZnCl_2(H_2O)$	79	114
H	Br	Br	$ZnCl_2(H_2O)$	88	114
HgCl	Cl	Cl	$HgCl_2$	88	109, 114
HgBr	Br	Br	$HgBr_2$	84	109, 114
$OCOCH_3$	$OCOCH_3$	Br	$(CH_3CO_2)_2Cu$	83	109, 115
Cl	Cl	Br	$CuCl_2$	—	114
$N(CO)_2C_6H_4$-*o*	Cl	Cl	$Cu[N(CO)_2C_6H_4\text{-}o]_2$	50	116
C. From 1,1′-Ferrocenediboronic Acid					
$N(CO)_2C_6H_4$-*o*	H	$B(OH)_2$	$Cu[N(CO)_2C_6H_4\text{-}o]_2$, C_5H_5N	29	111
$OCOCH_3$	$OCOCH_3$	$B(OH)_2$	$(CH_3CO_2)_2Cu$, H_2O	41	115
Cl	$B(OH)_2$	$B(OH)_2$	$CuCl_2$, CH_3OH	52	109, 114
Cl	Cl	$B(OH)_2$	$CuCl_2$, H_2O	75	108, 109
Br	$B(OH)_2$	$B(OH)_2$	$CuBr_2$, CH_3OH	65	109, 114
Br	Br	$B(OH)_2$	$CuBr_2$, H_2O	76	108, 109

Note: References 510–659 are on pp. 151–154.

* Acetoxyferrocene was also obtained in unstated yield.

† Diferrocenyl was also obtained in 20% yield.

TABLE VII. Preparation of Chloromercuriferrocene

Starting Material	Reagents	Solvent	Yield (%)*	Refs.
Ferrocene	$(CH_3CO_2)_2Hg$, LiCl	CH_3OH, C_6H_6	73 [18]	120
	$(CH_3CO_2)_2Hg$, LiCl	$(C_2H_5)_2O$, CH_3OH	14–53 [30–38]	118, 119
	$(CH_3CO_2)_2Hg$, KCl	CH_3CO_2H	14–50 [14–64]	118
	$(CH_3CO_2)_2Hg$, KCl	$(C_2H_5)_2O$, C_2H_5OH	—	117
	CH_3CO_2Na, $HgCl_2$	$(C_2H_5)_2O$, CH_3OH	47–53 [30–38]	119
Ferroceneboronic acid	$HgCl_2$	H_2O, CH_3COCH_3	76	108, 109
1′-Chloroferroceneboronic acid	$HgCl_2$	CH_3COCH_3	88†	114
Sodium ferrocenesulfonate	$HgCl_2$	80% alcohol	100	324

Note: References 510–659 are on pp. 151–154.

* The yield of 1,1′-disubstituted material is indicated in brackets.
† The product was 1′-chloro-1-chloromercuriferrocene.

TABLE VIII. SUBSTITUTED FERROCENES PREPARED FROM MERCURIFERROCENES

Substituent in Ferrocene	Reagent	Solvent/Temperature	Yield (%)	Refs.
A. From Chloromercuriferrocene				
Li	C_2H_5Li	$(C_2H_5)_2O$	64*	95, 98
	n-C_4H_9Li	$(C_2H_5)_2O$	43–50†	98
	C_6H_5Li	$(C_2H_5)_2O$	58*	98
$HgC_5H_4FeC_5H_5$	Na, then CO_2	C_6H_6	70	118
	NaI	C_2H_5OH	64	118
	Na_2SnO_2	C_2H_5OH-H_2O	70	118
	I_2, then $Na_2S_2O_3$	Xylene	—	117, 375
Br	N-Bromosuccinimide	$HCON(CH_3)_2$	57	120
	Pyridine-Br_2	$HCON(CH_3)_2$	58	120
	$CH_3CONHBr$	$HCON(CH_3)_2$	48	120
I	N-Iodosuccinimide	CH_2Cl_2	85	120
	I_2	Xylene	60–70	376
	I_2	C_2H_5OH	64	286
	I_2	CH_2Cl_2	70	110
B. From Diferrocenylmercury				
H	CH_3COCl	C_6H_6	94	326
	Pd	—	16–49‡	123
	Na	—	10	325
	$SnCl_2$	H_2O	15	325
	$CuCl_2$	H_2O	5–7	325
	CCl_4	CCl_4	12§	325
Li	n-C_4H_9Li	$(C_2H_5)_2O$, C_6H_6	29–80*	97
	n-C_4H_9Li	$(C_2H_5)_2O$	43*, ‖	98
C_6H_5	$(C_6H_5)_2Hg$, Ag	250–300°	45	122, 576

$C_6H_4C_6H_5$-*o*	(*o*-$C_6H_5C_6H_4)_2Hg$, Ag	235–300°	6	97, 576
$C_6H_4C_6H_5$-*m*	(*m*-$C_6H_5C_6H_4)_2Hg$, Ag	235–300°	22	97, 576
$C_6H_4C_6H_5$-*p*	(*p*-$C_6H_5C_6H_4)_2Hg$, Ag	235–300°	20	97, 576
$C(C_6H_5)_3$	$(C_6H_5)_3CCl$	C_6H_6	18	326
$C_5H_4FeC_5H_5$	Pd	—	1.5–6.3‡	123
	Ag	250–300°	47–61	122
$SC_5H_4FeC_5H_5$	S	C_2H_5OH	1	376
$SSC_5H_4FeC_5H_5$	$(CNS)_2$	C_2H_5OH	15†	326
$SO_2C_6H_5$	$C_6H_5SO_2I$	C_6H_6	22¶	326
	$C_6H_5SO_2Cl$	C_6H_6	6¶	326
$SO_2C_5H_4FeC_5H_5$	$C_5H_5FeC_5H_4SO_2I$	C_6H_6	27†	325
$SeC_5H_4FeC_5H_5$	$SeBr_4$	$CHCl_3$	21	326
Br	Br_2	$CHCl_3$	Low	375
I	I_2	Xylene	64	375
		C. From 1,1′-bis-(Chloromercuri)ferrocene		
Li	C_2H_5Li	$(C_2H_5)_2O$	—	98
Br	N-Bromosuccinimide	$HCON(CH_3)_2$	47	120
	Br_2	—	—	375
I	N-Iodosuccinimide	$HCON(CH_3)_2$	42	120
	I_2	—	25	375
$(HgC_5H_4FeC_5H_4)_n$-H	NaI	C_2H_5OH	—	124

Note: References 510–659 are on pp. 151–154.

* The yield was determined by carbonation to the corresponding acid.
† A 23% yield of ferrocene was obtained also.
‡ Another product was ferrocene or biferrocenyl.
§ No ferrocene was obtained in the absence of hydroquinone.
‖ A 19% yield of ferrocene was obtained also.
¶ A 35% yield of ferrocene was obtained also.

TABLE IX. PREPARATION OF DIALKYLAMINOMETHYLATED FERROCENES

Substituents in $R_1C_5H_4FeC_5H_3(R_2)N(R_3)_2$

R_1	R_2	R_3	Reactants	Solvent/Conditions	Yield (%)	Refs.
H	H	CH_3	$C_5H_5FeC_5H_5$, $[(CH_3)_2N]_2CH_2$	CH_3CO_2H, H_3PO_4	—	126, 127
			$C_5H_5FeC_5H_5$, $[(CH_3)_2N]_2CH_2$	CH_3CO_2H	51	125
			$C_5H_5FeC_5H_5$, $(CH_3)_2NH$, CH_2O	CH_3CO_2H	48	125, 130
H	H	CH_3	$C_5H_5FeC_5H_4CHO$, $(CH_3)_2NH$	CH_3OH, Ni, H_2, 120°, 165 atm.	76	139
H	H	C_2H_5	$C_5H_5FeC_5H_5$, $(C_2H_5)_2NH$, CH_2O	CH_3CO_2H, H_3PO_4	—	129
H	H	$—(CH_2)_5—$	$C_5H_5FeC_5H_5$, $(C_5H_{10}N)_2CH_2$	CH_3CO_2H, H_3PO_4	50	126
H	CH_3*	CH_3	$C_5H_5FeC_5H_4CH_3$, $[(CH_3)_2N]_2CH_2$	CH_3CO_2H, H_3PO_4	80†	129
C_6H_5†	H	CH_3	$C_5H_5FeC_5H_4C_6H_5$, $[(CH_3)_2N]_2CH_2$	CH_3CO_2H, H_3PO_4	54†	128
C_6H_5†	H	CH_3	$C_5H_5FeC_5H_4C_6H_5$, $[(CH_3)_2N]_2CH_2$	CH_3CO_2H, H_3PO_4	25†	577
$C_6H_4CH_3$-*p*†	H	CH_3	$C_5H_5FeC_5H_4C_6H_4CH_3$-*p*, $[(CH_3)_2N]_2CH_2$	CH_3CO_2H, H_3PO_4	62†	577
C_6H_4Cl-*p*†	H	CH_3	$C_5H_5FeC_5H_4C_6H_4Cl$-*p*, $[(CH_3)_2N]_2CH_2$	CH_3CO_2H, H_3PO_4	20†	577
SCH_3	SCH_3‡	CH_3	$CH_3SC_5H_4FeC_5H_4SCH_3$	—	—‡	578

Note: References 510–659 are on pp. 151–154.

* A mixture of positional isomers was probably obtained but not separated. Homoannular substitution was reportedly favored. Disubstitution (19%) was observed also.

† A mixture of positional isomers was obtained. Heteroannular substitution was strongly favored.

‡ A mixture of 2- and 3-isomers was obtained, with the former favored.

TABLE X. SUBSTITUTED FERROCENES PREPARED FROM FERROCENYLMETHYLTRIMETHYLAMMONIUM IODIDE

Substituent A in $C_5H_5FeC_5H_4CH_2A$	Displacing Agent or Base	Yield (%)	Refs.
H	Na(Hg)	89–94	133, 244
	Na(Hg)	71*	128
	Na(Hg)	75†	129
CH_3	CH_3MgX	59	243
C_2H_5	C_2H_5MgX	63	243
$CH{=}CH_2$	$CH_2{=}CHMgX$	34‡	243
$CH_2CH{=}CH_2$	$CH_2{=}CHCH_2MgX$	53	243
$CH_2C_6H_5$	$C_6H_5CH_2MgX$	60	243
C_6H_5	C_6H_5MgX	27	243
$C_{10}H_7$-1	1-$C_{10}H_7MgX$	51	243
C_5H_5	NaC_5H_5 §	41	131
$C_5H_4FeC_5H_4CH_2C_5H_4FeC_5H_5$	NaC_5H_5 ‖	53	131
CHO¶	—¶	37	155
CN	KCN	48–95	126, 579, 658
$CH_2N(CH_3)_2$	KNH_2 (base)	40–50	132, 580
2-Ketocyclopentyl	N-(1-Cyclopentenyl)piperidine	54	581
2-Ketocyclohexyl	N-(1-Cyclohexenyl)piperidine	88	581
$CH(CN)C_6H_5$	$C_6H_5CH_2CN$	35	582, 583
$C(C_6H_5)_2CN$	$(C_6H_5)_2CHCN$	76	582, 583
$CH(COCH_3)_2$	$CH_2(COCH_3)_2$	70	582, 583
$CH(COCH_3)CO_2C_2H_5$	$CH_3COCHNaCO_2C_2H_5$	85	583, 584
$CH(CN)CO_2C_2H_5$	$NCCH_2CO_2C_2H_5$	—	583
$CH(CO_2C_2H_5)_2$	$Na[CH(CO_2C_2H_5)_2]$	—	328
	$Na[CH(CO_2C_2H_5)_2]$	18	126
$C(COCO_2C_2H_5)(CO_2C_2H_5)C_5H_4FeC_5H_5$	$CH_2(CO_2C_2H_5)(COCO_2C_2H_5)$	—	583
$CCH_3(CO_2C_2H_5)_2$	$Na[C(CH_3)(CO_2C_2H_5)_2]$	—	585
$C(C_6H_5)(CO_2C_2H_5)_2$	$Na[C(C_6H_5)(CO_2C_2H_5)_2]$	—	585

Note: References 150–659 are on pp. 151–154.

* The product was 1-phenyl-1′-methylferrocene from the appropriately substituted quaternary salt.
† The starting material was (3-methylferrocenylmethyl)trimethylammonium iodide and the product was 1,3-dimethylferrocene.
‡ A trace of 1,2-diferrocenylethane was observed.
§ The product isolated after treatment with ferrous chloride which produced 12% 1,1′-bis(ferrocenylmethyl)ferrocene.
‖ The product was isolated after treatment with ferrous chloride.
¶ The aldehyde resulted from a Sommelet reaction of the quaternary salt with hexamethylenetetramine.

TABLE X. SUBSTITUTED FERROCENES PREPARED FROM FERROCENYLMETHYLTRIMETHYLAMMONIUM IODIDE (*Continued*)

Substituent A in $C_5H_5FeC_5H_4CH_2A$	Displacing Agent or Base	Yield %	Refs.
$C(CO_2C_2H_5)_2NHCHO$	$Na[C(NHCHO)(CO_2C_2H_5)_2]$	73	126
$C(CN)(CO_2C_2H_5)NHCOCH_3$	$Na[C(CN)(CO_2C_2H_5)NHCOCH_3]$	ca. 90	328
$N(C_5H_{10})$	Piperidine	94	587
NHC_6H_5	$C_6H_5NH_2$	75	587
NHC_6H_4Cl-*p*	$H_2NC_6H_4Cl$-*p*	ca. 50	588
$N(C_2H_4)_2O$	Morpholine	95	587
$N(CO)_2C_6H_4$-*o*	Potassium phthalimide	97	587
$P^{\oplus}(C_6H_5)_3I^{\ominus}$	$(C_6H_5)_3P$	97	225
OH	NaOH	90	125
	NaOH	44**	260
OCH_3	$NaOCH_3$	83	589
OC_2H_5	$NaOC_2H_5$	63	589
	$CH_3CO_2C_2H_5$	42	587
OC_4H_9-*n*	$CH_3CO_2C_4H_9$-*n*	35	587
OC_4H_9-*t*	$NaOC_4H_9$-*t*	9††	589
OC_9H_{19}-*n*‡‡	*n*-$C_9H_{19}OH$	28	589
$OC_{10}H_{21}$-*n*‡‡	*n*-$C_{10}H_{21}OH$	30	589
$OCH_2CH{=}CH_2$	$NaOCH_2CH{=}CH_2$	60	589
$OCH_2C_6H_5$	$NaOCH_2C_6H_5$	80	589
	$C_6H_5CH_2OH$	55	589
$OCH_2C_4H_3O$	NaO-(2-furyl)	58	589
OC_6H_5	$NaOC_6H_5$	80	129
$O\text{-}C_6H_4\text{-}N{=}NC_6H_5$	$NaO\text{-}C_6H_4\text{-}N{=}NC_6H_5$	80	129
$OCOCH_3$	CH_3CO_2Na, CH_3CO_2H	25	587
$SCH_2C_5H_4FeC_5H_5$	Na_2S	54	587
$SSCH_2C_5H_4FeC_5H_5$	NaSH	33	587
SCN	KCNS	46	129
SO_3Na	Na_2SO_3	77	129

Note: References 510–659 are on pp. 151–154.

** The product was 2-ethyl-1-hydroxymethylferrocene from the appropriately substituted quaternary salt.

†† Bis(ferrocenylmethyl) ether was obtained in 25% yield.

‡‡ The displacement was carried out on ethyl(ferrocenylmethyl)dimethylammonium iodide.

TABLE XI. PREPARATION OF FERROCENECARBOXALDEHYDE

Starting Material	Reagents/Method	Yield (%)	Refs.
$C_5H_5FeC_5H_5$	$HCON(CH_3)C_6H_5$, $POCl_3$	Up to 80	139–146
	$HCON(CH_3)_2$, $POCl_3$	17	139
	$CHCl_2OC_2H_5$, $AlCl_3$	ca. 50	131
	$2\text{-}CH_3C_6H_4N_2^{\oplus}Cl^{\ominus}$, CH_2Cl_2	16	280
	$2,4,6\text{-}(CH_3)_3C_6H_2N_2^{\oplus}Cl^{\ominus}$, CH_2Cl_2	23	280
$C_5H_5FeC_5H_4CH_3$	MnO_2	23	242
$C_5H_5FeC_5H_4CH_2N(CH_3)_3^{\oplus}\ I^{\ominus}$	Sommelet	37	141, 147, 155
$C_5H_5FeC_5H_4CH_2OH$	MnO_2	98–100	125, 154
$C_5H_5FeC_5H_4CONHNH_2$	McFadyen-Stevens	—	91
$CH_3C_5H_4FeC_5H_4CH_3$	MnO_2	7*	242
$HOCH_2C_5H_4FeC_5H_4CH_2OH$	MnO_2	10†	91

Note: References 510–659 are on pp. 151–154.

* The product was 1′-methylferrocenecarboxaldehyde.
† The product was 1,1′-ferrocenedicarboxaldehyde.

TABLE XII. EXAMPLES OF COMPOUNDS PREPARED FROM FERROCENECARBOXALDEHYDE

Substituent R in Product $C_5H_5FeC_5H_4R$	Reaction/Reagents	Yield (%)	Refs.
CH_3	Zn, HCl, CH_3CO_2H, 95°, 15 min.	66	91
$CH{=}CHCH{=}CH_2$	$(C_6H_5)_3P{=}CHCH{=}CH_2$	50	91
$CH{=}C_5H_4$ (cyclopentadienylidene)	C_5H_6, $NaOC_2H_5$, 95°	98	91
$CH{=}CHC_6H_5$	$(C_6H_5)_3P{=}CHC_6H_5$	45	261
$CH{=}CHC_5H_4FeC_5H_5$ (*trans*)	$(C_6H_5)_2PONa$	73	225
$CH{=}CH\text{-}C_6H_4\text{-}CH{=}CHC_5H_4FeC_5H_5$	$(C_6H_5)_3P{=}CH\text{-}C_6H_4\text{-}CH{=}P(C_6H_5)_3$	6	159
CH_2OH	$LiAlH_4$	90	139, 142
	$NaBH_4$	85	141
	Cannizzaro	24–47	125, 141
$CHOHCH_3$	CH_3MgI	81	222
$CHOHC_4H_9$-*t*	*t*-C_4H_9MgCl	ca. 30	257
$CHOHC{\equiv}CH$	$NaC{\equiv}CH$, liq. NH_3	70–89	154, 202
$CHOHC_6H_5$	C_6H_5MgBr	98	222
$CH_2OCH_2C_5H_4FeC_5H_5$	Raney Ni, H_2	57	139
$CHOHC{\equiv}CCHOHC_5H_4FeC_5H_5$	$LiC{\equiv}CLi$, $(C_2H_5)_2O$	44	202
CO_2H	Cannizzaro	10–27	125, 141

$CHOHCX_3$	CHX_3, KOH (X = Cl, Br, I)	7–32	194
CHOHCN	$NaHSO_3$, KCN	52	139
$CH(CN)N(CH_3)_2$	$NaHSO_3$, $(CH_3)_2NH$, NaCN	90	154
$COCHOHC_6H_5$	C_6H_5CHO, KCN	25	141
$CH{=}CHCO_2H$	$CH_2(CO_2H)_2$, $C_5H_{11}N$, 95°	68, 75	140–142
$CH{=}CHCOCH_3$	CH_3COCH_3, KOH	—	91
$CH{=}CHCOC_6H_5$	$CH_3COC_6H_5$, NaOH	92	154
$CH{=}C(C_6H_5)CHO$	$C_6H_5CH_2CHO$, $NaOC_2H_5$	14	160
$CH{=}C(NO_2)CH_3$	$C_2H_5NO_2$, $CH_3CO_2NH_4$	79	156
$CH{=}C(CN)CO_2C_2H_5$	$NCCH_2CO_2C_2H_5$, $C_5H_{11}N$, C_2H_5OH	58	157
$CH{=}\underline{CCO_2C(C_6H_5){=}N}$	(1) $LiCH_2CO_2C_4H_9$-*t*, (2) dil. HCl,	63	154
	$C_6H_5CONHCH_2CO_2H$, CH_3CO_2Na, $(CH_3CO)_2O$	58–64	126, 142
$CH[C_6H_4N(CH_3)_2\text{-}4]_2$	$C_6H_5N(CH_3)_2$, C_2H_5OH, HCl	ca. 50	590
$CH{=}CHCO_2C_2H_5$	Zn, $BrCH_2CH_2CO_2C_2H_5$, $C_6H_5CH_3$	74	162
	$(C_6H_5)_3P{=}CHCO_2C_2H_5$	82	162

Note: References 510–659 are on pp. 151–154.

TABLE XIII. MONOACYL FERROCENES

A. By Friedel-Crafts Reaction

R in Acyl Ferrocene $C_5H_5FeC_5H_4COR$*	Reagent	Lewis Acid	Solvent	Yield (%)	Refs.
CH_3	RCOCl	$AlCl_3$	$CHCl_3$	47–54	141
	$(RCO)_2O$	$AlCl_3$	CH_2Cl_2	65	182
	$(RCO)_2O$	BF_3	$ClCH_2CH_2Cl$, $CH_3CO_2C_2H_5$	88	163
	$(RCO)_2O$	BF_3	CH_2Cl_2	90	163
	$(RCO)_2O$	HF	HF	ca. 90	164, 591
	$(RCO)_2O$	H_3PO_4	$(CH_3CO)_2O$	71	139, 218
	Acetosilico anhydride	$SnCl_4$	C_6H_6	—	117
CH_2Cl	RCOCl	$AlCl_3$	CS_2	37	195
$CHCl_2$	RCOCl	$AlCl_3$	CS_2	30	195
C_2H_5	RCOCl	$AlCl_3$	CH_2Cl_2	48	210
	$(RCO)_2O$	$AlCl_3$	CH_2Cl_2	53	211
	$(RCO)_2O$	BF_3	$ClCH_2CH_2Cl$, $CH_3CO_2C_2H_5$	ca. 90	163
C_3H_7-*n*	RCOCl	$AlCl_3$	CH_2Cl_2	90	331
C_3H_7-*i*	RCOCl	$AlCl_3$†	$(C_2H_5)_2O$	—	592
C_4H_9-*n*	RCOCl	$AlCl_3$	CH_2Cl_2	55	201
$CH(CH_3)C_2H_5$	RCOCl	$BF_3\cdot(C_2H_5)_2O$	CH_2Cl_2	60–80	262
C_4H_9-*t*	RCO_2H	$FeCl_3$	$ClCH_2CH_2Cl$	—	594
	RCOCl	$AlCl_3$	$ClCH_2CH_2Cl$	—	595
	$(RCO)_2O$	$AlCl_3$	—	—	595
	RCOCl	HF	HF	—	595
C_5H_{11}-*n*	RCOCl	$AlCl_3$	CH_2Cl_2	88	596
$CH(C_2H_5)_2$	RCOCl	HF	HF	—	597
$CH_2CH_2Si(CH_3)_3$	RCOCl	$AlCl_3$	CH_2Cl_2	76	413
$(CH_2)_3Si(CH_3)_3$	RCOCl	$AlCl_3$	CH_2Cl_2	83	413
C_7H_{15}-*n*	RCOCl	$AlCl_3$	CH_2Cl_2	91	596

$C(C_2H_5)_3$	RCOCl	HF	HF	—	597
$CH(C_4H_9\text{-}n)C_3H_7\text{-}i$	RCOCl	HF	HF	—	597
$C_9H_{19}\text{-}n$	RCOCl	$AlCl_3$	CH_2Cl_2	84	596
$C_{11}H_{23}\text{-}n$	RCOCl	$AlCl_3$	CH_2Cl_2	79	596
$C_{15}H_{31}\text{-}n$	RCOCl	$AlCl_3$	CH_2Cl_2	46	599
$CH(CH_2C_6H_5)_2$	RCOCl	$AlCl_3$	CH_2Cl_2	42	212
Cyclohexyl	RCOCl	HF	HF	—	595
$CH_2C_6H_5$	RCOCl	$AlCl_3$	CH_2Cl_2	80	197
$CH_2CH_2C_6H_5$	RCOCl	$AlCl_3$	$ClCH_2CH_2Cl$	41	229
$CH_2C_5H_4FeC_5H_5$	RCOCl	$AlCl_3$	CH_2Cl_2	—	245, 262
$CH{=}CH(C_6H_5)$	RCOCl	$AlCl_3$	CH_2Cl_2	81	210
C_6H_5	RCOCl	$AlCl_3$	CS_2	67	200
	RCOCl	$AlCl_3$	CH_2Cl_2	70–75	206
$C_6H_4CH_3\text{-}2$	RCOCl	$AlCl_3$	CH_2Cl_2	—	197
$C_6H_4CH_3\text{-}3$	RCOCl	$AlCl_3$	CH_2Cl_2	—	197
$C_6H_4CH_3\text{-}4$	RCOCl	$AlCl_3$	CH_2Cl_2	—	197
$C_6H_4OH\text{-}2$	RCOCl	$AlCl_3$	CH_2Cl_2, $(C_2H_5)_2O$	22	199
$C_6H_4OCH_3\text{-}2$	RCOCl	$AlCl_3$	CH_2Cl_2	45	319
$C_6H_4OCH_3\text{-}4$	RCOCl	$AlCl_3$	CS_2	23	200
$C_6H_3(OCH_3)_2\text{-}2,4$	RCOCl	$AlCl_3$	CH_2Cl_2	41–57	319
$C_6H_3(OCH_3)_2\text{-}3,4$	RCOCl	$AlCl_3$	CH_2Cl_2	ca. 50	615
2-Pyrryl	RCOCl	$AlCl_3$	CH_2Cl_2	—	197
2-Furyl	RCOCl	$AlCl_3$	CH_2Cl_2	—	197
2-Thienyl	RCOCl	$AlCl_3$	CH_2Cl_2	63	201
$CH_2(C_4H_3S\text{-}2)$	RCOCl	$AlCl_3$	CH_2Cl_2	60–80	262

Note: References 510–659 are on pp. 151–154.

* When an acyl ferrocene has been prepared in the same way by several different groups of chemists, a representative preparation is listed. When more than one method of preparation has been used, an example of each is listed.

† Metallic aluminum was also present.

TABLE XIII. MONOACYL FERROCENES (*Continued*)

A. By Friedel-Crafts Reaction (*Continued*)

R in Acyl Ferrocene $C_5H_5FeC_5H_4COR$*	Reagent	Lewis Acid	Solvent	Yield (%)	Refs.
2-($C_4H_2SCOCH_3$-5)	RCOCl	$AlCl_3$	CH_2Cl_2	—	201
$C_5H_4FeC_5H_5$	$(COCl)_2$	$AlCl_3$	CH_2Cl_2	0.5	277
	RCOCl	$AlCl_3$	CH_2Cl_2	25	429
	$COCl_2$	$AlCl_3$	CH_2Cl_2	12	202
$C_5H_4Mn(CO)_3$	RCOCl	—	—	ca. 50	463
$CH_2CH_2CO_2H$	$(RCO)_2O$	$AlCl_3$	CH_2Cl_2	91	196
	Fumaryl chloride	$AlCl_3$	CH_2Cl_2	36	366
$CH_2CH_2CO_2CH_3$	Ester acid chloride	$AlCl_3$	CS_2	32	344
$(CH_2)_3CO_2H$	$(RCO)_2O$	$AlCl_3$	CH_2Cl_2	92	211
$CH_2CH(CH_3)CO_2H$	Citraconyl chloride	$AlCl_3$	CH_2Cl_2	ca. 30	366
	$(RCO)_2O$	$AlCl_3$	CH_2Cl_2	12	598
$CH(CH_3)CH_2CO_2H$	$(RCO)_2O$	$AlCl_3$	CH_2Cl_2	4	598
$(CH_2)_4CO_2C_2H_5$	Ester acid chloride	$AlCl_3$	CH_2Cl_2	71	211
$CH_2C(CH_3)_2CO_2H$	$(RCO)_2O$	$AlCl_3$	CH_2Cl_2	ca. 50	598
$CH_2CH(CH_3)CH_2CO_2H$	$(RCO)_2O$	$AlCl_3$	CH_2Cl_2	65	598
1-(2-Carboxycyclohexyl)	$(RCO)_2O$	$AlCl_3$	—	—	197
$CH{=}CHCO_2H$	Bromo-succinic anhydride	$AlCl_3$	—	—	197
$CH{=}CHCO_2CH_3$	Ester acid chloride	$AlCl_3$	$(C_2H_5)_2O$	—	593

$CH_2CH(C_6H_5)CO_2H$	$(RCO)_2O$	$AlCl_3$	—	—	348, 598
$CH(C_6H_5)CH_2CO_2H$	$(RCO)_2O$	$AlCl_3$	—	—	348, 598
$CH_2C(C_6H_5)_2CO_2H$	$(RCO)_2O$	$AlCl_3$	CH_2Cl_2	—	598
$CH_2CH_2COC_5H_4FeC_5H_5$	Fumaryl chloride	$AlCl_3$	CH_2Cl_2	ca. 10	366
	Succinyl chloride	$AlCl_3$	CH_2Cl_2	ca. 20	198
$CH_2(CH_2)_2COC_5H_4FeC_5H_5$	Citraconyl chloride	$AlCl_3$	CH_2Cl_2	ca. 7	366
$CH_2C_6H_4CO_2H$-2	$(RCO)_2O$	—	—	—	355
$C_6H_4CO_2H$-2	$(RCO)_2O$	—	—	—	63, 351
$C_6H_4CO_2CH_3$-2	RCOCl	$AlCl_3$	CS_2	83	344
$C_6H_4CH_2CO_2H$-2	$(RCO)_2O$	—	—	—	355

B. By Other Routes

R in Acyl Ferrocene $C_5H_5FeC_5H_4COR$*	Route	Solvent	Yield (%)	Refs.
CH_3	MnO_2 oxidation of $C_5H_5FeC_5H_4C_2H_5$	Methylcyclohexane	15	242
	CH_3MgI and $C_5H_5FeC_5H_4CN$	$(C_2H_5)_2O$	79	168
C_2H_5	C_2H_5MgBr and $C_5H_5FeC_5H_4CN$	$(C_2H_5)O_2$	70	168
$C_5H_4FeC_5H_5$	MnO_2 oxidation of $(C_5H_5FeC_5H_4)_2CH_2$	$CHCl_3$	72	242
	$C_5H_5FeC_5H_4Li$ and $C_5H_5FeC_5H_4CN$	THF	80	168
CO_2CH_3	MnO_2 oxidation of $C_5H_5FeC_5H_4CH_2CO_2CH_3$	$CHCl_3$	42	382
$C_5H_5FeC_5H_4CO$	MnO_2 oxidation of $C_5H_5FeC_5H_4COCH_2C_5H_4FeC_5H_5$	$CHCl_3$	66	242

Note: References 510–659 are on pp. 151–154.

* When an acyl ferrocene has been prepared in the same way by several different groups of chemists, a representative preparation is listed. When more than one method of preparation has been used, an example of each is listed.

TABLE XIV. HETEROANNULAR DIACYL FERROCENES FROM REACTION OF FERROCENES WITH ACID CHLORIDES AND ALUMINUM CHLORIDE

A. Symmetrical Diacyl Ferrocenes from Ferrocene

Substituents in $R_1COC_5H_4FeC_5H_4COR_2$	Yield (%)	Refs.
$R_1 = R_2 =$		
CH_3	60–74	63, 207
	85–89	141, 178, 599
C_2H_5	40	600
CH_2CH_2Cl	—	63
C_3H_7-*n*	40–77	600, 602
C_3H_7-*i*	—	595
C_4H_9-*t*	31	544, 594, 595
$CH_2C_4H_9$-*t*	—	601
$CH_2CH_2Si(CH_3)_3$	57	413
$(CH_2)_3Si(CH_3)_3$	73	413
C_7H_{15}-*n*	65	599
C_9H_{19}-*n*	41	599
$C_{11}H_{23}$-*n*	44	599
$C_{12}H_{25}$-*n*	15	599
$C_{15}H_{31}$-*n*	12	599
Cyclohexyl	—	311, 595
$CH{=}CH_2$	—	63
$CH_2C_6H_5$	60	197
C_6H_5	52–91	206, 207, 221
C_6H_4F-*p*	76	602
C_6H_4Cl-*p*	52	221
2-Thienyl	37	201
$C_5H_4Mn(CO)_3$	85	463, 603
x-$C_5H_3(CH_3)Mn(CO)_3$	—	603

$CH_2CH_2CO_2H$		18–38	139, 177, 351
$CH_2CH_2CO_2CH_3$		27–33	332, 351
$(CH_2)_3CO_2C_2H_5$		70	332
$CH{=}CHCO_2CH_3$		22	593
$C_6H_4CO_2H$-*o*		—	351
$C_6H_4CO_2CH_3$-*o*		—	63

B. Unsymmetrical Diacyl Ferrocenes from Monoacyl Ferrocenes

R_1 was present in the monoacyl ferrocene; R_2 was introduced in the reaction reported in this table.

R_1	R_2		
$R_1 = CH_3$	$R_2 = C_2H_5$	57	340
CH_3	C_3H_7-*n*	54	340
CH_3	$CH(CH_3)CH_2CH_3$	50	340
CH_3	$(CH_2)_7CH{=}CH(CH_2)_7CH_3$	53	340
CH_3	$CH_2C_6H_5$	55	340
CH_3	C_6H_5	48	178
$(CH_2)_2CO_2C_2H_5$	$(CH_2)_3CO_2C_2H_5$	65	332
$N(C_6H_5)_2$	CH_3	74	307
$N(C_6H_5)_2$	C_6H_5	71	307
OCH_3	CH_3	—	604
OCH_3	C_2H_5	—	604
OCH_3	C_3H_7-*n*	45	192
OCH_3	C_6H_5	—	604
$CH_2CH_2Si(CH_3)_3$	$(4\text{-}C_6H_4)COC_5H_4FeC_5H_4CO(CH_2)_2Si(CH_3)_3$	22	412

Note: References 510–659 are on pp. 151–154.

TABLE XV. Examples of Compounds Prepared from Acyl Ferrocenes

Substituent R in $C_5H_5FeC_5H_4R$	Reactants*	Yield (%)	Refs.
R =			
$C(CH_3)=CCH=CHCH=CH$ (ring closed between C and terminal CH)	C_5H_6, NaOH	—	230
$C(C_6H_5)=C(C_6H_5)C_5H_4FeC_5H_5$	$C_5H_5FeC_5H_4COC_6H_5$, $(C_6H_5)_2PONa$	47	225
$CH(NH_2)CH_3$	Liq. NH_3, Raney Ni, C_2H_5OH	Low	139, 606
$CHOHCH_3$	$LiAlH_4$	89	204
$CHOHCH_2Cl$	$C_5H_5FeC_5H_4COCH_2Cl$, $LiAlH_4$	100	195
$CH(OCH_3)CH_3$	$NaBH_4$, CH_3OH	48	226
$C(CH_3)_2OH$	CH_3MgI	58–75	55, 223
$C\equiv CH$	$POCl_3$, $HCON(CH_3)_2$	93	262
$C(Cl)=CH_2$	$POCl_3$, $HCON(CH_3)_2$	60	262
$C(OH)(CH_3)C\equiv CH$	$NaC\equiv CH$	16	202
$CHOHC_6H_5$	$C_5H_5FeC_5H_4COC_6H_5$, $LiAlH_4$	96	200
	$C_5H_5FeC_5H_4COC_6H_5$, $NaBH_4$	66	206
$CHOHCH_2C_6H_5$	$C_5H_5FeC_5H_4COCH_2C_6H_5$, $LiAlH_4$	93	198
$CHOHC_4H_3S$-2	$C_5H_5FeC_5H_4COC_4H_3S$-2, $LiAlH_4$	95	201
$C(OH)(C_5H_4FeC_5H_5)_2$	$(C_5H_5FeC_5H_4)_2CO$, $C_5H_5FeC_5H_4Li$	57–60	131, 168
$C(OH)C(OH)C_5H_4FeC_5H_5$ (with C_2H_5 on each C)	$C_5H_5FeC_5H_4COC_2H_5$, Mg, $HgCl_2$	31	605
$C(OCOCH_3)=CH_2$	$CH_2=C(CH_3)OCOCH_3$, p-$CH_3C_6H_4SO_3H$	25	140
CH_2CO_2H	(1) S, morpholine, (2) KOH, H_2O	10	126
$CH_2CO_2CH_3$	(1) S, morpholine, (2) KOH, CH_3OH	18	139
$COCH(CH_2C_6H_5)_2$	$C_6H_5CH_2Cl$, KNH_2, liq. NH_3	39	212
$COCH=CHC_6H_5$	C_6H_5CHO, NaOH	70	163
$COCH=C(CH_3)C_5H_4FeC_5H_5$	KOC_4H_9-t, C_6H_6	64	131
$COCH_2CH_2N(CH_3)_2$	CH_2O, $(CH_3)_2NH$, HCl	47–69	203, 210
$COCH_2COCH_3$	$CH_3CO_2C_2H_5$, $NaOCH_3$	63	215
	$CH_3CO_2C_2H_5$, KNH_2, liq. NH_3	60	232
$COCH_2COC_6H_5$	$C_6H_5CO_2CH_3$, KNH_2, liq. NH_3	63	163
$COCH_2COC_5H_4FeC_5H_5$	$C_5H_5FeC_5H_4CO_2CH_3$, $NaNH_2$	30	218
$COCH_2CO_2C_2H_5$	NaH, $CO(OC_2H_5)_2$	83	205
	KNH_2, $CO(OC_2H_5)_2$	65	163

Note: References 510–659 are on pp. 151–154.

* The acyl ferrocene employed was acetylferrocene unless otherwise specified.

TABLE XVI. ALKYL FERROCENES

A. Monoalkyl Ferrocenes Prepared by Alkylation of Ferrocene

Alkyl Group(s) in Product	Reagents/Method	Yield (%)	Refs.
CH_3	CH_3Cl, C_7H_{16}	ca. 4	235
C_2H_5	C_2H_5Br, C_7H_{16}	18	235
	C_2H_4	20	236
	C_2H_5Br, $LiAlH_4$	—	612, 618
C_3H_7-*i*	C_3H_6	30	236
	i-C_3H_7Cl, C_7H_{16}	ca. 5	235
C_4H_9-*t*	$(CH_3)_2C{=}CH_2(H_3PO_4, BF_3)$	ca. 15(45)	236, 607, 608
	t-C_4H_9Cl, C_7H_{16}	14	607
	t-C_4H_9OH, H_3PO_4	22	611
	t-C_4H_9Cl, $LiAlH_4$	—	612
$C(CH_3)_2C_2H_5$	$(CH_3)_2C(Cl)C_2H_5$	14	607
$CH(CH_3)C_3H_7$-*i*	$CH_3CH{=}C(CH_3)_2$, H_3PO_4, BF_3	25	608
$CH_2C(CH_3)_2CH_2CH(CH_3)_2$	Diisobutylene, H_3PO_4	ca. 90	613
$CH(CH_3)CH_2CH_2CH(CH_3)_2$	Isooctylene, H_3PO_4, BF_3	50	608
$CH_2[C(CH_3)_2CH_2]_2CH(CH_3)_2$	Triisobutylene, H_3PO_4	38	613
Cyclohexyl	Cyclohexene, H_3PO_4	51–74	613
$CH_2C_6H_5$	$C_6H_5CH_2Cl$	—	237
	$C_6H_5CH_3$, $C_6H_5CO_2OC_4H_9$-*t*	ca. 22	609, 610
$C(CH_3)(C_6H_5)_2$	$(C_6H_5)_2C{=}CH_2$	15	611
$C(C_6H_5)_3$	$(C_6H_5)_3COH$	72	611
$CH_2CH_2C_5H_4FeC_5H_5$	CH_2ClCH_2Cl	—	246
$CH(CH_3)C_5H_4FeC_5H_5$*	CH_2ClCH_2Cl	—	250
	CH_2ClCH_2Cl	27	237

Note: References 510–659 are on pp. 151–154.

* An incorrect structure was originally assigned this material; see ref. 245.

TABLE XVI. ALKYL FERROCENES (*Continued*)

B. Di- and Poly-alkylated Ferrocenes Prepared by Alkylation of Ferrocene

Alkyl Group(s) in Product	Reagents/Method	Yield (%)	Refs.
1,1′-$(CH_3)_2$†	CH_3Cl, C_7H_{16}	ca. 6	235
1,1′-$(C_2H_5)_2$	C_2H_5Br, C_7H_{16}	39	235
1,1′-$(C_3H_7\text{-}i)_2$†	$i\text{-}C_3H_7Cl$, C_7H_{16}	ca. 5	235
1,1′-$(C_4H_9\text{-}t)_2$†	$(CH_3)_2C{=}CH_2$, H_3PO_4, BF_3	26	607, 608
	$t\text{-}C_4H_9Cl$	10–26	544, 607
1,3-$(C_4H_9\text{-}t)_2$	$t\text{-}C_4H_9Cl$	—	544
1,1′,3-$(C_4H_9\text{-}t)_3$	$t\text{-}C_4H_9Cl$	—	544
1,1′,3,3′-$(C_4H_9\text{-}t)_4$	$t\text{-}C_4H_9Cl$	68	544
1,1′-$(C_5H_{11}\text{-}t)_2$†	$(CH_3)_2C(Cl)C_2H_5$	15	607
1,1′-$[CH(CH_3)C_3H_7\text{-}i]_2$	$(CH_3)_2C{=}CHCH_3$, H_3PO_4, BF_3	20	608
1,1′-$(C_6H_{11})_2$	Cyclohexene, H_3PO_4	26–36	613
1,1′-$[CH(C_6H_5)_2]_2$†	$(C_6H_5)_2CHOH$	43–55	611
1,1′-$[C(CH_3)(C_6H_5)_2]_2$	$(C_6H_5)_2C{=}CH_2$	27	611
1,1′-$[C(C_6H_5)_3]_2$	$(C_6H_5)_3COH$	95	611

C. Monoalkyl Ferrocenes Prepared by Reduction

Alkyl Group(s) in Product	Reagents/Method	Substituent in Ferrocene Reduced	Yield (%)	Refs.
CH_3	$LiAlH_4/AlCl_3$	CH_2OH	100	234
	$LiAlH_4/AlCl_3$	CHO	—‡	331
	Clemmensen	CHO	66	91
	Na(Hg), H_2O	$CH_2N(CH_3)_3I$	94	244
	$LiAlH_4/AlCl_3$	$CO_2H(CH_3)$	90–100	167, 234
	$LiAlH_4/AlCl_3$	$CH_2OCH_2C_5H_4FeC_5H_5$	Quant.	331
	$LiAlH_4$	$CH_2P(C_6H_5)_3I$	82	225

C_2H_5	Clemmensen	$COCH_3$	67	600
	H_2, PtO_2, C_2H_5OH	$COCH_3$	77	172
	$LiAlH_4/AlCl_3$	$COCH_3$	100	167, 331
C_3H_7-*n*	H_2, Pd/C, C_2H_5OH	$CH{=}CHCH_3$	85	202
	H_2, Pd/C, C_2H_5OH	$CH_2C{\equiv}CH$	100	269
	$LiAlH_4/AlCl_3$	COC_2H_5	—‡	331
	Clemmensen	COC_2H_5	—	65
C_3H_7-*i*	RMgX, then Clemmensen	$COCH_3$	95	55, 65
C_4H_9-*n*	Na, C_2H_5OH	$CH{=}CHC{\equiv}CH$	95	225
	Clemmensen	COC_3H_7-*n*	7	596
	$LiAlH_4/AlCl_3$	COC_3H_7-*n*	—	331
C_4H_9-*i*	Clemmensen	COC_3H_7-*i*	—	592
C_4H_9-*sec*	Hydrogenation	$CH(CH_3)C{\equiv}CH$	—	269
	H_2, Pd/C, C_2H_5OH	$C(OH)(CH_3)C{\equiv}CH$	95	202
	H_2, Pd/C, C_2H_5OH	$C(OH)(CH_3)C_2H_5$	—	202
	Clemmensen	$C(OH)(CH_3)C_2H_5$	—	65
$(CH_2)_3Si(CH_3)_3$	Clemmensen	$CO(CH_2)_2Si(CH_3)_3$	78	413
C_5H_{11}-*n*	H_2, Raney Ni	$CH_2C_4H_3S$-2	80	201
	$LiAlH_4/AlCl_3$	COC_4H_9-*n*	—	331
$CH_2C(CH_3)_3$	Clemmensen	COC_4H_9-*t*	—	592, 597
$(CH_2)_4Si(CH_3)_3$	Clemmensen	$CO(CH_2)_3Si(CH_3)_3$	91	413
C_6H_{13}-*n*	Clemmensen	COC_5H_{11}-*n*	93	596
$CH(CH_3)C_4H_9$-*t*	RMgX, then Clemmensen	$COCH_3$	—	65
$CH_2CH(C_2H_5)_2$	Clemmensen	$COCH(C_2H_5)_2$	—	597
C_8H_{17}-*n*	Clemmensen	COC_6H_{13}-*n*	71	596
$CH_2C(C_2H_5)_3$	Clemmensen	$COC(C_2H_5)_3$	—	597
$CH_2CH(C_3H_7$-*i*$)C_4H_9$-*n*	Clemmensen	$COCH(C_3H_7$-*i*$)C_4H_9$-*n*	—	597

Note: References 510–659 are on pp. 151–154.

† This product was also prepared directly from a substituted cyclopentadiene; see Table II.
‡ Yields were not reported; however, they are in general greater than 90%.

TABLE XVI. ALKYL FERROCENES (*Continued*)

C. Monoalkyl Ferrocenes Prepared by Reduction (*Continued*)

Alkyl Group(s) in Product	Reagents/Method	Substituent in Ferrocene Reduced	Yield (%)	Refs.
$C_{10}H_{21}$-*n*	Clemmensen	COC_9H_{19}-*n*	10	596
$C_{12}H_{25}$-*n*	Clemmensen	$COC_{11}H_{23}$-*n*	96	596
$C_{16}H_{33}$-*n*	Clemmensen	$COC_{15}H_{31}$-*n*	77	599
Cyclopentyl	H_2, PtO_2, C_6H_6	Cyclopentenyl	92	164, 614
$CH_2C_6H_5$	Clemmensen	COC_6H_5	49	200, 246
	H_2, Pt/C, *n*-C_4H_9OH	COC_6H_5	70	206
	Na, C_2H_5OH	COC_6H_5	98	206
	$LiAlH_4/AlCl_3$	COC_6H_5	90–100	167, 331
$CH_2CH_2C_6H_5$	Clemmensen	$COCH_2C_6H_5$	—	65, 197, 592, 601
$CH(CH_3)C_6H_5$	RMgX, then Clemmensen	Not stated	—	65
$CH(C_2H_5)C_6H_5$	RMgX, then $LiAlH_4/AlCl_3$	COC_6H_5	50	160
	RMgX, then Clemmensen	Not stated	—	65
$CH(CH_3)CH_2C_6H_5$	RMgX, then $LiAlH_4/AlCl_3$	$COCH_3$	68	160
$CH_2(CH_2)_3C_6H_5$	H_2, Pd/C, $CH_3CO_2C_2H_5$	$CH{=}CHCH{=}CHC_6H_5$	—	195
$CH_2CH(C_6H_5)_2$	Na, C_2H_5OH	$CH{=}C(C_6H_5)_2$	94	225
$CH_2C_6H_3(OCH_3)_2$-3,4	$LiAlH_4/AlCl_3$	$COC_6H_3(OCH_3)_2$-3,4	66	615
$CH_2C_4H_3S$-2	$LiAlH_4/AlCl_3$	COC_4H_3S-2	90	201
	Na, C_2H_5OH	COC_4H_3S-2	17	201
$CH_2C_5H_4FeC_5H_5$	Na, C_2H_5OH	$CHOHC_5H_4FeC_5H_5$	26	202
	H_2, Pd/C, C_2H_5OH	$CHOHC_5H_4FeC_5H_5$	30	202
$CH_2CH_2C_5H_4FeC_5H_5$	Na, C_2H_5OH	$CH{=}CHC_5H_4FeC_5H_5$	60	225
$CH(CH_3)C_5H_4FeC_5H_5$	Hydrogenation	$C(C_5H_4FeC_5H_5){=}CH_2$	—	250
$(CH_2)_3C_5H_4FeC_5H_5$	$LiAlH_4/AlCl_3$	$COCH_2CH_2C_5H_4FeC_5H_5$	—	195

$(CH_2)_4C_5H_4FeC_5H_5$	H_2, Pd/C, C_2H_5OH	$CH_2C{\equiv}CCH_2C_5H_4FeC_5H_5$	—	269
	H_2, Pd/C, cyclohexane	$CH{=}CHCH{=}CHC_5H_4FeC_5H_5$	—	202
$(CH_2)_6C_5H_4FeC_5H_5$	H_2, Pt, cyclohexane	$CH_2(C{\equiv}C)_2CH_2C_5H_4FeC_5H_5$	86	269
	H_2, Pd/C, C_2H_5OH	$CH{=}CH(CH_2)_2CH{=}CHC_5H_4$-$FeC_5H_5$	—	202
$CH(CH_3)(CH_2)_4CH(CH_3)$-$C_5H_4FeC_5H_5$	H_2, Pd/C, C_2H_5OH	$C(OH)(CH_3)C{\equiv}CC{\equiv}CC(OH)$-$(CH_3)C_5H_4FeC_5H_5$	61	202
$(CH_2)_8C_5H_4FeC_5H_5$	Na, C_2H_5OH	$CH{=}CH(C{\equiv}C)_2CH{=}CHC_5H_4$-$FeC_5H_5$	79	225
$CH(C_6H_5)C_5H_4FeC_5H_5$	Zn, C_2H_5OH, $TiCl_3$	$C(OH)(C_6H_5)C_5H_4FeC_5H_5$	87	131
$C(C_5H_4FeC_5H_5)(C_6H_5)$ $CH_2C_6H_5$	$LiAlH_4$	$C(C_5H_4FeC_5H_5)(C_6H_5)$ COC_6H_5	—	246
$CH(C_5H_4FeC_5H_5)_2$	Zn, C_2H_5OH, $TiCl_3$	$C(OH)(C_5H_4FeC_5H_5)_2$	64	131
	Zn, CH_3CO_2H	$C(OH)(C_5H_4FeC_5H_5)_2$	59	168
$CH(C_6H_5)CH(C_6H_5)C_5H_4$-$FeC_5H_5$	Na, C_2H_5OH	$C(C_6H_5){=}C(C_6H_5)C_5H_4FeC_5H_5$	80	225
$CH_2C_6H_4CH_2CO_2H$-2	Clemmensen	$COC_6H_4CH_2CO_2H$-2	—	224, 355
$CH_2CH_2C_6H_4CO_2H$-2	Clemmensen	$COCH_2C_6H_4CO_2H$-2	—	224, 355
$CH_2CH(C_6H_5)CH_2CO_2H$	Clemmensen	$COCH(C_6H_5)CH_2CO_2H$	—	348
$(CH_2)_2CH(C_6H_5)CO_2H$	Clemmensen	$COCH_2CH(C_6H_5)CO_2H$	—	348
$(CH_2)_2CO_2H$	Clemmensen	$COCH_2CO_2C_2H_5$	30	258
$(CH_2)_3CO_2C_2H_5$	Clemmensen	$CO(CH_2)_2CO_2C_2H_5$	100	258
		D. Di- and Poly-alkyl Ferrocenes Prepared by Reduction		
1,1′-$(CH_3)_2$	$LiAlH_4$	1,1′-$(CO_2CH_3)_2$	89	104
	$LiAlH_4/AlCl_3$	1,1′-$(CO_2CH_3)_2$	69	176, 234
1,2-$(CH_3)_2$	Na(Hg), H_2O	1-$CH_2N(CH_3)_3I$-2-CH_3	—	129
1,1′-$(C_2H_5)_2$	$LiAlH_4/AlCl_3$	1,1′-$(COCH_3)_2$	—‡	331
	Clemmensen	1,1′-$(COCH_3)_2$	40–60	599, 600

Note: References 510–659 are on pp. 151–154.

‡ Yields were not reported; however, they are in general greater than 90%.

TABLE XVI. ALKYL FERROCENES (*Continued*)

D. Di- and Poly-alkyl Ferrocenes Prepared by Reduction (*Continued*)

Alkyl Group(s) in Product	Reagents/Method	Substituent in Ferrocene Reduced	Yield (%)	Refs.
1,1′-(C_3H_7-*n*)$_2$	$LiAlH_4/AlCl_3$	1,1′-(COC_2H_5)$_2$	‡	331
	Clemmensen	1,1′-(COC_2H_5)$_2$	70	600
1,1′-(C_3H_7-*i*)$_2$	RMgX, then $LiAlH_4/AlCl_3$	1,1′-($COCH_3$)$_2$	88	55
1,1′-(CH_2)$_3$Si(CH_3)$_3$	Clemmensen	1,1′[$COCH_2CH_2$Si(CH_3)$_3$]$_2$	82	413
1,1-(C_4H_9-*n*)$_2$	$LiAlH_4/AlCl_3$	1,1′-(COC_3H_7-*n*)$_2$	‡	331
	Clemmensen	1,1′-(COC_3H_7-*n*)$_2$	60	600
1,1′-(CH_2)$_4$Si(CH_3)$_3$	Clemmensen	1,1′-[CO(CH_2)$_3$Si(CH_3)$_3$]$_2$	72	413
1,1′-(C_5H_{11}-*n*)$_2$	$LiAlH_4/AlCl_3$	1,1′-(COC_4H_9-*n*)$_2$	‡	331
	H_2, Raney Ni	1,1′-(C_4H_3S-2)$_2$	70	201
1,1′-($CH_2C_4H_9$-*t*)$_2$	Clemmensen	1,1′-(COC_4H_9-*t*)$_2$	—	179, 595
1,1′-(C_5H_{11}-*i*)$_2$	Clemmensen	1,1′-($COCH_2C_3H_7$-*i*)$_2$	—	601
1,1′-($CH_2CH_2C_4H_9$-*t*)$_2$	Clemmensen	1,1′-($COCH_2C_4H_9$-*t*)$_2$	—	601
1,1′-(C_8H_{17}-*n*)$_2$	Clemmensen	1,1-(COC_7H_{15}-*n*)$_2$	58	599
1,1′-[(CH_2)$_2$CH(CH_3)-$CH_2C_4H_9$-*t*]$_2$	Clemmensen	1,1′-[$COCH_2$CH(CH_3)CH_2-C_4H_9-*t*]$_2$	—	601
1,1′-($C_{10}H_{21}$-*n*)$_2$	Clemmensen	1,1′-(COC_9H_{19}-*n*)$_2$	50	599
1,1′-($C_{12}H_{25}$-*n*)$_2$	Clemmensen	1,1′-($COC_{11}H_{23}$-*n*)$_2$	—	599
1,1′-($C_{13}H_{27}$-*n*)$_2$	Clemmensen	1,1′-($COC_{12}H_{25}$-*n*)$_2$	32	599
1,1′-($C_{16}H_{33}$-*n*)$_2$	Clemmensen	1,1′-($COC_{15}H_{31}$-*n*)$_2$	33	599
1,1′-($CH_2CH_2C_6H_{11}$)$_2$	Clemmensen	1,1′-($COCH_2C_6H_{11}$)$_2$	—	601
1,1′-($CH_2C_6H_5$)$_2$†	$LiAlH_4/AlCl_3$	1,1′-(COC_6H_5)$_2$	‡	331
	Clemmensen	1,1′-(COC_6H_5)$_2$	60	600
	H_2, Pt/C, *n*-C_4H_9OH	1,1′-(COC_1H_5)$_2$	68	206
	Na, C_2H_5OH	1,1′-(COC_6H_5)$_2$	80	206
	$LiAlH_4$	1,1′-(COC_6H_5)$_2$	—	104

1,1′-$(CH_2CH_2C_6H_5)_2$	Clemmensen	1,1′-$(COCH_2C_6H_5)_2$	—	197
1,1′-$[CH(C_6H_5)_2]_2$†§	$LiAlH_4$	1,1′-$[C(OH)(C_6H_5)_2]_2$	87	104
1,1′-$[CH_2CH_2C_6H_3$-$(CH_3)_2$-3,4$]_2$	Clemmensen	1,1′-$[COCH_2C_6H_3(CH_3)_2$-3,4$]_2$	—	601
1,1′-$(CH_2C_4H_3S$-2$)_2$	$LiAlH_4/AlCl_3$	1,1′-$(COC_4H_3S$-2$)_2$	95	201
1-CH_3-1′-C_6H_5	Na(Hg), H_2O	1-C_6H_5-1′-$CH_2N(CH_3)_3I$	71	128
1-C_5H_{11}-*n*-1′-$CH_2C_4H_3S$-2	Clemmensen	1,1′-$(COC_4H_3S$-2$)_2$	48	201
1-$CH_2C_4H_9$-*t*-1′-C_3H_7-*n*	Clemmensen	—	—	179
1-$CH_2C_4H_9$-*t*-1′-C_4H_9-*i*	Clemmensen	—	—	179
1-$CH_2C_4H_9$-*t*-1′-$(CH_2)_2$ $CH(CH_3)CH_2C_4H_9$-*t*	Clemmensen	—	—	179
1-$CH_2C_4H_9$-*t*-1′-$CH_2C_6H_{11}$	Clemmensen	—	—	179
1-$CH_2C_4H_9$-*t*-1′-$CH_2CH_2C_6H_5$	Clemmensen	—	—	179
1-$CH_2C_4H_9$-*t*-1′-$CH_2C_6H_4Cl$	Clemmensen	—	—	179
1,1′-$(CH_3)_2$-2-C_2H_5	H_2, PtO_2	1,1′-$(CH_3)_2$-2-$COCH_3$	—	182
1,1′-$(CH_3)_2$-3-C_2H_5	H_2, PtO_2	1,1′-$(CH_3)_2$-3-$COCH_3$	—	182
1,1′-$(CH_3)_2$-2,2′-$(C_2H_5)_2$	H_2, catalyst	1,1′-$(CH_3)_2$-2,2′-$(COCH_3)_2$	—	62

E. Alkyl Ferrocenes Prepared by Other Reactions

Alkyl Group(s) in Product	Reagents/Method	Substituent in Starting Ferrocene	Yield (%)	Refs.
CH_3	(1) Li, THF, (2) NH_4Cl	CH_2OCH_3	82	616
C_2H_5	RMgX	$CH_2N(CH_3)_3I$	59	243
C_3H_7-*n*	RMgX	$CH_2N(CH_3)_3I$	63	243

Note: References 510–659 are on pp. 151–154.

† This product was also prepared directly from a substituted cyclopentadiene; see Table II.
‡ Yields were not reported; however, they are in general greater than 90%.
§ This product was also prepared from the appropriately substituted fulvene; see Table III.

TABLE XVI. ALKYL FERROCENES (*Continued*)

E. Alkyl Ferrocenes Prepared by Other Reactions (*Continued*)

Alkyl Group(s) in Product	Reagents/Method	Substituent in Starting Ferrocene	Yield (%)	Refs.
$CH_2CH{=}CH_2$	RMgX	$CH_2N(CH_3)_3I$	34	243
$CH_2CH_2CH{=}CH_2$	RMgX	$CH_2N(CH_3)_3I$	53	243
$CH_2C_5H_5$	C_5H_5Na, THF	$CH_2N(CH_3)_3I$	65	91
	RMgX	$CH_2N(CH_3)_3I$	27	243
$CH_2CH_2C_6H_5$	RMgX	$CH_2N(CH_3)_3I$	60	243
$CH_2C(C_6H_5)_3$	$(C_6H_5)_3CCl$	$HgC_5H_4FeC_5H_5$	18	326
$CH_2C_{10}H_7$-1	RMgX	$CH_2N(CH_3)_3I$	51	243
$CH_2C_5H_4FeC_5H_5$	CH_2O, H_2SO_4	H	58	131
	Ferrocene, H_2SO_4	CH_2OH	29	131
$CH_2CH_2C_5H_4FeC_5H_5$	CH_2O, HF	H	50	164
	CH_2O, H_2SO_4	H	65–75	247
	$FeCl_3$	CH_2Li	34	616
$CH(C_6H_5)CH(C_6H_5)$-$C_5H_4FeC_5H_5$	$SOCl_2$ or $KHSO_4$	$CHOHC_6H_5$	—	246
	H_2SO_4, C_6H_5CHO	H	—	245–248
$CH_2Si(CH_3)_3$	$(CH_3)_3SiCl$	CH_2Li	68	616

Note: References 510–659 are on pp. 151–154.

TABLE XVII. SUBSTITUTED ALKYL FERROCENES PREPARED FROM REDUCTION OF SUBSTITUTED ACYL FERROCENES

Substituents in Product	Acyl Group Reduced	Reagents	Yield (%)	Refs.
1,1′-$(CH_3)_2$	1-CHO	$LiAlH_4/AlCl_3$	—	55
1,2-$(CH_3)_2$	1-CHO	$LiAlH_4/AlCl_3$	—	55
1,3-$(CH_3)_2$	1-CHO	$LiAlH_4/AlCl_3$	—	55
1,2-$(C_2H_5)_2$	1-$COCH_3$	$LiAlH_4/AlCl_3$	>90	239
1,1′-$(CH_3)_2$-2-C_2H_5	$COCH_3$	H_2, PtO_2	—	182
1,1′-$(CH_3)_2$-3-C_2H_5	$COCH_3$	H_2, PtO_2	—	182
1,1′,2-$(C_2H_5)_3$	1-$COCH_3$	$LiAlH_4/AlCl_3$	>90	239
1,1′,3-$(C_2H_5)_3$	1-$COCH_3$	$LiAlH_4/AlCl_3$	>90	239
1-C_2H_5-1′,3-$(C_6H_5)_2$	$COCH_3$	Zn(Hg), CH_3CO_2H	40	185
1-C_2H_5-1′-Br	$COCH_3$	Zn(Hg), HCl	77	188
1′-C_2H_5-1-CO_2H	$COCH_3$	Zn(Hg), CH_3CO_2H/HCl	70	192
1′-C_4H_9-*n*-1-CO_2H	$CO(CH_2)_2CH_3$	Zn(Hg), CH_3CO_2H/HCl	74	192

Note: References 510–659 are on pp. 151–154.

TABLE XVIII. ALKENYL FERROCENES

A. Monosubstituted Alkenyl Ferrocenes

Alkenyl Group(s) in Product	Substituent(s) in Starting Ferrocene	Reagent(s)	Yield (%)	Refs.
$CH{=}CH_2$	$CHOHCH_3$	Al_2O_3	21–50	202, 204, 259
	$CH_2CH_2N(CH_3)_3I$	KNH_2, liq. NH_3	ca. 60	132
	$CH_2CH_2N(CH_3)_3I$	KOC_4H_9-*t*	81	225
	$CH(OCOCH_3)CH_3$	Pyrolysis	56–85	204, 255
	$CHOHCH_3$	$(CH_3CO)_2O$, C_5H_5N	—	617
	$CHOHCH_3$	$CuSO_4$, $C_6H_5CH_3$, diamylhydroquinone	97	618
$CH{=}CHCl$	$CHOHCH_2Cl$	Al_2O_3	77	195
$CH{=}CCl_2$	$CHOHCHCl_2$	Al_2O_3	21	195
$CH{=}CHCH_3$	$CHOHCH_2CH_3$	Al_2O_3	81	202
$CH{=}CHCH{=}CH_2$	CHO	Phosphorus ylide	50	91
$CH{=}CHC{\equiv}CH$	$CH{=}P(C_6H_5)_3$	$HC{\equiv}CCHO$	71	225
$C(CH_3){=}CHCH_3$	$C(OH)(CH_3)CH_2CH_3$	Al_2O_3	80	202
	$C(OH)(CH_3)CH_2CH_3$	Al_2O_3	—	259
$CH{=}CHC_6H_5$	$CHOHCH_2C_6H_5$	PCl_5	80	198
	$CHOHCH_2C_6H_5$	Al_2O_3	70	222
	$CH{=}P(C_6H_5)_3$	C_6H_5CHO	48	225
	CHO	Phosphorus ylide	45	261
	$CHOHCH_2C_6H_5$	$(CH_3CO)_2O$, CH_3CO_2K	99	619
	$CH_2CH(OCOCH_3)C_6H_5$	Heat	60	619
	$CH(CH_2C_6H_5)N(CH_3)_3I$	KCN	84	134
	$CH_2CH(C_6H_5)N(CH_3)_3I$	Base	43	135
$CH{=}CHCH{=}CHC_6H_5$	CHO	Phosphorus ylide	35	261
	$C{\equiv}CH$	$C_6H_5C{\equiv}CH$, CH_3OH, C_5H_5N, $(CH_3CO)_2Cu$	50	195

$C(C_6H_5)=CHCH_3$	$C(OH)(C_6H_5)CH_2CH_3$	Heat in vacuum	53	222
$CH=C(C_6H_5)_2$	$CH_2P(C_6H_5)_2$	$(C_6H_5)_2CO$	67	225
$CH=CHC_6H_4CH_3$-*p*	CHO	Phosphorus ylide	53	261
$CH=CH(C_6H_4)CH=CHC_6H_5$	CHO	Phosphorus ylide	33	261
$CH=CHC_{10}H_7$-1	CHO	Phosphorus ylide	23	261
$CH=CHC_5H_4FeC_5H_5$	CHO	Phosphorus ylide	26	225
	CHO	$(C_6H_5)_3PO$, NaH	73	225
	$C\equiv CC_5H_4FeC_5H_5$	H_2, Lindlar catalyst	94	225
	$CHOHCH_2C_5H_4FeC_5H_5$	$(CH_3CO)_2O$, CH_3CO_2K	70	620
$CH=CHCH=CH$-$C_5H_4FeC_5H_5$	$C\equiv CH$	CH_3OH, C_5H_5N, $(CH_3CO_2)_2Cu$	90	195
	$CHOHC\equiv CCHOHC_5H_4FeC_5H_5$	$LiAlH_4$	90	202
$CH=CHCH_2CH_2CH=CH$-$C_5H_4FeC_5H_5$	$CHOH(CH_2)_3CH=CHC_5H_4$-FeC_5H_5	Al_2O_3	30	202, 259
$CH=CH(C_6H_4)CH=CHC_5H_4$-FeC_5H_5	CHO	Phosphorus ylide	31	261
$C(C_6H_5)=C(C_6H_5)C_5H_4$-$FeC_5H_5$	COC_6H_5	$(C_6H_5)_3PO$, NaH	47	225

B. 1,1′-Dialkenyl Ferrocenes

$CH=CH_2$	1,1′-$(CHOHCH_3)_2$	Al_2O_3	—	259
$C(CH_3)=CHCH_3$	1,1′-$[C(OH)(CH_3)CH_2CH_3]_2$	$KHSO_4$	—	207
$C(C_6H_5)=CHC_6H_5$	1,1′-$[C(OH)(C_6H_5)CH_2C_6H_5]_2$	$KHSO_4$	50	221
$C(C_6H_4Cl$-*p*$)=CHC_6H_5$	1,1′$[C(OH)(C_6H_4Cl$-*p*$)CH_2C_6H_5]_2$	$KHSO_4$	—	221

Note: References 510–659 are on pp. 151–154.

TABLE XIX. ARYL FERROCENES

A. Monoaryl Ferrocenes from Arylation of Ferrocene

Aryl Group(s) in Product	Solvent	Yield (%)	Refs.
C_6H_5	H_2O	17	271
	CH_3CO_2H	ca. 40–66	66, 186, 275*
	—	56†	621
	H_2O	ca. 5‡	272
$C_6H_4CH_3$-*o*	CH_3CO_2H	43	275
	CH_2Cl_2	40	621
$C_6H_4CH_3$-*p*	$(C_2H_5)_2O$, H_2O	57	274
	H_2SO_4	62	276
	—	53‡	268
$C_6H_4CF_3$-*m*	CH_2Cl_2	—	621
$C_6H_4C_6H_5$-*o*	CH_2Cl_2	28	621
$C_6H_4C_6H_5$-*p*	CH_2Cl_2	13	622
$C_{10}H_7$-1	H_2O	50	623
$C_{10}H_7$-2	H_2O	18	623
$C_6H_4C_5H_4FeC_5H_5$-*m*	CH_2Cl_2	16	622
$C_6H_4C_6H_4C_5H_4FeC_5H_5$-*p*	CH_2Cl_2	7	622
1,8-Naphtho	CH_2Cl_2	—	622
$C_6H_4CO_2H$-*o*	CH_3CO_2H	7	275
$C_6H_4CO_2CH_3$-*o*	CH_2Cl_2	23	621
C_6H_4Cl-*m*	CH_3CO_2H	34	275
$C_6H_4CO_2CH_3$-*m*	CH_2Cl_2	15	621
$C_6H_4CO_2C_2H_5$-*m*	CH_2Cl_2	20	621
C_6H_4CHO-*p*	CH_3CO_2H, H_2O	6	267, 624
$C_6H_4COCH_3$-*p*	CH_3CO_2H	<1–20	66, 186, 272*
C_6H_4CN-*p*	CH_2Cl_2	10	621
$C_6H_4CH_2C(NHCHO)(CO_2C_2H_5)_2$-*p*	CH_3CO_2H	64	142
$C_6H_4NO_2$-*o*	CH_3CO_2H	5, 8§	275, 288

$C_6H_4NO_2$-*m*	$(C_2H_5)_2O$, H_2O	—	278
$C_6H_4NO_2$-*p*	$(C_2H_5)_2O$, H_2O	64	117, 274
	CH_3CO_2H	18	186, 275*
	H_2O	10	271
$C_6H_4N{=}NC_6H_5$-*p*	H_2O, H_2SO_4, CH_3CO_2H	21	279
C_6H_4OH-*p*	$(C_2H_5)_2O$, H_2O	39	278
	CH_3CO_2H	14	275
	H_2O	60	271
$C_6H_4OCH_3$-*o*	H_2O	ca. 30	615
$C_6H_4OCH_3$-*p*	$(C_2H_5)_2O$, H_2O	40	274
	CH_3CO_2H	35	66, 275*
	H_2O, CH_3COCH_3	8	272
$C_6H_4OC_2H_5$-*o*	CH_2Cl_2	4	621
$C_6H_4OC_6H_5$-*p*	$(C_2H_5)_2O$, H_2O	31	277
$C_6H_4SO_3H$-*p*	CH_3CO_2H	—	275
C_6H_4F-*o*	CH_2Cl_2	32	621
C_6H_4Cl-*o*	CH_2Cl_2	22	621
C_6H_4Cl-*p*	CH_3CO_2H	—	66
	H_2O	—	271
C_6H_4Br-*p*	CH_3CO_2H	—	66
C_6H_4Br-*o*	CH_2Cl_2	22	621
C_6H_4I-*o*	CH_2Cl_2	33	621
$C_6H_3CH_3$-2-NO_2-4	CH_2Cl_2	90	621
$C_6H_3CH_3$-2-NO_2-5	CH_2Cl_2	43	621
$C_6H_3CH_3$-2-NO_2-6	CH_2Cl_2	11	621

Note: References 510–659 are on pp. 151–154.

* The highest yield was reported in this reference.

† This product was prepared from bromoferrocene and the appropriate tetraarylboron.

‡ This product was prepared from ferricenium chloride and the appropriate aryldiazonium salt.

§ This product was prepared by Ullmann coupling of iodoferrocene and *o*-iodonitrobenzene.

TABLE XIX. ARYL FERROCENES (*Continued*)

Aryl Group(s) in Product	Solvent	Yield (%)	Refs.
$C_6H_2(CH_3)_2$-2,6-NO_2-4	CH_2Cl_2	39	621
$C_6H_3(OH)_2$-3,4	H_2O	17	615
$C_{10}H_6CO_2H$-8	H_2O	6	271
B. Di- and Poly-aryl Ferrocenes from Arylation of Ferrocene			
1,1′-$(C_6H_5)_2$	H_2O	4–20	66, 186, 271, 272
	H_2O, CH_3CO_2H	52‖	275
	—	40¶	88
1,2-$(C_6H_5)_2$	H_2O	—	272
1,3-$(C_6H_5)_2$	H_2O	—	272
1,1′-$(C_6H_4CH_3$-$p)_2$	H_2SO_4	9‖	276
1,1′-$(C_5H_4FeC_5H_5)_2$	—	14**	291
1,1′-$(C_6H_4COCH_3$-$p)_2$	H_2O, CH_3CO_2H	ca. 8	186
1,2-$(C_6H_4COCH_3$-$p)_2$	H_2O, CH_3CO_2H	ca. 1	272
1,3-$(C_6H_4COCH_3$-$p)_2$	H_2O, CH_3CO_2H	—	272
1,1′-$(C_6H_4CO_2H$-$o)_2$	H_2O	15	271
1,1′-$(C_6H_4NO_2$-$p)_2$	H_2O	60–67	121, 271
1,1′-$(C_6H_4OCH_3$-$o)_2$	H_2O	ca. 5	615
1,1′-$(C_6H_4OCH_3$-$p)_2$	H_2O, CH_3COCH_3	7	272, 275
1,1′-$(C_6H_4CHO$-$p)_2$	—	—	624
1,2-$(C_6H_4OCH_3$-$p)_2$	H_2O, CH_3COCH_3	ca. 2	272
1,3-$(C_6H_4OCH_3$-$p)_2$	H_2O, CH_3COCH_3	—	272

1,1′-$(C_6H_4OC_6H_5\text{-}m)_2$	H_2O	9	277
1,1′-$(C_6H_4Cl\text{-}m)_2$	H_2O, CH_3CO_2H	0.4	275
1,1′-$(C_6H_4Cl\text{-}p)_2$	H_2O	—	66, 271
1,1′-$(C_6H_4Br\text{-}p)_2$	CH_3CO_2H	17	121
1,1′-$[C_6H_3(OH)_2\text{-}3,4]_2$	H_2O	3	615
1-$C_6H_4OCH_3$-*p*-1′-$C_6H_4NO_2$-*p*	H_2O	—	625
1-$C_6H_4OCH_3$-*p*-2-$C_6H_4NO_2$-*p*	H_2O	—	625
1,1′,*x*-$(C_6H_5)_3$	H_2O, CH_3CO_2H	10‖	275
1,*x*,*y*-$(C_6H_4C_6H_5\text{-}p)_3$	H_2O	50	271
1,*x*,*y*-$(C_6H_4CO_2H\text{-}o)_3$	H_2O	—	271
1,*x*,*y*-$(C_{10}H_6CO_2H\text{-}8)_3$	H_2O	30	271
1,*x*,*y*,*z*-$(C_6H_4CO_2H\text{-}o)_4$	H_2O	—	271
1,1′,*x*,*y*,*z*-$(C_6H_5)_5$	$(C_2H_5)_2O$, H_2O	42	274
C. Monoaryl Ferrocenes from Diferrocenylmercury, a Diarylmercury, and Silver			
C_6H_5	—	45	97, 122
$C_6H_4C_6H_5$-*o*	—	6	97
$C_6H_4C_6H_5$-*m*	—	22	97
$C_6H_4C_6H_5$-*p*	—	20	97

Note: References 510–659 are on pp. 151–154.

‖ This product was prepared by arylation of the monoaryl ferrocene.
¶ This product was prepared by dehydrogenation of 1,1′-bis-(1-cyclohexenyl)ferrocene with Pd/C.
** This product was prepared by Ullmann coupling of bromo- and 1,1′-dibromo-ferrocene.

TABLE XX. PREPARATION OF DIFERROCENYL

Substituent X in Starting Material $C_5H_5FeC_5H_4X$	Reagents/Conditions	Yield (%)	Refs.
Li	$CoBr_2$, n-C_4H_9Br	11	289
	(n-$C_6H_{13})_3SiBr$	—	284
	(n-$C_6H_{13})_2SiBr$	0.11	285
MgBr	$CoCl_2$, $BrCH_2CH_2Br$	80	110
$HgC_5H_4FeC_5H_5$	Ag, 265°	54–61	97, 122*
	Pd (black)	1.5–6	123
$B(OH)_2$	NH_3, Ag_2O	52	108, 109
	$(CH_3CO_2)_2Cu$	21	109, 112, 113
Cl	Cu bronze, 140–150°	60–80	287, 288*
	Zn	17	288
Br	Cu bronze, 140–150°	60–97	287, 288*
I	Cu bronze, 150–160°	96–100	122, 287, 288*
	n-C_4H_9Li, CO_2	20	110
	Cu bronze, $IC_6H_4NO_2$-o	71	288
	Cu, 150°	79	286
	Cu bronze, 90°	63	285

Note: References 510–659 are on pp. 151–154.

* The highest yield is reported in this reference.

TABLE XXI. CYANO FERROCENES

Substituent(s) in Product	Substituent(s) in Starting Ferrocene	Method	Reagents	Yield (%)	Refs.
CN	CH=NOH	Dehydration	$(CH_3CO)_2O$*	46	139, 145
	CH=NOH	Dehydration	PCl_5	25	141
	CH=NOH	Dehydration	Raney Ni	6	141
	CH=NOH	Dehydration	$C_6H_{11}N=C=NC_6H_{11}$	50–78	141
	—	Cyanation	BrCN, $AlCl_3$, CS_2	Trace	141
	—	Cyanation	$FeCl_3$, HCN, THF	86	169, 170
	$FeCl_4^{\ominus}$	Cyanation	$FeCl_3$, HCN, HCl, NaOH†	52	169, 170
	$FeBr_4^{\ominus}$	Cyanation	$FeBr_3$, KCN	3	169, 170
	Cl	Cyanation	CuCN	42	109, 297
	Br	Cyanation	CuCN	84	109, 297
	$CONH_2$	Dehydration	$(CH_3CO)_2O$	—	300
1-CN-(1′,2 or 3)-CH_3	CH_3	Cyanation	$FeCl_3$, HCN, THF	36–43	169, 170, 299‡
1-CN-1′-C_2H_5	1-Br-1′-C_2H_5	Cyanation	CuCN	80	188
1-CN-(1′,2 or 3)-C_2H_5	C_2H_5	Cyanation	$FeCl_3$, HCN, THF	35–53	169, 170,‡ 299
1-CN-1′-$C_6H_4NO_2$-*p*	$C_6H_4NO_2$-*p*	Cyanation	$FeCl_3$, HCN, THF	100	301
1-CN-1′-CO_2H	1-CN-1′-$CONH_2$	Dehydration	$(CH_3CO)_2O$	55	300
1-CN-1′-CO_2CH_3	CO_2CH_3	Cyanation	$FeCl_3$, HCN, THF	90	298
1,1′-$(CN)_2$	—	Cyanation	$FeCl_3$, HCN, THF	23–62	169, 170, 301‡
	1,1′-$(CONH_2)_2$	Dehydration	$(CH_3CO)_2O$	30	300
	1,1′-$(CH=NOH)_2$	″	$C_6H_{11}N=C=NC_6H_{11}$	98	91
1-CN-1′-$COCH_3$	$COCH_3$	Cyanation	$FeCl_3$, HCN, THF	18	301
1-CN-1′-NO_2	NO_2	Cyanation	$FeCl_3$, HCN, THF	3	301
1-CN-1′-$SO_2C_6H_5$	$SO_2C_6H_5$	Cyanation	$FeCl_3$, HCN, THF	52	301
1-CN-1′-Cl	Cl	Cyanation	$FeCl_3$, HCN, THF	73-79	169, 170, 298‡
1-CN-1′-Br	Br	Cyanation	$FeCl_3$, HCN, THF	78	298, 301

Note: References 510-659 are on pp. 151–154.

* The oxime acetate on storage slowly decomposes to nitrile.

† Omission of the treatment with sodium hydroxide lowered the yield to 31%.

‡ The highest yield is reported in this reference.

TABLE XXII. Preparation of Ferrocene Mono- and Di-carboxylic Acids

Starting Material	Reagents	Yield (%) of Monocarboxylic Acid*	Refs.
$C_5H_5FeC_5H_5$	n-C_4H_9Li, CO_2	30 [50]	117
	n-C_4H_9Li, CO_2	57 [19]	118
	n-C_4H_9Li, CO_2	45 [19]	93
	n-C_4H_9Li, CO_2	45 [20]	94
	n-C_4H_0Li, CO_2	25	96
	n-C_4H_9Li, CO_2	60 [40]	302
	$C_5H_{11}Na$, CO_2	5 [54]	98, 104–106, 175†
	C_6H_5Na, CO_2	— [36–38]	104, 105
$C_5H_5FeC_5H_4MgBr$	CO_2	84	110
$C_5H_5FeC_5H_4HgCl$	C_2H_5Li, CO_2	64	98
$(C_5H_5FeC_5H_4)_2Hg$	n-C_4H_9Li, CO_2	30–73	97
	n-C_4H_9Li, CO_2	43‡	98
$C_5H_5FeC_5H_4CHO$	50% KOH	10	141
	40% KOH	27	125
$C_5H_5FeC_5H_4CN$	Hydrolysis	80	309
$C_5H_5FeC_5H_4COSCH_3$	Hydrolysis	80	308
$C_5H_5FeC_5H_4CONH_2$	Hydrolysis	40	307
$C_5H_5FeC_5H_4CON(C_6H_5)_2$	Hydrolysis	72	307
$C_5H_5FeC_5H_4COCH_3$	Pyridine, I_2	35–50	164,† 182, 303
	KOCl	50–60	304
	NaOCl	14	303
$C_5H_5FeC_5H_4I$	n-C_4H_9Li, CO_2	17§	110
$Fe(C_5H_4MgBr)_2$	CO_2	16 [59]	110
	,,	Trace [30]	98
$Fe(C_5H_4CN)_2$	Hydrolysis	— [34]	309
$Fe(C_5H_4COSCH_3)_2$	Hydrolysis	— [ca. 90]	308
$Fe(C_5H_4COCH_3)_2$	NaOI	— [56–87]	63, 228,† 320
$CH_3COC_5H_4FeC_5H_4CO_2H$	Pyridine, I_2	— [ca. 10]	192

Note: References 510–659 are on pp. 151–154.

* Figures in brackets refer to the yield of 1,1′-dicarboxylic acid, where obtained.
† The highest yield is reported in this reference.
‡ In addition, ferrocene was obtained in 19% yield.
In addition, ferrocene and diferrocenyl were obtained in 60% and 20% yields, respectively.

TABLE XXIII. SUBSTITUTED FERROCENOIC ACIDS

Substituent(s) in Ferrocenoic Acid	Substituent(s) in Starting Ferrocene	Method/Reagents	Yield (%)	Refs.
1′-CH_3	1′-CH_3-1-$CONH_2$	Hydrolysis	73	55
2-CH_3	2-CH_3-1-$CONH_2$	Hydrolysis	12	55
1′-C_2H_5	1′-$COCH_3$-1-CO_2CH_3	Clemmensen	70	192
2-(or 3)-C_2H_5	2-(or 3)-C_2H_5-1-CN	Hydrolysis	61	170
1′,2-$(CH_3)_2$	1′,2-$(CH_3)_2$-1-$COCH_3$	NaOI	—	182
	1,1′-$(CH_3)_2$	n-C_4H_9Li, CO_2	—	53
1′,3-$(CH_3)_2$	1′,3-$(CH_3)_2$-1-$COCH_3$	NaOI	—	182
	1,1′-$(CH_3)_2$	n-C_4H_9Li, CO_2	—	53
1′-$CH_2C_6H_5$	1′-$CHOHC_6H_5$-1-CO_2H	Catalytic reduction	92	209
1′-$CHOHC_6H_5$	1′-$CHOHC_6H_5$-1-$CON(C_6H_5)_2$	Hydrolysis	81	209
1′-CO_2CH_3	1,1′-$(CO_2CH_3)_2$	Partial saponification	25	192
2-CO_2CH_3	1,2-$(COCH_3)_2$	NaOCl, CH_3OH	—	173
1′-CN	1′-$CONH_2$-1-CO_2H	Dehydration	55	300
1′-$COCH_3$	1′-$COCH_3$-1-$CON(C_6H_5)_2$	Hydrolysis	86	307
1′-COC_6H_5	1′-COC_6H_5-1-$CON(C_6H_5)_2$	Hydrolysis	74	307
1′-$OCOCH_3$	1′-Br-1-CO_2H	$(CH_3CO_2)_2Cu$	60	305
1′-SO_3H	CO_2H	Sulfonation	ca. 90	322
1′-SO_2NH_2	1′-SO_2NH_2-1-CO_2CH_3	Saponification	45–75	322
1′-SO_2F	1-COCl-1′-SO_2Cl	KHF_2	—	322
1′-Cl	1′-Cl-1-CN	Hydrolysis	78	298
1′-Br	1′-Br-1-$COCH_3$	Pyridine, I_2	41	305
	1′-Br-1-CN	Hydrolysis	89	298

Note: References 510–659 are on pp. 151–154.

TABLE XXIV. FERROCENECARBOXAMIDES FROM FRIEDEL-CRAFTS ACYLATION
Aluminum chloride was the catalyst in all reactions listed. The solvent was methylene chloride unless otherwise specified.

Substituent(s) in Product	Reagent	Yield (%)	Refs.
$CONH_2$	H_2NCOCl*	70	307
$CONHC_2H_5$	C_2H_5NCO	30 (54)†	306
$CONHC_{18}H_{37}$-*n*	*n*-$C_{18}H_{37}NCO$	58	306
$CONHC_6H_5$	C_6H_5NCO	57	306
$CON(C_6H_5)_2$	$(C_6H_5)_2NCOCl$*	64	307
$CONHC_6H_4Br$-*p*	*p*-BrC_6H_4NCO	47	306
$CONHC_6H_4C_6H_5$-*p*	*p*-$C_6H_5C_6H_4NCO$	23	306
$CONHC_{10}H_7$-1	1-$C_{10}H_7NCO$	39	306
1-$CONH_2$-1′-CH_3	H_2NCOCl	—	55
1-$CONH_2$-2-CH_3	H_2NCOCl	12	55
1-$CONH_2$-3-CH_3	H_2NCOCl	—	55
1-$CONH_2$-1′-C_3H_7-*i*	H_2NCOCl	35	55
1-$CONH_2$-2-C_3H_7-*i*	H_2NCOCl	12	55
1-$CONH_2$-3-C_3H_7-*i*	H_2NCOCl	28	55

Note: References 510–659 are on pp. 151–154.

* 1,2-Dichloroethane was the solvent.

† Two and one half moles of ethyl isocyanate was used per mole of ferrocene.

TABLE XXV. BRIDGED FERROCENES

Bridge between 1,1′ Positions in Product	Substituent(s) in Starting Ferrocene	Yield (%)	Refs.
$—(CH_2)_3—$*	Bridge: $—(CH_2)_2CO—$	70–99	140, 176, 205,† 331, 332
$—(CH_2)_4—$*	Bridge: $—(CH_2)_3CO—$	75	140
$—CH_2C(CH_3)_2CH_2—$	Bridge: $—CH_2C(CH_3)_2CO—$	77	140
$—CH_2CH{=}CH—$	Bridge: $—CH_2CH_2C(=NNHSO_2C_6H_4CH_3\text{-}p)—$	22	330
$—CH_2OCH_2—$	1,1′-$(CH_2OH)_2$	—	89
$—CH_2OCH(CH_3)—$	1-$COCH_3$-1′-CO_2CH_3	83	150
$—CH(CH_3)OCH(CH_3)—$	1,1′-$(CHOHCH_3)_2$	60–98	226–228†
$—(CH_2)_2CO—$	1-$CH_2CH_2CO_2H$	69–81	176, 205†
$—CH_2COCH_2—$	Bridge: $—CH_2COCH(CO_2CH_3)—$	85	338
$—CH_2CO(CH_2)_2—$	Bridge: $—(CH_2)_2CO—$	ca. 50	140
$—(CH_2)_3CO—$	Bridge: $—(CH_2)_2CO—$	Trace	140
	1-$(CH_2)_3CO_2H$	7	56, 57
$—CH_2CH(CH_3)CO—$	1-$CH_2CH(CH_3)CO_2H$	—	585
$—CH(CH_3)CH_2CO—$	1-$CH(CH_3)CH_2CO_2H$	75	586
$—CH_2C(CH_3)_2CO—$	1-$CH_2C(CH_3)_2CO_2H$	ca. 30	140
$—CH_2CH(C_6H_5)CO—$	1-$CH_2CH(C_6H_5)CO_2H$	—	582, 585
$—CH(CH_3)CH(CH_3)CO—$	1-$CH(CH_3)CH(CH_3)CO_2H$	41	586
$—CH(C_6H_5)CH_2CO—$	1-$CH(C_6H_5)CH_2CO_2H$	21	586
$—CH_2CHOHCOCH_2—$	1,1′-$(CH_2CO_2C_2H_5)_2$	20	332
$—(CH_2)_2CHOHCO(CH_2)_2—$	1,1′-$(CH_2CH_2CO_2C_2H_5)_2$	50	332
$—(CH_2)_3CHOHCO(CH_2)_3—$	1,1′-$(CH_2CH_2CH_2CO_2C_2H_5)_2$	58	332
$—(CH_2)_3CHOHCO(CH_2)_4—$	1-$(CH_2)_3CO_2C_2H_5$-1′-$(CH_2)_4CO_2C_2H_5$	55	332
$—CH_2COCH(CO_2CH_3)—$	1,1′-$(CH_2CO_2CH_3)_2$	85	338
$—Si(CH_3)_2OSi(CH_3)_2—$	1,1′-$[Si(CH_3)_2OC_2H_5]_2$	90	339, 554
	1,1′-$[Si(CH_3)_2Si(CH_3)_3]_2$	—	556
$—Si(CH_3)(OH)OSi(CH_3)(OH)—$	1,1′-$[SiCH_3(OC_2H_5)_2]_2$	21	339
$—Si(CH_3)(OC_2H_5)OSi(CH_3)(OC_2H_5)—$	1,1′-$[SiCH_3(OC_2H_5)_2]_2$	77	339

Note: References 510–659 are on pp. 151–154.

* This compound was also prepared directly from a substituted cyclopentadiene and an iron source; see Table II.

† The highest yield is reported in this reference.

TABLE XXVI. BRIDGED FERROCENES FROM 1,1′-DIACYL FERROCENES AND ALDEHYDES

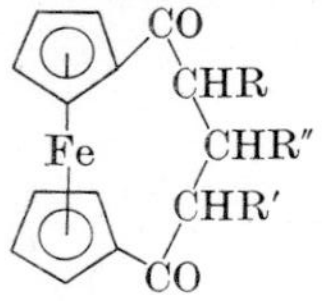

Substituents R and R′ are from the diacyl ferrocene; R″ is from the aldehyde.

Substituents in Product			Yield (%)	Refs.
R	R′	R″		
CH_3	CH_3	H	12	214
CH_3	CH_3	CH_3	10	214
CH_3	CH_3	C_2H_5	13	214
CH_3	CH_3	C_3H_7-*n*	15	214
CH_3	CH_3	C_3H_7-*i*	18	214
CH_3	CH_3	C_4H_9-*i*	9	214
CH_3	CH_3	C_6H_5	68	341
CH_3	CH_3	C_6H_4Cl-*o*	78 (25)	341, 342
CH_3	CH_3	$C_6H_4NO_2$-*m*	10	341
CH_3	CH_3	$C_6H_3O_2CH_2$-3,4	59	341
CH_3	CH_3	C_4H_3O-2	67	341
CH_3	CH_3	$C_5H_4FeC_5H_5$	41	343
C_2H_5	C_2H_5	C_6H_4Cl-*o*	30	340
C_3H_7-*n*	C_3H_7-*n*	C_6H_4Cl-*o*	—	340
CH_3	C_2H_5	C_6H_4Cl-*o*	47	340
CH_3	C_3H_7-*n*	C_6H_4Cl-*o*	48	340
CH_3	C_3H_7-*i*	C_6H_4Cl-*o*	40	340
CH_3	$(CH_2)_7CH{=}CH(CH_2)_7CH_3$	C_6H_4Cl-*o*	—	340
CH_3	$CH_2C_6H_5$	C_6H_4Cl-*o*	5	340

Note: References 510–659 are on pp. 151–154.

TABLE XXVII. Ferroçocarbocyclic Compounds

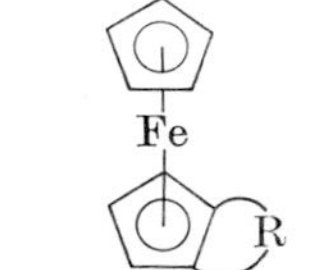

Substituent —R— in Product	Substituent in Ferrocene Precursor	Yield (%)	Refs.
$—(CH_2)_4—$	Carbocycle: $—(CH_2)_3CO—$	—	345, 358
	$—CH{=}CHCH{=}CH—$	—	358
$—CH_2CH_2CH{=}CH—$	$—(CH_2)_3CHOH—$	—	57
$—(CH_2)_3CHOH—$	$—(CH_2)_3CO—$	—*	176, 208
$—(CH_2)_3CO—$	$(CH_2)_3CO_2H$	15–86	176, 196, 205,† 329, 344, 626‡
$—CH_2SCH_2CO—$	CH_2SCH_2COCl	—	627
$—(CH_2)_4CO—$	$(CH_2)_4CO_2H$	14–28	205,† 329, 627
$—CH_2C(CH_3)_2CO—$	$CH_2C(CH_3)_2CO_2H$	ca. 30	140
$—CH_2CH(C_6H_5)CH_2CO—$	$CH_2CH(C_6H_5)CH_2CO_2H$	—	60, 348, 349
$—(CH_2)_2CH(C_6H_5)CO—$	$(CH_2)_2CH(C_6H_5)CO_2H$	—	60, 348, 349
$—CH_2CH(CH_2C_5H_4FeC_5H_5)COCO—$	$CH_2CH(CH_2C_5H_4FeC_5H_5)COCO_2H$	—	582, 583
$—COCH{=}CHCO—$	$—(CH_2)_3CO—$	11	242
$—C_6H_4CO$-*o*	$C_6H_4CO_2H$-*o*	11–20	187, 354
$—C_6H_4CO$-*o* §	$C_6H_4CO_2H$-*o* §	—	186

Note: References 510–659 are on pp. 151–154.

* A mixture of *endo* and *exo* products in a ratio of 9:1 was obtained.
† The highest yield is reported in this reference.
‡ The isomeric monoethylated derivatives are reported in this reference.
§ This material also contains a phenyl group in the 1′ position.

TABLE XXVIII. FERROCENYL AND FERROÇO HETEROCYCLIC COMPOUNDS

Product	Reactants	Yield (%)	Refs.
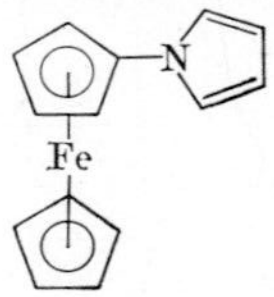	$C_5H_5FeC_5H_4Br$, NaC_4H_4N, Cu_2Br_2	17	268
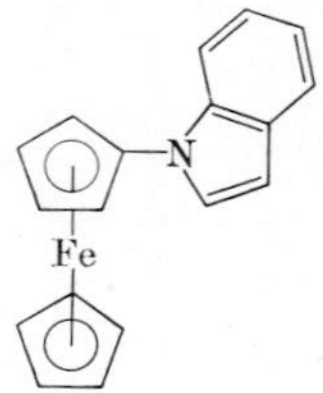	$C_5H_5FeC_5H_4Br$, $KB(C_8H_6N)_4$, Cu_2Br_2	35	268
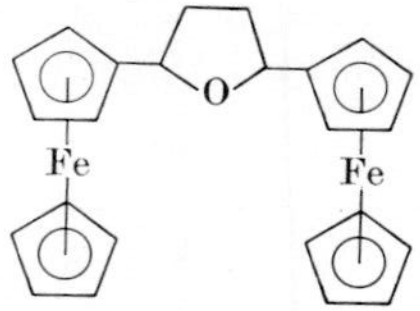	$C_5H_5FeC_5H_4CHOH(CH_2)_2CHOHC_5H_4FeC_5H_5$, $(CH_3CO)_2O$	—	198
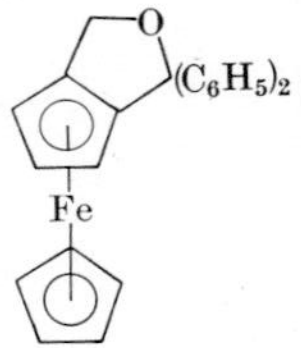	$C_5H_5FeC_5H_3$-$[C(OH)(C_6H_5)_2]$-1-$[CH_2N(CH_3)_3I$-2$]$, KNH_2, NH_3	79	102

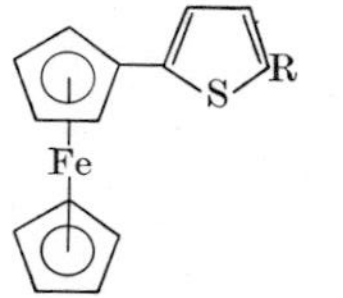	$C_5H_5FeC_5H_4Br$, $KB(C_4H_3S)_4$, Cu_2Br_2	81	268
	$C_5H_5FeC_5H_4C{\equiv}CC{\equiv}CCH_2OH$, Na, C_2H_5OH, H_2S	70	267
	$C_5H_5FeC_5H_4C{\equiv}CC{\equiv}CC_6H_5$, Na, C_2H_5OH, H_2S	51	267
	$C_5H_5FeC_5H_4C{\equiv}CC{\equiv}CC_5H_4FeC_5H_5$, Na, C_2H_5OH, H_2S	80	267
R = H, CH_2OH, C_6H_5, $C_5H_4FeC_5H_5$			
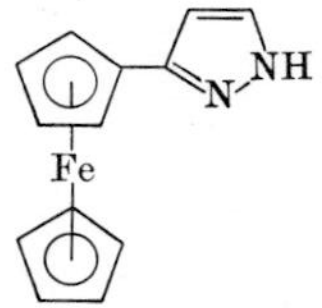	$C_5H_5FeC_5H_4COC{\equiv}CH$, N_2H_4	85	269, 623
	$C_5H_5FeC_5H_4COCH_2CHO$, N_2H_4	80	623
	$C_5H_5FeC_5H_4COCH{=}CHCl$, N_2H_4	82	623
	$C_5H_5FeC_5H_4C(OC_2H_5){=}CHCHO$, N_2H_4	70	262
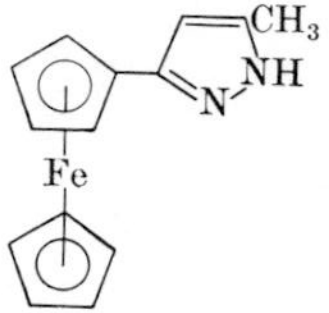	$C_5H_5FeC_5H_4COCH_2COCH_3$, N_2H_4	68	628
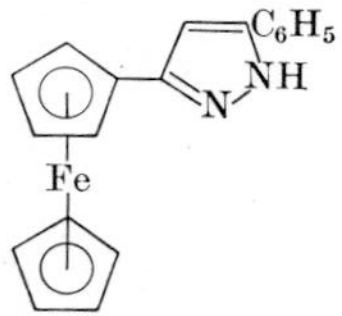	$C_5H_5FeC_5H_4COCH_2COC_6H_5$, N_2H_4	61	163

Note: References 510–659 are on pp. 151–154.

TABLE XXVIII. FERROCENYL AND FERROÇO HETEROCYCLIC COMPOUNDS (*Continued*)

Product	Reactants	Yield (%)	Refs.
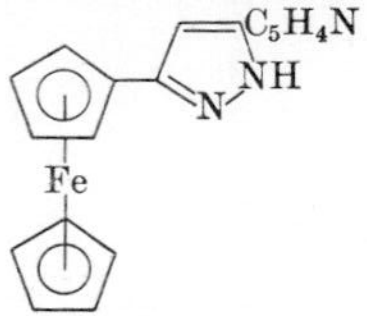 C_5H_4N, NH, N, Fe	$C_5H_5FeC_5H_4COCH_2COC_5H_4N$, N_2H_4	ca. 70*	629
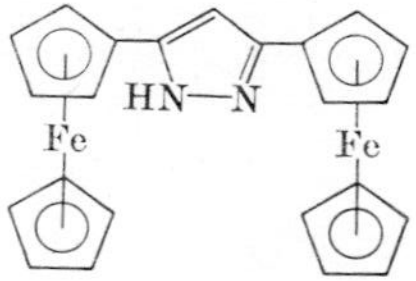 HN—N, Fe, Fe	$(C_5H_5FeC_5H_4CO)_2CH_2$, N_2H_4	—	218
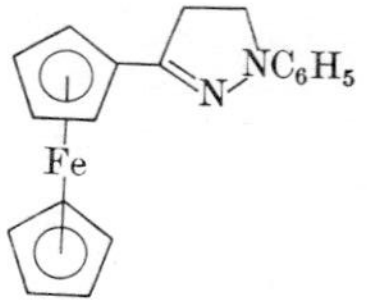 N, NC_6H_5, Fe	$C_5H_5FeC_5H_4CO(CH_2)_2N(C_2H_5)_2 \cdot HCl$, $C_6H_5NHNH_2$	49	203
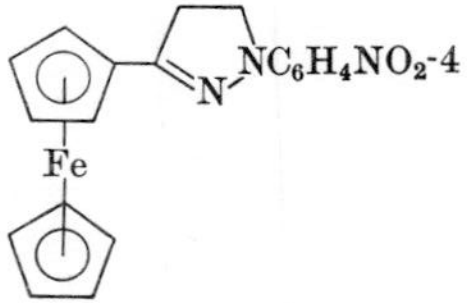 N, $NC_6H_4NO_2$-4, Fe	$C_5H_5FeC_5H_4CO(CH_2)_2N(C_2H_5)_2 \cdot HCl$, $O_2NC_6H_4NHNH_2$-4	—	630
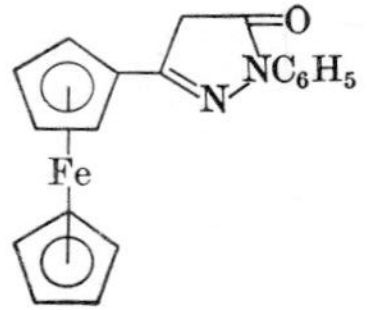 O, N, NC_6H_5, Fe	$C_5H_5FeC_5H_4COCH_2CO_2C_2H_5$, $C_6H_5NHNH_2$	89	163

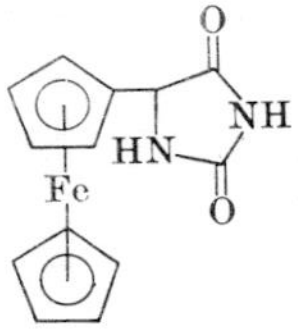

	$C_5H_5FeC_5H_4CHO$, KCN, $(NH_4)_2CO_3$	42–58	142, 156
	$C_5H_5FeC_5H_4CHO$, $NaHSO_3$, KCN, $(NH_4)_2CO_3$	32	139
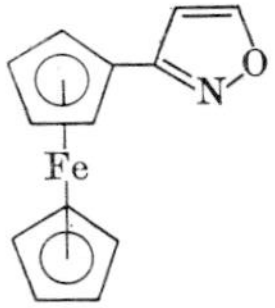	$C_5H_5FeC_5H_4COC{\equiv}CH$, NH_2OH	50	269
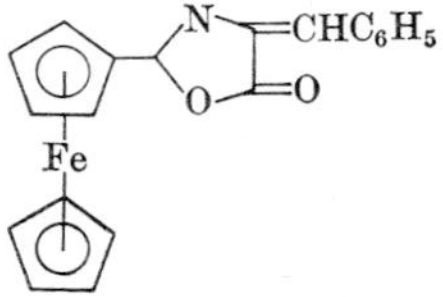	$C_5H_5FeC_5H_4CONHCH_2CO_2H$, $(CH_3CO)_2O$, C_6H_5CHO	70	142
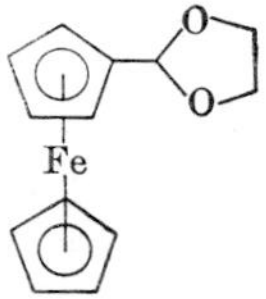	$C_5H_5FeC_5H_4CHO$, $HOCH_2CH_2OH$	94	141

Note: References 510–659 are on pp. 151–154.

* Each of the three isomeric pyridyl ketones was utilized to provide the corresponding substituted pyrazole.

TABLE XXVIII. FERROCENYL AND FERROÇO HETEROCYCLIC COMPOUNDS (*Continued*)

Product	Reactants	Yield (%)	Refs.
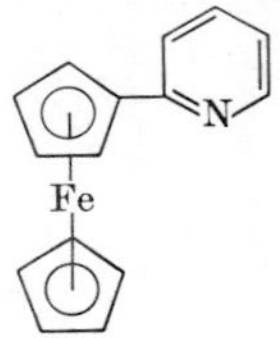 N, Fe	$C_5H_5FeC_5H_4Li$, C_5H_5N	24–32	364, 365
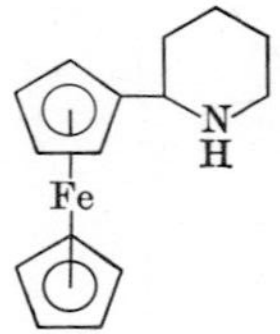 NH, Fe	$C_5H_5FeC_5H_4C_5H_4N$-2, H_2, PtO_2	59	365
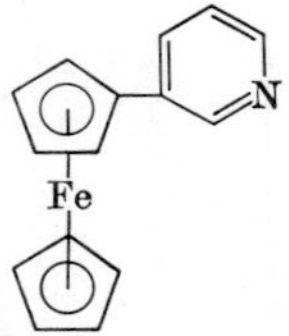 N, Fe	$C_5H_5FeC_5H_5$, 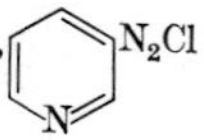 N, N_2Cl	27	365
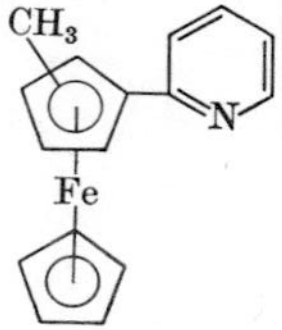 CH_3, N, Fe †	$FeC_{10}H_8(CH_3)Li$ (mixture), C_5H_5N	7†	365

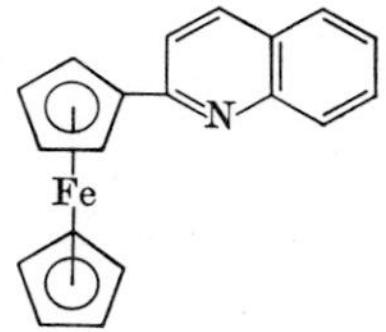	$C_5H_5FeC_5H_4Li$, C_9H_7N		365, 572
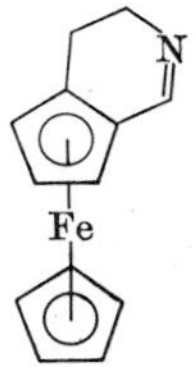	$C_5H_5FeC_5H_4(CH_2)_2NHCHO$, $POCl_3$, C_6H_6	50‡	56, 362, 363
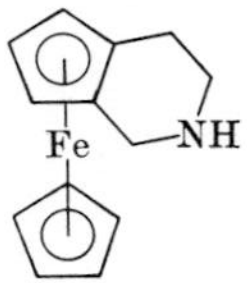	$C_5H_5FeC_5H_4(CH_2)_2N{=}CH_2$, HCl, H_2O	63	56
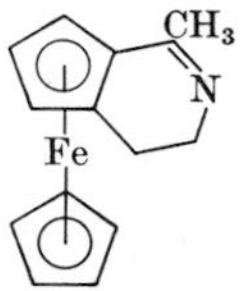	$C_5H_5FeC_5H_4(CH_2)_2NHCOCH_3$, $POCl_3$, C_6H_6	85‡	362, 363

Note: References 510–659 are on pp. 151–154.

† The product is a mixture of the 2-, 3-, and 1′-methyl isomers in the ratio 2:70:28.
‡ Reduction with lithium aluminum hydride gives the corresponding tetrahydropyridoferrocene.

TABLE XXVIII. FERROCENYL AND FERROÇO HETEROCYCLIC COMPOUNDS (*Continued*)

Product	Reactants	Yield (%)	Refs.
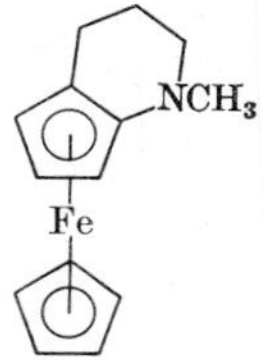	$C_5H_5FeC_5H_4(CH_2)_2NH_2$, HCO_2H, CH_2O	70	260, 361, 362
	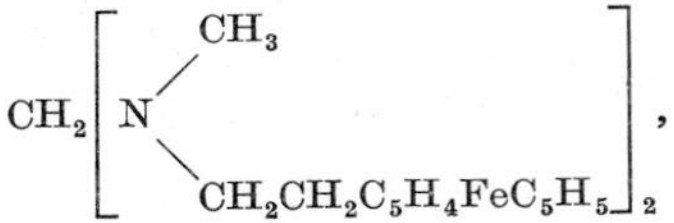 $CH_2[N(CH_3)CH_2CH_2C_5H_4FeC_5H_5]_2$, H_3PO_4, CH_3CO_2H	17	362, 363
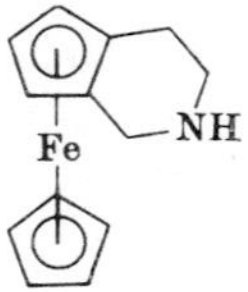	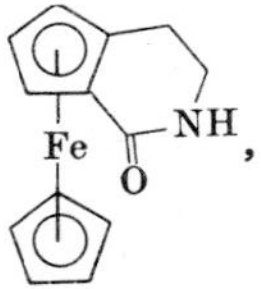 , $LiAlH_4$	60	56
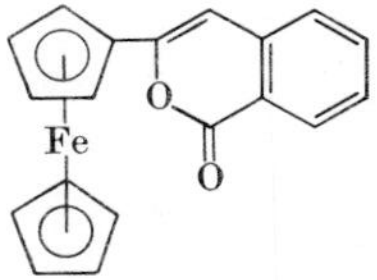	$C_5H_5FeC_5H_4CH_2C_6H_4CO_2H$-2, polyphosphoric acid	—	224, 355
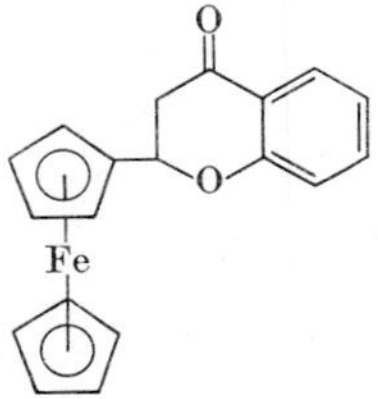	$C_5H_5FeC_5H_4CH{=}CHCOC_6H_4OH$-2, CH_3CO_2Na	—	158

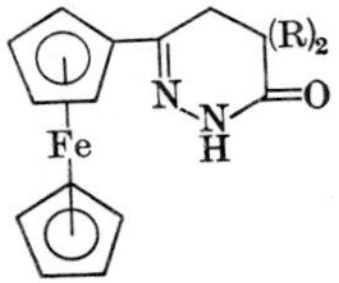 $R = CH_3, C_6H_5$	$C_5H_5FeC_5H_4COCH_2CR_2CO_2H$, N_2H_4	—	598
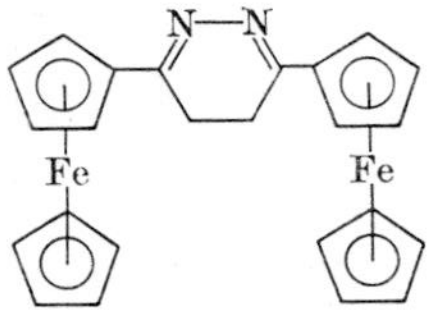	$(C_5H_5FeC_5H_4COCH_2)_2$, N_2H_4	—	366
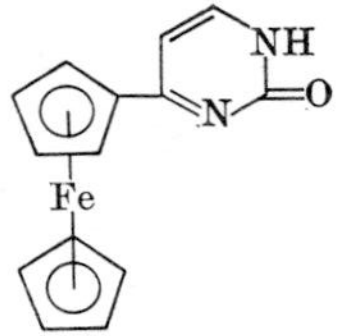	$C_5H_5FeC_5H_4COCH_2CHO$, H_2NCONH_2	—	623
Ferrocene			
1,1′-Di-(3-pyrazolyl)-	$Fe(C_5H_4COCH_2CHO)_2$, H_2NNH_2	16	623
1,1′-Di-(5-methyl-3-pyrazolyl)-	$Fe(C_5H_4COCH_2COCH_3)_2$, H_2NNH_2	89	216
1,1′-Di-(5-ethyl-3-pyrazolyl)-	$Fe(C_5H_4COCH_2CO_2C_2H_5)_2$, H_2NNH_2	99	216
1,1′-Di-(5-phenyl-3-pyrazolyl)-	$Fe(C_5H_4COCH_2C_6H_5)_2$, H_2NNH_2	82	216, 231
1,1′-Di-(2-piperidinyl)-	$Fe(C_5H_4C_5H_4N\text{-}2)_2$, H_2, Pt, CH_3CO_2H	79	365
1,1′-Di-(2-pyridyl)-	$Fe(C_5H_4Li)_2$, C_5H_5N	3	364, 365
1,1′-Di-(2-quinolyl)-	$Fe(C_5H_4Li)_2$, C_9H_7N	1–7	365, 572
1,1′-Di-(3-pyridyl)-	$C_{10}H_{10}Fe$, $3\text{-}C_5H_4NNH_2$, HONO, CH_3CO_2H	26	365

Note: References 510–659 are on pp. 151–154.

TABLE XXIX. HALO FERROCENES

Substituent(s) in Product	Substituent(s) in Starting Ferrocene	Reagent	Yield (%)	Refs.
A. Monohaloferrocenes				
Cl	$B(OH)_2$	$CuCl_2$	80	108, 109
	1-$B(OH)_2$-1′-Cl	$ZnCl_2$, H_2O	79	114
	$N{=}NNHC_6H_5$	HCl	76	378
	$N{=}NNHC_5H_4FeC_5H_5$	HCl	72	378
Br	HgCl	N-Bromosuccinimide	57	120
	HgCl	N-Bromoacetamide	48	120
	HgCl	Pyridine·Br_2	58	120
	$HgC_5H_4FeC_5H_5$	$CHCl_3$, Br_2	—	375
	$B(OH)_2$	$CuBr_2$	80	108, 109
	$N{=}NNHC_6H_5$	HBr	70	378
I	HgCl	I_2, xylene	60–70	286, 376
	HgCl	N-Iodosuccinimide	85	120
	$HgC_5H_4FeC_5H_5$	I_2	64	375
	$N{=}NNHC_6H_5$	HI	72	378
B. Dihaloferrocenes				
1,1′-Cl_2	1,1′-$[B(OH)_2]_2$	$CuCl_2$	75	108, 109
	1-$B(OH)_2$-1′-Br	$CuCl_2$	—	109, 114
1,1′-Br_2	1,1′-$(HgCl)_2$	Br_2	—	375
	1,1′-$(HgCl)_2$	N-Bromosuccinimide	47	120
	1,1′-$[B(OH)_2]_2$	$CuBr_2$	76	108, 109
1,1′-I_2	1,1′-$(HgCl)_2$	I_2	25	375
	1,1′-$(HgCl)_2$	N-Iodosuccinimide	42	120
1-I-1′-Cl	1-HgCl-1′-Cl	I_2	94	114
	$(ClC_5H_4FeC_5H_4)_2Hg$*	I_2	64	109
1-I-1′-Br	1-HgCl-1′-Br	I_2	76	114
	$(BrC_5H_4FeC_5H_4)_2Hg$*	I_2	76	109

Note: References 510–659 are on pp. 151–154.

* This is the formula of the starting ferrocene.

TABLE XXX. ADDITIONAL FERROCENES SUBSTITUTED BY HETEROATOMS

A. Monosubstituted Ferrocenes

Substituent(s) in Product	Substituent(s) in Starting Ferrocene	Reagent(s)	Yield (%)	Refs.
$Si(CH_3)_3$	Na	$(CH_3)_3SiCl$	10	98
	Li	"	19	118
$Si(C_2H_5)_3$	Na	$(C_2H_5)_3SiCl$	8	107
$Si(C_6H_{13}\text{-}n)_3$	Li	$(n\text{-}C_6H_{13})_3SiCl$	—	285
	Li	$(n\text{-}C_6H_{13})_3SiBr$	32	103
	Na	"	<1.0	103
$Si(CH_3)_2C_6H_5$	$C_6H_5(CH_3)_2SiC_5H_4Fe\text{-}C_5H_4Si(CH_3)_2OLi$*	HCl	60	391
$Si(C_6H_5)_3$	Li	$(C_6H_5)_3SiCl$	27–65	93, 98,† 575
$Si(CH_3)_2OSi(CH_3)_2C_6H_5$	$H_2N(CH_3)_2SiC_5H_4Fe\text{-}C_5H_4Si(CH_3)_2ONa$*	HCl	49	391
$Ge(C_6H_5)_3$	Na	$(C_6H_5)_3GeBr$	4	98
	$(MgBr)_2$	"	32	98
$P(C_6H_5)_2$	—	$(C_6H_5)_2PCl$, $AlCl_3$	53	387
$P(C_6H_5)C_5H_4FeC_5H_5$	—	$C_6H_5PCl_2$, $AlCl_3$	67	387
$P(C_5H_4FeC_5H_5)_2$	—	$(C_2H_5)_2NPCl_2$, $AlCl_3$	—	389
$P(O)(C_6H_5)_2$	—	$(C_6H_5)_2PCl$, $AlCl_3$	5	387
$P(O)(C_6H_5)C_5H_4FeC_5H_5$	—	$C_6H_5PCl_2$, $AlCl_3$	8	387
$P(O)(C_5H_4FeC_5H_5)_2$	—	PCl_3, $AlCl_3$	5–100	388, 631†
		$(C_2H_5)_2NPCl_2$, $AlCl_3$	53	389
$P(O)(OH)H$	—	PCl_3, $AlCl_3$	4	388
$P(O)(OH)C_5H_4FeC_5H_5$	—	PCl_3, $AlCl_3$	4	388
	$(C_5H_5FeC_5H_4)_3PO$*	H_2SO_4	—	632

Note: References 510–659 are on pp. 151–154.

* This is the complete formula of the starting ferrocene.

† The highest yield is reported in this reference.

TABLE XXX. ADDITIONAL FERROCENES SUBSTITUTED BY HETEROATOMS (*Continued*)

Substituent(s) in Product	Substituent(s) in Starting Ferrocene	Reagent(s)	Yield (%)	Refs.
$P(Se)(C_5H_4FeC_5H_5)_2$	$(C_5H_5FeC_5H_4)_3P$*	Se	—	389
$P(S)(C_5H_4FeC_5H_5)_2$	$(C_5H_5FeC_5H_4)_3PO$*	P_2S_5	—	631
		S_8	100	389
PCl_2	—	$(C_2H_5)_2NPCl_2$, $AlCl_3$	5	389
$P(Cl)C_5H_4FeC_5H_5$	—	$(C_2H_5)_2NPCl_2$, $AlCl_3$	9	389
$P(OH)C_5H_4FeC_5H_5$	—	$(C_2H_5)_2NPCl_2$, $AlCl_3$	15	389
$P[N(C_2H_5)_2]C_5H_4FeC_5H_5$	—	$(C_2H_5)_2NPCl_2$, $AlCl_3$	13	389
$AsO_2AsC_5H_4FeC_5H_5$	—	$AsCl_3$, $AlCl_3$	22	392
$SeC_5H_4FeC_5H_5$	$(C_5H_5FeC_5H_4)_2Hg$*	$SeBr_4$	21	326
		B. 1,1′-bis-Trialkylsilyl-ferrocenes		
$[Si(CH_3)_3]_2$	Li_2	$(CH_3)_3SiCl$	27	118
	Na_2	$(CH_3)_3SiCl$	—	633
$[Si(C_2H_5)_3]_2$	Na_2	$(C_2H_5)_3SiCl$	12	107, 633
$[Si(C_3H_7\text{-}n)_3]_2$	Na_2	$(n\text{-}C_3H_7)_3SiCl$	—	633
$[Si(C_6H_{13}\text{-}n)_3]_2$	Li_2	$(n\text{-}C_6H_{13})_3SiBr$	35	103
	Na_2	$(n\text{-}C_6H_{13})_3SiBr$	8	103
$[Si(C_6H_5)_3]_2$	Li_2	$(C_6H_5)_3SiCl$	7–50	93, 575†

Note: References 510–659 are on pp. 151–154.

* This is the complete formula of the starting ferrocene.

† The highest yield is reported in this reference.

TABLE XXXI. SUBSTITUTION PRODUCTS OF RUTHENOCENE AND OSMOCENE

Substituent(s) in Product	Reagent(s)	Yield (%)	Refs.
A. Derivatives of Ruthenocene			
HgCl	$(CH_3CO_2)_2Hg$, KCl	21–25*	429
CHO	$C_6H_5N(CH_3)CHO$, $POCl_3$	9	187
CO_2H	n-C_4H_9Li, CO_2	24	429
$CONHC_6H_5$	C_6H_5NCO, $AlCl_3$	20	429
$COSCH_3$	CH_3SCOCl, $AlCl_3$	—	308
$COCH_3$	CH_3COCl, $AlCl_3$	45	429
COC_6H_5	C_6H_5COCl, $AlCl_3$	64	425, 429
$COC_5H_4FeC_5H_5$	$C_5H_5FeC_5H_4COCl$, $AlCl_3$	45	429
1,1′-$(HgCl)_2$	$(CH_3CO_2)_2Hg$, KCl	75–79*	429
1,1′-$(CO_2H)_2$	n-C_4H_9Li, CO_2	24	429
1,1′-$(COCH_3)_2$	CH_3COCl, $AlCl_3$	5	429
1,1′-$(COC_6H_5)_2$	C_6H_5COCl, $AlCl_3$	19–24	425, 429
1,2-$(COC_6H_5)_2$	C_6H_5COCl, $AlCl_3$	—	187
$Si(CH_3)_3$	n-C_4H_9Li, $(CH_3)_3SiCl$	—	390
B. Derivatives of Osmocene			
CO_2H	n-C_4H_9Li, CO_2	42*	429
$COCH_3$	CH_3COCl, $AlCl_3$	58–69	429,† 440
	$(CH_3CO)_2O$, H_3PO_4	85–91	431,† 438
COC_6H_5	C_6H_5COCl, $AlCl_3$	60–62	429, 440
1,1′-$(CO_2H)_2$	n-C_4H_9Li, CO_2	58*	429
1,1′-$(COCH_3)_2$	$(CH_3CO)_2O$, H_3PO_4	0.1	431
$Si(CH_3)_3$	n-C_4H_9Li, $(CH_3)_3SiCl$	—	390

Note: References 510–659 are on pp. 151–154.

* The mixture of mono- and di-substituted products was not separated.

† The highest yield is reported in this reference.

TABLE XXXII. Derivatives of Cyclopentadienylmanganese Tricarbonyl Prepared Directly from Cyclopentadienes

Substituent R in $RMn(CO)_3$	Reactants/Solvents	Pressure (p.s.i.)	Temp. (°C.)	Yield (%)	Refs.
C_5H_5	$(C_5H_5)_2Mn$, CO	3000–3700	90–150	80	443, 445
	$(C_5H_5)_2Mn$, CO, $(C_2H_5)_2O$	1975	158	75	442
	$(C_5H_5)_2Mn$, CO, THF	800	250	40	441
	C_5H_5Na, MnX_2, CO, THF	800	167	40–70	441, 444
	C_5H_5Na, $(C_5H_5N)_2Mn(CO)_3Br$, t-C_4H_9OH	—	83	—	634
	C_5H_5Na, $Mn(CO)_5X$, THF	—	—	72–85	448, 635
	C_5H_5Tl, $Mn(CO)_5X$	—	—	93	636
	$(C_5H_5)_2Mn$, $Fe(CO)_5$, CO, diglyme	400–750	165	78	637
	$(C_5H_5)_3Al$, $Mn[OCH(CH_3)_2]_2$, CO, $C_6H_5CH_3$	700	165	48	638
	C_5H_6, $Mn(CO)_5$	—	100	ca. 60	447
	C_5H_6, Mg, $MnCl_2$, C_5H_5N, $HCON(CH_3)_2$, CO	3600	200	—	639
	C_5H_6, Mg, $Mn(C_5H_5N)_2Cl_2$, CO, $HCON(CH_3)_2$	4500	180	70	640
$C_5H_4CH_3$	$(C_5H_4CH_3)_2Mn$, Na, CO, $MnCl_2$, diglyme	300–500	165	50–80	641
	$(C_5H_4CH_3)_2Mn$, CO	1000	200	70	540
	$(C_5H_4CH_3)_2Mn$, CO, $(C_2H_5)_2O$	1975	158	—	442
	$(C_5H_4CH_3)_2Mn$, $Fe(CO)_5$	—	—	—	449
	$(C_5H_4CH_3)_2Mn$, $MnCl_2$, NaH, diglyme	500	165	53	642
	$(C_5H_4CH_3)_2Mn$, $Mn(CO)_5$	—	175	—	643
	$(C_5H_4CH_3)Na$, $Mn(CO)_5Cl$	—	—	—	448
	$(C_5H_4CH_3)Na$, $Mn(CO)_5Br$, THF	—	—	50	635
	$(C_6H_4CH_3)Na$, $(C_5H_5N)_2Mn(CO)_3Br$, $C_6H_5N(C_2H_5)_2$	—	—	—	634
	$(C_5H_4CH_3)Na$, CO, mineral oil, THF, $MnCl_2$	2100–3100	160	—	644
	$C_5H_5CH_3$, Mg, $MnCl_2$, $Fe(CO)_5$, $HCON(CH_3)_2$	3500	195	28	656, 657
	$C_5H_5CH_3$, $MnCl_2$, DMF, $Fe(CO)_5$, CO, 25–30 volts	100	195	7*	645
	$C_5H_5CH_3$, $MnSO_4$, Al, allyl tolyl ether, CO	0.1–5.5	210	—	646
$C_5H_4C_2H_5$	$(C_5H_4C_2H_5)_2Mn$, CO, $(C_2H_5)_2O$	1975	158	—	442
	$(C_5H_4C_2H_5)_2Mn$, $Fe(CO)_5$	—	—	—	449

	$(C_5H_4C_2H_5)Li$, $C_6H_5NH_2$, CH_3OH, $(C_5H_5N)_2Mn(CO)_3Br$	—	—	—	634
	$(C_5H_4C_2H_5)Na$, $Mn(CO)_5I$	—	—	—	448
	$C_5H_4C_2H_5$, $MnCl_2$, CO	1470	—	55	455
$C_5H_4C_3H_7$-*i*	$(CH_3)_2CHC_5H_4Li$, THF, *n*-$C_4H_9NH_2$, $(C_5H_5N)_2Mn(CO)_3Br$	—	—	—	634
$C_5H_4C_8H_{17}$-*n*†	$(C_5H_4C_8H_{17}\text{-}n)_2Mn$, CO, $(C_2H_5)_2O$	—	—	—	442†
	$(C_5H_4C_8H_{17}\text{-}n)Na$, *i*-$C_3H_7OH$, $(C_5H_5N)_2Mn(CO)_3Br$	—	—	—	634
$C_5H_4C_{10}H_{21}$-*n*	$(C_5H_4C_{10}H_{21}\text{-}n)Na$, $MnCl_2$, diglyme, CO	300	—	—	641
$C_5(CH_3)_5$	$[C_5(CH_3)_5]_2Mn$, CO, $(C_2H_5)_2O$	1975	158	—	442
$C_5H_4CH{=}CHCH_3$	$(C_5H_4CH{=}CHCH_3)_2Mn$, CO, $(C_2H_5)_2O$	1975	158	—	442
$C_5H_4CH_2CH{=}CH_2$	$C_5H_4CH_2CH{=}CH_2$, Mn, CO	—	40	—	646
$C_5H_4C_6H_5$	$(C_5H_4C_6H_5)_2Mn$, $Fe(CO)_5$	—	—	—	449
	$(C_5H_4C_6H_5)_2Mn$, CO, $(C_2H_5)_2O$	1975	158	—	442, 446
Indenyl	$(C_9H_7)_2Mn$, CO, $(C_2H_5)_2O$	1975	158	—	442, 446
	C_9H_7K, $(C_5H_5N)_2Mn(CO)_3Br$, $(C_2H_5)_2O$	—	—	—	634
Fluorenyl	$(C_{13}H_9)_2Mn$, $Fe(CO)_5$	—	—	—	449
Dihydropentalenyl	C_8H_8, $Mn_2(CO)_{10}$	500	150	—	450
	$CH_3C_5H_4Mn(CO)_3$, C_2H_2, $Mn_2(CO)_{10}$, THF	500	150	27	450
	C_2H_2, $Mn(CO)_6$, THF	600	150	40	450, 451
	C_8H_8, $[Mn(CO)_6]_2$, THF, $(C_2H_5)_2O$	500	150	—	647
	C_2H_2, $Mn(CO)_5Br$, CH_3COCH_3	—	150	—	648
	C_2H_2, $[Mn(CO)_5]_2$	600	—	—	649
	C_2H_2, $[XMn(CO)_4]_2$‡	—	140–160	—	650

Note: References 510–659 are on pp. 151–154.

* In a similar fashion the ethylcyclopentadienyl-, phenylcyclopentadienyl-, indenyl-, and fluorenyl-manganese tricarbonyls were prepared in unstated yields.

† In similar reactions tetrahydroindenyl-, 1,3,4,7-tetramethylindenyl-, 3-cyclohexylindenyl-, 1,2,3,4,5,6,7,8-octahydrofluorenyl-, 1,8-diethylfluorenyl-, and 2-ethyl-1-naphthylcyclopentadienyl- manganese tricarbonyls have been prepared in unstated yields.

‡ X is a Lewis base such as PH_3, AsH_3, SbH_3, BiR_3, $(C_6H_5)_3P$, an amine, or NO.

TABLE XXXIII. Derivatives of Cyclopentadienylmanganese Tricarbonyl Prepared by Substitution Reactions

Substituent in $(OC)_3MnC_5H_4R$	Reagents	Yield (%)	Refs.
HgCl	$(CH_3CO_2)_2Hg$, $CaCl_2$, C_2H_5OH	37*	452
CH_2Cl	CH_2O, HCl, $ZnCl_2$, CCl_4	73	473
C_2H_5	C_2H_5Br, $AlCl_3$	63	455
$C(CH_3)_3$	$(CH_3)_3CCl$, $AlCl_3$	69	471
$C_2H_4C_5H_4Mn(CO)_3$	$ClCH_2CH_2Cl$, $AlCl_3$	—	481
$COSCH_3$	CH_3SCOCl, $AlCl_3$	—	308
$COCH_3$	CH_3COCl, $AlCl_3$, CS_2	46–98	443, 454–456, 459†
	$(CH_3CO)_2O$, $AlCl_3$	76	443, 456
$COCH_2C_6H_5$	$C_6H_5CH_2COCl$, $AlCl_3$	60–80	651
COC_6H_5	C_6H_5COCl, $AlCl_3$	82–95	443, 455–457, 459†
$COC_6H_4CH_3$-*o*	*o*-$CH_3C_6H_4COCl$, $AlCl_3$	60–80	651
$COC_6H_4CH_3$-*p*	*p*-$CH_3C_6H_4COCl$, $AlCl_3$	60–80	651
COC_4H_3O-2	2-C_4H_3OCOCl, $AlCl_3$	60–80	651
COC_4H_3S-2	2-C_4H_3SCOCl, $AlCl_3$	60–80	651
$COC_5H_4FeC_5H_5$	$ClCOC_5H_4Mn(CO)_3$, $C_{10}H_{10}Fe$, $AlCl_3$	—	463
$COC_5H_4Mn(CO)_3$	$ClCOC_5H_4Mn(CO)_3$, $AlCl_3$	28	463
$CO(CH_2)_2CO_2H$	$(CH_2CO)_2O$, $AlCl_3$	85–86	443, 460
$CO(CH_2)_3CO_2H$	$CH_2(CH_2CO)_2O$, $AlCl_3$	95	460
$SO_2NHC_6H_4CH_3$-*p*	H_2SO_4, *p*-$H_2NC_6H_4CH_3$, $(CH_3CO)_2O$	93	453

Note: References 510–659 are on pp. 151–154.

* A bis-chloromercurated derivative of unspecified structure was also obtained in 14% yield. In similar reactions with methylcyclopentadienylmanganese tricarbonyl a monochloromercurio derivative was obtained in 37% yield and a bis-chloromercurio derivative in 18% yield; structures of both were unspecified.

† The highest yield is reported in this reference.

TABLE XXXIV. LIGAND EXCHANGE PRODUCTS OF CYCLOPENTADIENYLMANGANESE TRICARBONYL

A. Products of Type $RC_5H_4Mn(CO)_2D$
(R is hydrogen unless otherwise specified.)

Displacing Reagent (D in Product)	Yield (%)	Refs.
$CH_2{=}CH_2$	63	652
$CH_2{=}CHCH{=}CH_2$	—	653
Cyclopentene	38	652
Cycloheptene	40	652
Cyclooctene	62	652
$C_6H_5C{\equiv}CH$	33	654
$C_6H_5C{\equiv}CC_6H_5$	34	484
	11*	486
$CH_2{=}CHCN$	10	653
Cyclohexylisonitrile	14–43	489,† 652
$C_6H_5NH_2$	34	489
NH_3	15	491
$(CH_3)_2NH$	64	652
$(CH_3)_3N$	52	491
n-$C_3H_7NH_2$	34	491
i-$C_3H_7NH_2$	31	491
n-$C_6H_{13}NH_2$	29	491
Pyrrolidine	41	491
2,5-Dimethylpyrrolidine	36	491
$N(CH_2CH_2)_3N$	45	490
$(CH_2)_6N_4$	66	490
C_5H_5N	53	483
	—*	483
Quinoline	50	491
Piperazine	40	490
PCl_3	28‡	493
$(C_6H_5)_3P$	25–50	485, 487,† 488*
$(C_6H_5)_2PCH_2CH_2P(C_6H_5)_2$	—	488
	—*	488
$(C_6H_5)_3As$	—	488, 489
	—*	488
$(C_6H_5)_3Sb$	23	489
o-$C_6H_4[As(CH_3)_2]_2$	ca. 10	488
Tetrahydrofuran	—	652
$(C_2H_5)_2S$	44	492
Tetrahydrothiophene	49	492
$(CH_3)_2SO$	32	489, 655
Tetramethylene sulfoxide	35	492
$(C_6H_5)_2SO$	38	492
Ethylene sulfite	77	489, 492
SO_2	11	489, 492

Note: References 510–659 are on pp. 151–154.

* R in the cyclopentadiene and the product is CH_3.

† The best yield is reported in this reference.

‡ The product is $C_5H_5Mn(CO)_2PCl_2C_5H_4Mn(CO)_3$.

TABLE XXXIV. LIGAND EXCHANGE PRODUCTS OF CYCLOPENTADIENYLMANGANESE TRICARBONYL (*Continued*)

Displacing Reagent (D in Product)	Yield (%)	Refs.
B. Products of Type $RC_5H_4Mn(CO)D_2$		
Cyclononene	59	652
Bicyclo[2.2.1]heptadiene	30–35	652
1,3-Cyclooctadiene	16–17	652
Dicyclopentadiene	25	652
$N(CH_2CH_2)_3N$	39	490
Piperazine	20	490
$(C_6H_5)_3P$	48–89	485, 652†
$(CH_3)_2SO$	50	655
$[(CH_3)_2PCH_2]_2$	31	659
$[(C_6H_5)_2PCH_2]_2$	34	659
$[(C_6H_5)_2P]_2CH_2$	32	659
C. Products of Type $RC_5H_4Mn(CO)_2D(CO)_2MnC_5H_4R$		
$CH_2{=}CHCH{=}CH_2$	—	653
$(CH_3)_2PP(CH_3)_2$	21	659
$(C_6H_5)_2PP(C_6H_5)_2$	30	659
$(CH_3)_2PCH_2CH_2P(CH_3)_2$	20	659
$(C_6H_5)_2PCH_2P(C_6H_5)_2$	29	659
$(C_6H_5)_2PCH_2CH_2P(C_6H_5)_2$	21	488, 659
	—*	488
o-$C_6H_4[As(CH_3)_2]_2$	ca. 20*	488
$(CH_3)_2AsAs(CH_3)_2$	10	659

Note: References 510–659 are on pp. 151–154.

* R in the cyclopentadiene and the product is CH_3.

† The best yield is reported in this reference.

REFERENCES FOR TABLES

[510] C. L. Hobbs, Jr., Brit. pat. 733,129, [*C.A.*, **50**, 7146 (1956)].
[511] G. Wilkinson, F. A. Cotton, and J. M. Birmingham, *J. Inorg. Nucl. Chem.*, **2**, 95 (1956).
[512] J. F. Cordes, Fr. pat. 1,341,880 [*C.A.*, **60**, 6873 (1964)].
[513] W. F. Little, R. C. Koestler, and R. Eisenthal, *J. Org. Chem.*, **25**, 1435 (1960).
[514] D. B. Clapp, Brit. pat. 763,047 [*C.A.*, **51**, 10588 (1957)].
[515] G. O. Schenk and E. K. von Gustorf, Fr. pat. 1,343,770 [*C.A.*, **60**, 8062 (1964).
[516] A. N. Nesmeyanov, R. B. Materikova, and N. S. Kochetkova, *Izv. Akad. Nauk SSSR, Ser. Khim.*, 1334 (1963) [*C.A.*, **59**, 12841 (1963)].
[517] L. I. Zakharkin and V. V. Gavrilenko, *Zh. Obshch. Khim.*, **33**, 3112 (1963) [*C.A.*, 60, 1781 (1964)].
[518] C. L. Hobbs, Jr., U.S. pat. 2,763,700 [*C.A.*, **51**, 8806 (1957)].
[519] J. C. Brantley, U.S. pat. 3,028,406 [*C.A.*, **57**, 9883 (1962)].
[520] L. Kaplan, W. L. Kester, and J. J. Katz, *J. Am. Chem. Soc.*, **74**, 5531 (1952).
[521] G. Wilkinson and J. M. Birmingham, *J. Am. Chem. Soc.*, **76**, 6210 (1954).
[522] G. J. Pedersen, Ger. pat. 935,467 [*C.A.*, **52**, 19109 (1958)].
[523] J. Hartley, Brit. pat. 737,109 [*C.A.*, **50**, 13086 (1956)].
[524] E. G. Lindstrom and M. R. Barusch, U.S. pat. 3,122,577 [*C.A.*, **60**, 12057 (1964)].
[525] E. L. Morehouse, Brit. pat. 797,151 [*C.A.*, **53**, 4297 (1959)].
[526] A. Ekemark and K. Skagius, *Acta Chem. Scand.*, **16**, 1136 (1962).
[527] M. Cais, Israeli pat. 17,013 [*C.A.*, **60**, 9314 (1964).]
[528] E. L. Morehouse, U.S. pat. 3,071,605 [*C.A.*, **58**, 12602 (1963)].
[529] E. B. Sokolova, M. P. Shebanova, and N. F. Nikolaeva, *Zh. Obschch. Khim.*, **31**, 332 (1961) [*C.A.*, **55**, 22271 (1961)].
[530] J. M. Birmingham, D. Seyferth, and G. Wilkinson, *J. Am. Chem. Soc.*, **76**, 4179 (1954).
[531] E. G. Lindstrom, U.S. pat. 3,057,899 [*C.A.*, **58**, 3460 (1963)].
[532] G. Wilkinson, P. L. Pauson, and F. A. Cotton, *J. Am. Chem. Soc.*, **76**, 1970 (1954).
[533] W. F. Anzilotti and V. Weinmayr, U.S. pat. 2,791,597 [*C.A.*, **51**, 15560 (1957)].
[534] California Research Corp., Brit. pat. 744,450 [*C.A.*, **51**, 491 (1957)].
[535] F. S. Arimoto, U.S. pat. 2,804,468 [*C.A.*, **52**, 2086 (1958)].
[536] E. I. du Pont de Nemours and Co., Brit. pat. 737,780 [*C.A.*, **50**, 14000 (1956)].
[537] R. Riemschneider and D. Helm, *Z. Naturforsch.*, **16b**, 234 (1961).
[538] R. Riemschneider and D. Helm, *Z. Naturforsch.*, **14b**, 811 (1959).
[539] K. Issleib and A. Brack, *Z. Naturforsch.*, **11b**, 420 (1956).
[540] L. T. Reynolds and G. Wilkinson, *J. Inorg. Nucl. Chem.*, **9**, 86 (1959).
[541] D. S. Breslow, U.S. pat. 2,848,506 [*C.A.*, **53**, 2250 (1959)].
[542] M. A. Lynch, Jr., and J. C. Brantley, Brit. pat. 785, 760 [*C.A.*, **52**, 11126 (1958)].
[543] R. Riemschneider and E. B. Grabitz, *Monatsh. Chem.*, **89**, 748 (1958).
[544] T. Leigh, *J. Chem. Soc.*, 3294 (1964).
[545] R. Riemschneider and R. Nehring, *Monatsh. Chem.*, **90**, 568 (1959).
[546] T. Leigh, Brit. pat. 870,949 [*C.A.*, **56**, 3517 (1962)].
[547] T. Leigh, Brit. pat. 869,058 [*C.A.*, **56**, 3516 (1962)].
[548] W. F. Little and R. C. Koestler, *J. Org. Chem.*, **26**, 3247 (1961).
[549] R. L. Schaaf and C. T. Lenk, *J. Org. Chem.*, **29**, 3430 (1964).
[550] H. Rosenberg and M. D. Rausch, U.S. pat. 3,060,215 [*C.A.*, **58**, 6865 (1963)].
[551] N. S. Nametkin, T. I. Chernysheva, and L. V. Babare, *Zh. Obshch. Khim.*, **34**, 2258 (1964) [*C.A.*, **61**, 9525 (1964)].
[552] E. A. Mailey, C. R. Dickey, G. M. Goodale, and V. E. Matthews, *J. Org. Chem.*, **27**, 616 (1962).
[553] C. R. Dickey, E. A. Mailey, and V. E. Matthews, U. S. pat 3,062,854 [*C.A.*, **58**, 9145 (1963)].
[554] R. L. Schaaf, U.S. pat. 3,036,105 [*C.A.*, **57**, 16656 (1962)].
[555] R. L. Schaaf, P. T. Kan, C. T. Lenk, and E. P. Dick, *J. Org. Chem.*, **25**, 1986 (1960).
[556] M. Kumada, K. Mimura, M. Ishikawa, and K. Shiina, *Tetrahedron Letters*, 83 (1965).
[557] R. L. Schaaf, C. T. Lenk, and H. Rosenberg, U.S. pat. 3,010,982 [*C.A.*, **57**, 2253 (1962)].

[558] S. McVey and P. L. Pauson, *J. Chem. Soc.*, 4312 (1965).
[559] P. L. Pauson and G. Wilkinson, *J. Am. Chem. Soc.*, **76**, 2024 (1954).
[560] E. O. Fischer and D. Seus, *Z. Naturforsch.*, **8b**, 694 (1953).
[561] E. B. Sokolova, M. P. Shebanova, and V. D. Sheludyakov, *Zh. Obshch. Khim.*, **31**, 3379 (1961) [*C.A.*, **57**, 863 (1962)].
[562] E. B. Sokolova, M. P. Shebanova, and H. C. Chou, *Zh. Obshch. Khim.*, **33**, 217 (1963) [*C.A.*, **59**, 658 (1963)].
[563] R. Riemschneider, *Z. Naturforsch.*, **17b**, 133 (1962).
[564] W. M. Sweeney, U.S. pat. 2,912,449 [*C.A.*, **54**, 4616 (1960)].
[565] E. B. Sokolova, M. P. Shebanova, H. C. Chou, and S. A. Pisareva, *Zh. Obshch. Khim.*, **34**, 2693 (1964) [*C.A.*, **61**, 14708 (1964)].
[566] R. L. Pruett and E. L. Morehouse, Ger. pat. 1,052,401 [*C.A.*, **55**, 18770 (1961)].
[567] W. M. Sweeney, U.S. pat. 3,035,075 [*C.A.*, **58**, 10242 (1963)].
[568] R. L. Pruett and E. L. Morehouse, U.S. pat. 3,063,974 [*C.A.*, **58**, 11404 (1962)].
[569] A. D. Petrov, E. B. Sokolova, and G. P. Bakunchik, *Dokl. Akad. Nauk SSSR*, **148**, 598 (1963) [*C.A.*, **59**, 2855 (1963)].
[570] R. L. Schaaf and C. T. Lenk, *J. Chem. Eng. Data*, **9**, 103 (1964).
[571] K. Hata, I. Motoyama, and H. Watanabe, *Bull. Chem. Soc. Japan*, **36**, 1698 (1963).
[572] A. N. Nesmeyanov, V. I. Sazonova, V. A. Romanenko, N. A. Rodionova, and G. P. Zol'nikova, *Dokl. Akad. Nauk. SSSR*, **155**, 1130 (1964) [*C.A.*, **61**, 1891 (1964)].
[573] K. M. Berdichevskaya, V. S. Chugunov, and A. D. Petrov, *Dokl. Akad. Nauk SSSR*, **151**, 1319 (1963) [*C.A.*, **59**, 14022 (1963)].
[574] A. D. Petrov, E. B. Sokolova, M. P. Shebanova, and N. I. Golovina, *Dokl. Akad. Nauk SSSR*, **152**, 1118 (1963) [*C.A.*, **60**, 1792 (1964)].
[575] R. A. Benkeser, U.S. pat. 2,831,880 [*C.A.*, **52**, 14694 (1958)].
[576] M. D. Rausch, U.S. pat. 3,098,864 [*C.A.*, **60**, 561 (1964)].
[577] A. N. Nesmeyanov, E. G. Perevalova, and L. S. Shilovtseva, *Izv. Akad. SSSR, Otd. Khim. Nauk*, 1767 (1962) [*C.A.*, **58**, 7972 (1963)].
[578] G. R. Knox, P. L. Pauson, and G. V. D. Tiers, *Chem. Ind.* (*London*), 1046 (1959).
[579] D. Lednicer, J. K. Lindsay, and C. R. Hauser, *J. Org. Chem.*, **23**, 653 (1958).
[580] C. R. Hauser, J. K. Lindsay, D. Lednicer, and C. E. Cain, *J. Org. Chem.*, **22**, 717 (1957).
[581] T. I. Bieber and M. T. Dorsett, *J. Org. Chem.*, **29**, 2028 (1964).
[582] J. Decombe, A. Dormond, and J. Ravoux, *Compt. Rend.*, **259**, 4289 (1964).
[583] J. Decombe, J. P. Ravoux, and A. Dormond, *Bull. Soc. Chim. France*, 1261 (1965).
[584] J. Decombe, J. Ravoux, and A. Dormond, *Compt. Rend.*, **258**, 2348 (1964).
[585] B. Gautheron and J. Tirouflet, *Compt. Rend.*, **258**, 6443 (1964).
[586] J. W. Huffman and R. L. Asbury, *J. Org. Chem.*, **30**, 3941 (1965).
[587] A. N. Nesmeyanov, E. G. Perevalova, L. S. Shilovtseva, and V. D. Tyurin, *Izv. Akad. Nauk SSSR, Otd. Khim. Nauk*, 1997 (1962) [*C.A.*, **58**, 9132 (1963)].
[588] H. J. Lorkowski, *J. Prakt. Chem.*, **23**, 98 (1964).
[589] A. N. Nesmeyanov, E. G. Perevalova, Yu. A. Ustynyuk, and L. S. Shilovtseva, *Izv. Akad. Nauk SSSR, Otd. Khim. Nauk*, 554 (1960) [*C.A.*, **54**, 22540 (1960)].
[590] C. C. Barker, G. Hallus, and M. N. Thornber, *J. Chem. Soc.*, 5759 (1965).
[591] V. Weinmayr, U.S. pat. 2,988,562 [*C.A.*, **55**, 22338 (1961)].
[592] R. J. Stephenson, Brit. pat., 864,197 [*C.A.*, **55**, 17647 (1961)].
[593] A. N. Nesmeyanov, V. D. Vil'chevskaya, and N. S. Kochetkova, *Dokl. Akad. Nauk SSSR*, **152**, 627 (1963) [*C.A.*, **60**, 1793 (1964)].
[594] R. J. Stephenson, Brit. pat. 861,833 [*C.A.*, **55**, 25981 (1961)].
[595] T. Leigh, Brit. pat. 819,108 [*C.A.*, **54**, 7732 (1960)].
[596] E. I. DeYoung, *J. Org. Chem.*, **26**, 1312 (1961).
[597] T. Leigh, Brit. pat. 898,633 [*C.A.*, **57**, 15156 (1962)].
[598] J. Tirouflet, B. Gautheron, and R. Dabard, *Bull. Soc. Chim. France*, 96 (1965).
[599] M. Vogel, M. D. Rausch, and H. Rosenberg, *J. Org. Chem.*, **22**, 1016 (1957).
[600] A. N. Nesmeyanov and N. A. Vol'kenau, *Dokl. Akad. Nauk SSSR*, **107**, 262 (1956) [*C.A.*, **50**, 15519 (1956)].
[601] T. Leigh, Brit. pat. 869,504 [*C.A.*, **55**, 24790 (1961)].

[602] R. L. Schaaf and C. T. Lenk, *J. Org. Chem.*, **28,** 3238 (1963).

[603] M. Cais, Israeli pat. 15,293 [*C.A.*, **57,** 11234 (1962)].

[604] E. G. Perevalova, M. D. Reshetova, I. I. Grandberg, and A. N. Nesmeyanov, *Izv. Akad. Nauk SSSR, Ser. Khim.*, 1901 (1964) [*C.A.*, **62,** 2792 (1965)].

[605] L. R. Moffett, Jr., *J. Org. Chem.*, **29,** 3726 (1964).

[606] P. J. Graham and G. M. Whitman, U.S. pat. 2,859,233 [*C.A.*, **53,** 8161 (1959)].

[607] A. N. Nesmeyanov and N. S. Kochetkova, *Dokl. Akad. Nauk SSSR*, **117,** 92 (1957) [*C.A.*, **52,** 7295 (1958)].

[608] Y. M. Paushkin, T. P. Vishnyakova, T. A. Sokolinskaya, K. I. Zimina, and G. G. Kotova, *Neftekhimiya*, **3,** 280 (1963) [*C.A.*, **59,** 6438 (1963)].

[609] A. L. J. Beckwith and R. J. Leydon, *Tetrahedron*, **20,** 791 (1964).

[610] A. L. J. Beckwith and R. J. Leydon, *Tetrahedron Letters*, 385 (1963).

[611] E. W. Neuse and D. S. Trifan, *J. Am. Chem. Soc.*, **84,** 1850 (1962).

[612] A. N. Nesmeyanov, N. S. Kochetkova, S. V. Vitt, V. B. Bondarev, and E. I. Kovshov, *Dokl. Akad. Nauk SSSR*, **156,** 99 (1964) [*C.A.*, **61,** 3145 (1964)].

[613] T. P. Vishnyakova, Ya. M. Paushkin, and T. A. Sokolinskaya, *Zh. Obshch. Khim.*, **33,** 3685 (1963) [*C.A.*, **60,** 8060 (1964)].

[614] V. Weinmayr, U.S. pat. 2,831,879 [*C.A.*, **52,** 16367 (1958)].

[615] H. J. Lorkowski, *J. Prakt. Chem.*, **27,** 6 (1965).

[616] A. N. Nesmeyanov, E. G. Perevalova, and Yu. A. Ustynyuk, *Dokl. Akad. Nauk SSSR*, **133,** 1105 (1960) [*C.A.*, **54,** 24616 (1960)].

[617] A. C. Haven, Jr., U.S. pat. 2,821,512 [*C.A.*, **52,** 6842 (1958)].

[618] I. Pascal and W. J. Borecki, U.S. pat. 3,132,165 [*C.A.*, **61,** 4396 (1964)].

[619] E. G. Perevalova, Yu. A. Ustynyuk, and A. N. Nesmeyanov, *Izv. Akad. Nauk SSSR, Ser. Khim.*, 1967 (1963) [*C.A.*, **60,** 6864 (1964)].

[620] Yu. A. Ustynyuk, E. G. Perevalova, and A. N. Nesmeyanov, *Izv. Akad. Nauk SSSR, Ser. Khim.*, 70 (1964) [*C.A.*, **60,** 9310 (1964)].

[621] W. F. Little, C. N. Reilley, J. D. Johnson, K. N. Lynn, and A. P. Sanders, *J. Am. Chem. Soc.*, **86,** 1376 (1964).

[622] W. F. Little, R. Nielsen, and R. Williams, *Chem. Ind.* (*London*) 195 (1964).

[623] K. Schlögl and H. Egger, *Monatsh. Chem.*, **94,** 1054 (1963).

[624] S. C. Shih, H. T. Sung, and F. C. Li, *K'o Hsueh T'ung Pao*, 78 (1965) [*C.A.*, **63,** 13314 (1965)].

[625] M. Rosenblum, J. O. Santer, and W. G. Howells, *J. Am. Chem. Soc.*, **85,** 1450 (1963).

[626] J. Tirouflet and G. Tainturier, *Tetrahedron Letters*, 4177 (1965).

[627] A. N. Nesmeyanov, E. G. Perevalova, L. I. Leonteva, and Yu. A. Ustynyuk, *Izv. Akad. Nauk SSSR, Ser. Khim.*, 1882 (1965) [*C.A.*, **64,** 2123 (1965)].

[628] I. I. Grandberg and A. N. Kost, *Zh. Obshch. Khim.*, **32,** 3025 (1962) [*C.A.*, **58,** 3412 (1963)].

[629] L. Wolf and H. Hennig, *Z. Anorg. Allgem. Chem.*, **341,** 1 (1965).

[630] M. Furdik, P. Elecko, and S. Kovac, *Chem. Zvesti*, **19,** 371 (1965) [*C.A.*, **63,** 7041 (1965)].

[631] A. N. Nesmeyanov, V. D. Vil'chevskaya, N. S. Kochetkova, and N. P. Palitsyn, *Izv. Akad. Nauk. SSSR, Ser. Khim.*, 2051 (1963) [*C.A.*, **60,** 5548 (1964)].

[632] A. N. Nesmeyanov, D. N. Kursanov, V. D. Vil'chevskaya, N. S. Kochetkova, V. N. Setkina, and Yu. N. Novikov, *Dokl. Akad. Nauk SSSR*, **160,** 1090 (1965) [*C.A.*, **62,** 14725 (1965)].

[633] Midland Silicones, Ltd., British pat. 799,067 [*C.A.*, **53,** 15094 (1959)].

[634] T. H. Coffield and H. Hebert, U.S. pat. 2,927,935 [*C.A.*, **54,** 12157 (1960)].

[635] A. N. Nesmeyanov, K. N. Anisimov, and N. E. Kolobova, *Izv. Akad. Nauk SSSR, Ser. Khim.*, 1880 (1963) [*C.A.*, **60,** 2998 (1964)].

[636] A. N. Nesmeyanov, K. N. Anisimov, and N. E. Kolobova, USSR pat. 159,519 [C.A., **60,** 14538 (1963)]; see also *Izv. Akad. Nauk SSSR*, 2220 (1964) [*C.A.*, **62,** 7788 (1965)].

[637] L. L. Sims, U.S. pat. 2,987,529 [*C.A.*, **55,** 22340 (1961)].

[638] H. Shapiro, E. G. DeWitt, and J. E. Brown, U.S. pat. 2,987,531 [*C.A.*, **55,** 22340 (1961)].

[639] E. Brodkorb and H. Cordes, Brit. pat. 921,031 [*C.A.*, **59,** 10125 (1963)].
[640] J. F. Cordes and D. Neubauer, *Z. Naturforsch.*, **17b**, 791 (1962).
[641] C. R. Bergeron and A. F. Limper, U.S. pat. 2,915,539 [*C.A.*, **54,** 17416 (1960)].
[642] H. Shapiro, U.S. pat. 2,916,504 [*C.A.*, **54,** 15402 (1960)].
[643] V. Hnizda, U.S. pat. 2,948,744 [*C.A.*, **55,** 1480 (1961)].
[644] H. E. Petree, U.S. pat. 2,868,816 [*C.A.*, **53,** 11407 (1959)].
[645] T. H. Pearson, U.S. pat. 2,915,440 [*C.A.*, **54,** 17415 (1960)].
[646] E. G. DeWitt, H. Shapiro, and J. E. Brown, U.S. pat. 2,964,547 [*C.A.*, **55,** 11434 (1961)].
[647] T. H. Coffield, U.S. pat. 3,100,214 [*C.A.*, **60,** 550 (1964)].
[648] T. H. Coffield, U.S. pat. 3,100,213 [*C.A.*, **60,** 550 (1964)].
[649] T. H. Coffield, U.S. pat. 3,100,211 [*C.A.*, **60,** 549 (1964)].
[650] T. H. Coffield, U.S. pat. 3,100,212 [*C.A.*, **60,** 549 (1964)].
[651] J. Tirouflet, R. Dabard, and E. Laviron, *Bull. Soc. Chim. France*, 1655 (1963).
[652] E. O. Fischer and M. Gerberhold, *Experientia Suppl.*, **9,** 259 (1964) [*C.A.*, **62,** 579 (1965)].
[653] M. L. Ziegler and R. K. Sheline, *Inorg. Chem.*, **4,** 1230 (1965).
[654] W. Strohmeier and H. Hellmann, *Chem. Ber.*, **98,** 1598 (1965).
[655] W. Strohmeier and J. Guttenberger, *Z. Naturforsch.*, **18b,** 667 (1963).
[656] T. H. Pearson and J. K. Presswood, U.S. pat. 2,987,530 [*C.A.*, **55,** 22340 (1961)].
[657] Ethyl Corp., Brit. pat. 861,371 [*C.A.*, **56,** 15547 (1962)].
[658] D. Lednicer and C. R. Hauser, *Org. Syntheses*, **40,** 45 (1960).
[659] R. G. Hayter and L. F. Williams, *J. Inorg. Nucl. Chem.*, **26,** 1977 (1964).

CHAPTER 2

THE γ-ALKYLATION AND γ-ARYLATION OF DIANIONS OF β-DICARBONYL COMPOUNDS

Thomas M. Harris and Constance M. Harris
Vanderbilt University

CONTENTS

INTRODUCTION

It is well known that β-diketones can undergo condensations at the α-methylene group with alkyl halides and certain other reagents through the intermediate formation of monoanions. For example, acetylacetone on treatment with an alkali metal or an alkali metal alkoxide or carbonate forms a monoanion which can be alkylated with alkyl halides. With benzyl chloride the α-benzyl derivative is formed.[1]

$$CH_3COCH_2COCH_3 \xrightarrow{\text{Base}} CH_3CO\overset{\ominus}{C}HCOCH_3 \xrightarrow{C_6H_5CH_2Cl} CH_3COCH(CH_2C_6H_5)COCH_3$$

These reactions are often carried out in boiling acetone, benzene, or toluene and in general proceed rather slowly. A second alkylation at the α position is frequently a side reaction.[1]

$$CH_3COCH(CH_2C_6H_5)COCH_3 \xrightarrow[C_6H_5CH_2Cl]{\text{Base}} CH_3COC(CH_2C_6H_5)_2COCH_3$$

In 1958 Hauser and Harris reported that, if benzoylacetone or acetylacetone is first converted to the dipotassium salt, the salt undergoes alkylation and other carbon-carbon condensations at the terminal methyl group rather than at the methylene group.[2] The dipotassium salts were prepared by treatment of the diketones with 2 molecular equivalents of potassium amide in liquid ammonia. Rapid alkylation occurred when the salts were treated with 1 equivalent of benzyl chloride and, after acidification, the terminal monobenzylation derivatives were obtained in good

[1] G. T. Morgan and C. J. A. Taylor, *J. Chem. Soc.*, **127**, 797 (1925).
[2] C. R. Hauser and T. M. Harris, *J. Am. Chem. Soc.*, **80**, 6360 (1958).

yield. Dibenzylation was not observed. In this chapter the alkylations of these and other β-diketones via their dianions will be surveyed.

$$C_6H_5COCH_2COCH_3 \xrightarrow[\text{Liq. } NH_3]{2\ KNH_2} C_6H_5COCHKCOCH_2K$$

$$\downarrow C_6H_5CH_2Cl$$

$$C_6H_5COCH_2COCH_2CH_2C_6H_5 \xleftarrow{H^{\oplus}} C_6H_5COCHKCOCH_2CH_2C_6H_5$$

β-Keto aldehydes and β-keto esters have also been converted to dianions by similar procedures; the dianions undergo alkylation reactions at the γ positions.[3, 4] These alkylations will also be discussed.

$$\overset{\ominus}{C}H_2CO\overset{\ominus}{C}HCHO \xrightarrow{RX} RCH_2CO\overset{\ominus}{C}HCHO$$

$$\overset{\ominus}{C}H_2CO\overset{\ominus}{C}HCO_2C_2H_5 \xrightarrow{RX} RCH_2CO\overset{\ominus}{C}HCO_2C_2H_5$$

Alkylation can be considered a typical reaction of these carbanions. The successful alkylation of a dicarbanion is an indication that acylation, carboxylation, and aldol-type condensations can also be effected. Thus the results obtained with alkylating agents are useful in predicting and understanding the reactions of the dianions of β-dicarbonyl compounds with many other electrophilic condensing agents. Reactions that have been observed include carboxylation with carbon dioxide,[5] acylation with aromatic and aliphatic esters,[4, 6–8] carbonyl addition reactions with aldehydes and ketones[4, 8–10] and 1,4-additions to α,β-unsaturated compounds.[6, 8, 10] Some of these are illustrated by the accompanying equations.

$$C_6H_5COCHKCOCH_2K$$

$$\xrightarrow[2.\ H^{\oplus}]{1.\ CO_2} C_6H_5COCH_2COCH_2CO_2H$$

$$\xrightarrow[2.\ H^{\oplus}]{1.\ C_6H_5CO_2CH_3} C_6H_5COCH_2COCH_2COC_6H_5$$

$$\xrightarrow[2.\ H^{\oplus}]{1.\ (C_6H_5)_2CO} C_6H_5COCH_2COCH_2C(OH)(C_6H_5)_2$$

$$\xrightarrow[2.\ H^{\oplus}]{1.\ C_6H_5CH{=}CHCO(2,4,6\text{-}(CH_3)_3C_6H_2)} C_6H_5COCH_2COCH_2CH(C_6H_5)CH_2CO(2,4,6\text{-}(CH_3)_3C_6H_2)$$

[3] T. M. Harris and C. R. Hauser, *J. Am. Chem. Soc.*, **84**, 1750 (1962).

[4] J. F. Wolfe, T. M. Harris, and C. R. Hauser, *J. Org. Chem.*, **29**, 3249 (1964).

[5] T. M. Harris and C. M. Harris, *J. Org. Chem.*, **31**, 1032 (1966).

[6] F. B. Kirby, T. M. Harris, and C. R. Hauser, *J. Org. Chem.*, **28**, 2266 (1963).

[7] S. D. Work and C. R. Hauser, *J. Org. Chem.*, **28**, 725 (1963).

[8] T. M. Harris, S. Boatman, and C. R. Hauser, *J. Am. Chem. Soc.*, **87**, 3186 (1965).

[9] R. J. Light and C. R. Hauser, *J. Org. Chem.*, **26**, 1716 (1961).

[10] R. J. Light, T. M. Harris, and C. R. Hauser, *J. Org. Chem.*, **26**, 1344 (1961).

Arylations, although apparently mechanistically different from alkylations,[11] are included in this chapter because of their formal similarity to alkylations and because alkylation and arylation are methods for directly converting β-diketones into higher homologs without increasing their functionality.

MECHANISM

Chemical evidence indicates that the intermediates in the reaction of β-dicarbonyl compounds at the γ position are the *bis*-enolate anions resulting from proton abstraction at both the α and the γ positions. Although canonical resonance forms having one or both negative charges on oxygen are undoubtedly important contributors, the dialkali metal salts will henceforth be pictured as dicarbanions to emphasize the high reactivity at the γ carbon atom.

$$\mathrm{R\overset{O^{\ominus}}{\overset{|}{C}}{=}CH\overset{O^{\ominus}}{\overset{|}{C}}{=}CH_2} \leftrightarrow \mathrm{R\overset{O^{\ominus}}{\overset{|}{C}}{=}CH\overset{O}{\overset{\|}{C}}{-}\overset{\ominus}{C}H_2}$$

$$\updownarrow \qquad\qquad \updownarrow$$

$$\mathrm{R\overset{O}{\overset{\|}{C}}\overset{\ominus}{C}H\overset{O^{\ominus}}{\overset{|}{C}}{=}CH_2} \leftrightarrow \mathrm{R\overset{O}{\overset{\|}{C}}\overset{\ominus}{C}H\overset{O}{\overset{\|}{C}}\overset{\ominus}{C}H_2}$$

The nuclear magnetic resonance spectrum of the dipotassium salt of benzoylacetone provides physical evidence that the species present is an α,γ-dianion.[12] The spectrum shows, in addition to the aromatic multiplet, a singlet and two doublets. The singlet has been assigned to proton *a* and the doublets to protons *b* and *c*.[12] Protons *b* and *c* are coupled by approximately 2 cps. The facts that protons *b* and *c* have different chemical shifts and that the coupling constant is quite small[13] support the contention that the terminal anion is more accurately represented as a planar enolate anion than as a tetrahedral carbanion.

$$\mathrm{C_6H_5{-}\overset{O^{\ominus}K^{\oplus}}{\overset{|}{C}}{=}\underset{H_a}{\underset{|}{C}}{-}\overset{O^{\ominus}K^{\oplus}}{\overset{|}{C}}{=}C\genfrac{}{}{0pt}{}{\diagup H_b}{\diagdown H_c}}$$

Chemical evidence for dianion formation is that a β-diketone such as dibenzoylmethane, having only α-hydrogen atoms, has been found to react with only 1 molecular equivalent of potassium amide in liquid

[11] K. G. Hampton, T. M. Harris, and C. R. Hauser, *J. Org. Chem.*, **29**, 3511 (1964).

[12] M. L. Miles, C. G. Moreland, D. M. von Schriltz, and C. R. Hauser, *Chem. & Ind.* (*London*), 2098 (1966).

[13] H. S. Gutowsky, M. Karplus, and D. M. Grant, *J. Chem. Phys.*, **31**, 1278 (1959).

ammonia,[14] whereas a β-diketone such as benzoylacetone, having both α- and γ-hydrogen atoms, reacts with 2 molecular equivalents of the base. The fact that alkylation and other condensations occur at the γ position of dicarbanions is *prima facie* evidence that this was the site of secondary ionization, since alkylations at this position are not observed when 1:1 ratios of dicarbonyl compound and base are employed. Additional evidence has been obtained from titration of disodioacetylacetone with acid. The salt was found to be dibasic and to be essentially free of ammonia.[15]

The possibility remains that acetylacetone, which has two terminal methyl groups as well as an α-methylene group, might form a 1,5-dicarbanion, $\overset{\ominus}{C}H_2COCH_2CO\overset{\ominus}{C}H_2$, instead of a 1,3-dicarbanion,

$$CH_3CO\overset{\ominus}{C}HCO\overset{\ominus}{C}H_2$$

There is no evidence to support this proposal. Some 1,5-dialkylation has been observed in the alkylation of dipotassioacetylacetone. However, this has been accounted for in another manner (p. 163).[16]

Evidently the γ-hydrogen atoms of singly ionized β-diketones or β-keto aldehydes are sufficiently acidic that treatment with a stoichiometric amount of alkali metal amide affords essentially complete conversion to the dicarbanions; that is, these dianions are weaker bases in ammonia than is amide ion. If the base strengths of amide ion and the dianion had been similar, significant amounts of amide ion and monoanion of the dicarbonyl compound would have remained in equilibrium with the dianion. The evidence that such a situation does not exist is that stilbene formation has usually not been observed during the alkylation

$$C_6H_5COCHKCOCH_3 + KNH_2 \rightleftharpoons C_6H_5COCHKCOCH_2K + NH_3$$

$$\downarrow C_6H_5CH_2Cl$$

$$\underset{\text{[None observed]}}{[C_6H_5CH{=}CHC_6H_5]} + C_6H_5COCHKCOCH_2CH_2C_6H_5$$

of these dianions, for example, dipotassiobenzoylacetone, with benzyl chloride.[2] Benzyl chloride is known to undergo displacement reactions with carbanions but exclusively self-condensation on treatment with amide ion.[17] The latter reaction pathway would be able to compete with

[14] T. M. Harris and C. R. Hauser, unpublished results.

[15] T. M. Harris, unpublished results.

[16] K. G. Hampton, T. M. Harris, and C. R. Hauser, *J. Org. Chem.*, **30**, 61 (1965).

[17] C. R. Hauser, W. R. Brasen, P. S. Skell, S. W. Kantor, and A. E. Brodhag, *J. Am. Chem. Soc.*, **78**, 1653 (1956).

alkylation if significant amounts of amide ion were present.* The self-condensation of benzyl chloride is characterized by a vivid, transient violet coloration of the reaction mixture which permits visual detection of this side reaction.[18] The inability to obtain stilbene or to observe this violet coloration during alkylations of dianions is evidence that essentially no amide ion remains in the dianion reaction mixture.

In general, good results have been obtained with β-keto aldehydes, and there is no indication that dianion formation is not complete. However, on the basis of limited study of ethyl acetoacetate, it is possible that this β-keto ester is not completely converted to the dianion by potassium amide.[4]

Alkylation at the α position of diketones has not been observed during the dianion reactions, even when excess alkyl halide is employed. Not only is the nucleophilicity of the monocarbanion too low for observable reaction to occur under these conditions, but also apparently the nucleophilicity of the α position of the dianion is substantially less than that of the γ position. Alkylations are presumably simple displacement reactions, although the details of the mechanism have not been investigated.

Alkylation reactions appear to have a high degree of stereospecificity. The alkylations of the disodium salts of 1-phenyl-2,4-pentanedione and 1,4-diphenyl-1,3-butanedione with α-phenethyl chloride gave exclusively the corresponding erythro products.[18a]

$$RCO\overset{\ominus}{C}HCO\overset{\ominus}{C}HC_6H_5 \xrightarrow[R\,=\,CH_3 \text{ or } C_6H_5]{C_6H_5CHClCH_3} \underset{\text{Erythro}}{RCOCH_2COCH(C_6H_5)CH(C_6H_5)CH_3}$$

In the first case ($R = CH_3$), the threo isomer was demonstrated to be stable to the alkylation conditions. Therefore it can be concluded that the more stable erythro isomer was the direct product of the alkylation reaction.[18a]

Alkylations of the dianion of 2-formyl-1-decalone occur at the 9 position to generate an additional asymmetric center (see p. 176). Considerable stereospecificity exists in these reactions also.

The arylation of dicarbanions with diaryliodonium salts appears to be at least partially a free-radical reaction. It has been postulated[11] that one electron is transferred from the dicarbanion to the iodonium ion to afford the anion radical of the diketone. The resulting diphenyliodine decomposes to iodobenzene and phenyl radical. The phenyl radical couples primarily at the γ position of the diketone anion radical. The

* The self-condensation of benzyl chloride by amide ion and the benzylation of a carbanion, that of diphenylmethane, have been shown to occur at comparable rates.[18]

[18] C. R. Hauser and P. J. Hamrick, *J. Am. Chem. Soc.*, **79**, 3142 (1957).

[18a] D. M. von Schriltz and C. R. Hauser, unpublished results.

sequence is illustrated with acetylacetone dicarbanion and diphenyliodonium ion in the accompanying scheme.

$$CH_3CO\overset{\ominus}{C}HCO\overset{\ominus}{C}H_2 + (C_6H_5)_2I^{\oplus} \rightarrow CH_3CO\overset{\ominus}{C}HCO\dot{C}H_2 + (C_6H_5)_2I\cdot$$

$$(C_6H_5)_2I\cdot \rightarrow C_6H_5I + C_6H_5\cdot$$

$$CH_3CO\overset{\ominus}{C}HCO\dot{C}H_2 \xrightarrow[\text{2. Neutralization}]{\text{1. } C_6H_5\cdot} CH_3COCH_2COCH_2C_6H_5$$

In support of this mechanism, both biphenyl and α-phenylacetylacetone have been detected in trace amounts.[11] The α-phenylated diketone was accounted for by resonance contributors to the intermediate anion radical in which the free electron is on the α- as well as on the γ-carbon atom. Apparently the difference in reactivity of the α and γ positions of the anion radical is less than the difference in reactivity of the two positions of the dianion.

$$CH_3\overset{O}{\overset{\|}{C}}\overset{\ominus}{C}H\overset{O}{\overset{\|}{C}}\dot{C}H_2 \leftrightarrow CH_3\overset{O}{\overset{\|}{C}}CH{=}\overset{O^{\ominus}}{\overset{|}{C}}\dot{C}H_2 \leftrightarrow CH_3\overset{O^{\ominus}}{\overset{|}{C}}{=}CH\overset{O}{\overset{\|}{C}}\dot{C}H_2$$

$$\updownarrow \qquad\qquad \updownarrow$$

$$CH_3\overset{O}{\overset{\|}{C}}\dot{C}H\overset{O}{\overset{\|}{C}}\overset{\ominus}{C}H_2 \leftrightarrow CH_3\overset{O}{\overset{\|}{C}}\dot{C}H\overset{O^{\ominus}}{\overset{|}{C}}{=}CH_2$$

Additional support for the free-radical mechanism of phenylation is the observation that, although dipotassioacetylacetone is inert toward iodobenzene, in the presence of diphenyliodonium chloride a significant amount of iodobenzene does react to form the γ-phenylation product. When disodioacetylacetone (2 equivalents) was treated with diphenyliodonium chloride, the yield was 92%. However, if an equivalent amount of iodobenzene was added simultaneously with the iodonium salt, a yield in excess of quantitative (109% based on iodonium salt) was obtained.[11]

The possibility that the arylation reaction is partly direct displacement cannot be eliminated. Direct displacement has been postulated in naphthylations of other anions by 1- and 2-naphthyl methyl sulfone and 1-fluoronaphthalene.[19] On the other hand, it is possible that certain alkylations such as those with benzyl chloride may proceed by a single electron transfer similar to the phenylation reaction.

Finally, several arylations have been effected with halobenzenes and excess amide ion.[2, 11] These reactions undoubtedly involve arylation by benzyne intermediates.

$$RCO\overset{\ominus}{C}HCO\overset{\ominus}{C}H_2 + C_6H_5Br \xrightarrow[\text{2. } H^{\oplus}]{\text{1. } NH_2^{\ominus}} RCOCH_2COCH_2C_6H_5$$

[19] J. F. Bunnett and T. K. Brotherton, *J. Am. Chem. Soc.*, **78**, 6265 (1956).

SCOPE AND LIMITATIONS

Preparation of Dianions

The dicarbanions of β-diketones are prepared by addition of the diketone to 2 equivalents of alkali amide in liquid ammonia. The amides of lithium, sodium, and potassium have been employed in this reaction. Conversion to the dianions is apparently rapid.

The solubility of the salts in liquid ammonia cannot be readily predicted. Although potassium amide is much more soluble in liquid ammonia than is sodium amide or lithium amide, disodioacetylacetone is more soluble in this medium than either the dipotassium or the dilithium salt.[16] The greater solubility of the disodium salt in ammonia has also been observed with other aliphatic diketones.[20]

The dianions of β-keto aldehydes have usually been prepared by addition of the monosodium salt of the keto aldehyde to potassium amide in liquid ammonia.[3, 21] Such a procedure is necessary with acetoacetaldehyde and other keto aldehydes lacking substituents at the α position because these compounds readily undergo aldol condensations leading to 1,3,5-triacylbenzenes.[22] The monosodium salts of many β-keto aldehydes are conveniently prepared by acylation of the corresponding ketone with formate esters and sodium alkoxides.[22] The condensations occur in high yield, and the sodium salts have been used without subsequent purification. The formation of sodioformylacetone and its conversion to the dianion are illustrated in the accompanying scheme.

$$CH_3COCH_3 + HCO_2C_2H_5 \xrightarrow[(C_2H_5)_2O]{NaOCH_3} CH_3COCHNaCHO \xrightarrow[\text{Liq. } NH_3]{KNH_2} K^{\oplus}, Na^{\oplus}[CH_2COCHCHO]^{\ominus}$$

With α-substituted β-keto aldehydes the procedure involving the monosodium salt is not necessary because these compounds do not readily undergo self-condensation.[22] However, the monosodium salts have commonly been employed since these keto aldehydes are also prepared by formylation of the corresponding ketones and are conveniently isolated as their sodium salts.[23]

Keto ester dianions have been prepared principally by means of potassium amide,[4, 24] although sodium amide may sometimes be satisfactory.[25]

[20] K. G. Hampton, T. M. Harris, and C. R. Hauser, *J. Org. Chem.*, **31,** 1035 (1966).

[21] T. M. Harris, S. Boatman, and C. R. Hauser, *J. Am. Chem. Soc.*, **85,** 3273 (1963).

[22] C. R. Hauser, F. W. Swamer, and J. T. Adams, *Org. Reactions*, **8,** 87–89 (1954).

[23] S. Boatman, T. M. Harris, and C. R. Hauser, *J. Am. Chem. Soc.*, **87,** 82 (1965).

[24] V. I. Gunar, L. F. Kudryavtseva, and S. I. Zav'yalov, *Bull. Acad. Sci. USSR, Div. Chem. Sci.*, 1343 (1962) [*Izv. Akad. Nauk SSSR, Otd. Khim. Nauk*, 1431 (1962)] [*C.A.*, **58,** 2378 (1963)].

[25] R. L. Carney and T. M. Harris, unpublished results.

Alkylations of Acetylacetone (Table III)

The dialkali metal salts of acetylacetone show certain distinct differences in reactivity. These differences cannot be correlated with solubility in ammonia but can be correlated with the size of the alkali metal ion. The dilithium salt has more covalent character and is markedly less reactive in alkylation reactions than is the disodium or dipotassium salt. The alkylation of dilithioacetylacetone with *n*-butyl bromide is reported to be only 25% complete after twice the reaction period employed to obtain 67–73% alkylation of disodioacetylacetone.[16] This lack of reactivity does not extend to all classes of reactions of dilithium salts because these salts have been found to be uniquely useful for condensations with aliphatic esters and with ketones having relatively acidic α-hydrogen atoms.[7,9] For example, acylation of dilithiobenzoylacetone with ethyl acetate proceeds rapidly and at a rate much greater than proton abstraction from the ester.[7] On the other hand, disodio- and dipotassio-benzoylacetone selectively ionize the ester.[7]

$$C_6H_5COCHLiCOCH_2Li \xrightarrow{CH_3CO_2C_2H_5} C_6H_5COCHLiCOCHLiCOCH_3$$

$$C_6H_5COCHKCOCH_2K \xrightarrow{CH_3CO_2C_2H_5} KCH_2CO_2C_2H_5 + C_6H_5COCHKCOCH_3$$

Disodium and dipotassium salts of acetylacetone were initially considered to have essentially the same properties.[26] However, a fundamental difference in the chemistry of the two salts of acetylacetone was later detected by gas chromatographic analysis of alkylation products.[16,27] Alkylation of dipotassioacetylacetone was found to afford significant amounts (14–26%) of the 1,5-dialkylation product in addition to the normal terminal alkylation product.[16] Although this might have occurred through a 1,5-dianion, $\overset{\ominus}{C}H_2COCH_2CO\overset{\ominus}{C}H_2$, it is more likely that as monoalkylation products are formed they react with dipotassioacetylacetone to form a new dianion that subsequently undergoes further alkylation.

$$CH_3COCHKCOCH_2K \xrightarrow{RX} CH_3COCHKCOCH_2R$$

$$CH_3COCHKCOCH_2R \xrightarrow{CH_3COCHKCOCH_2K} KCH_2COCHKCOCH_2R$$

$$KCH_2COCHKCOCH_2R \xrightarrow{RX} RCH_2COCHKCOCH_2R$$

The amount of dialkylation is somewhat dependent on the addition rate of alkyl halide to dipotassioacetylacetone; very slow addition lengthens

[26] R. B. Meyer and C. R. Hauser, *J. Org. Chem.*, **25**, 158 (1960).

[27] K. G. Hampton, T. M. Harris, and C. R. Hauser, *J. Org. Chem.*, **28**, 1946 (1963).

the time in which proton abstraction can occur and hence increases the extent of dialkylation.

On the other hand, disodioacetylacetone undergoes alkylation with alkyl halides without observable formation of dialkylation products.[16] Apparently proton interchange between a monoanion and a dianion is much slower with sodium salts than with potassium salts. However, the difference in alkylation rates is much less, if any. Similarly alkylation of dilithioacetylacetone, although impractical because of the low reaction rate, does not afford dialkylation products.[16]

The difference between the dipotassium salt of acetylacetone and the more covalent disodium salt was clearly demonstrated by treatment of mixtures of the dianion of acetylacetone (**1**) and the monoanion of another diketone (**2**) with an alkyl halide.[16] The yield of terminal alkylation product arising from the initially monoionized diketone was a measure of the extent to which dianion interchange had occurred. In addition, the monoanion of acetylacetone (**3**) and the dianion (**4**) of the other diketone were employed in the reverse experiment.

$$\underset{\mathbf{1}}{CH_3COCHMCOCH_2M} + \underset{\mathbf{2}}{n\text{-}C_5H_{11}COCHMCOCH_3} \underset{\text{Fast, M = K}}{\overset{\text{Very slow, M = Na}}{\rightleftharpoons}} \underset{\mathbf{3}}{CH_3COCHMCOCH_3} + \underset{\mathbf{4}}{n\text{-}C_5H_{11}COCHMCOCH_2M}$$

$$\xrightarrow[\text{2. } NH_4Cl]{\text{1. } C_2H_5Br} \underset{\mathbf{5}}{CH_3COCH_2COCH_2C_2H_5} + \underset{\mathbf{6}}{n\text{-}C_5H_{11}COCH_2COCH_2C_2H_5}$$

In the case of the potassium salts, when equal molar mixtures either of **1** and **2** or of **3** and **4** were stirred for 2 hours before alkylation, an approximate 4:1 mixture of alkylation products **5** and **6** was obtained.[16] Even after only 6 minutes of mixing, nearly the same results were obtained, indicating that equilibration was largely complete.

With the sodium salts, dianion interchange was not observed.[16] Alkylation of a mixture of sodium salts **1** and **2** afforded only **5**, and alkylation of **3** and **4** afforded only **6**.[16] This result was obtained even after the mono- and di-sodium salts were stirred together for 2 hours.

The metallic cation effect is made more striking by the fact that rapid dianion interchange occurred with the sparingly soluble potassium salts rather than with the highly soluble sodium salts.

One additional conclusion to be drawn from this study is that the disodium salts are not significantly involved in proton transfers with ammonia to form sodium amide and the monoanion of the diketone. This fact has been useful in interpretation of the results obtained in the alkylation of unsymmetrical diketones (see next section).

$$RCOCHNaCOCH_2Na + NH_3 \rightleftharpoons RCOCHNaCOCH_3 + NaNH_2$$

Many examples of alkylation of the disodium and dipotassium salts of acetylacetone have been reported (see Table III). The results obtained with the disodium salts are better in most cases, largely because dialkylation products are not produced.

Primary and secondary aliphatic halides have been used with almost uniform success; yields with disodioacetylacetone typically range from 55% to 80%. The alkylation of dipotassioacetylacetone with isopropyl bromide was effected in only 27% yield,[26] but a 66% yield was achieved in a similar alkylation of disodioacetylacetone.[16] Alkylation with *sec*-butyl bromide has been accomplished in 78% yield.[16]

Among alkylations with simple halides, failures have been reported with *t*-butyl chloride, β-phenethyl chloride, and benzhydryl chloride.[26] The cause of failure with the first of these is probably the inherent lack of reactivity of tertiary alkyl halides in displacement reactions. The failures with β-phenethyl chloride and benzhydryl chloride stem from the alkaline properties of dianions; high yields of styrene and tetraphenylethylene, respectively, were obtained from the reactions.[26] Tetraphenylethylene can arise by self-alkylation of benzhydryl chloride by means of its anion.[17]

$$CH_3COCHKCOCH_2K + C_6H_5CH_2CH_2Cl \rightarrow CH_3COCHKCOCH_3 + C_6H_5CH{=}CH_2$$

$$CH_3COCHKCOCH_2K + (C_6H_5)_2CHCl \rightarrow CH_3COCHKCOCH_3 + (C_6H_5)_2CKCl$$

$$(C_6H_5)_2CKCl \xrightarrow{(C_6H_5)_2CHCl} (C_6H_5)_2CClCH(C_6H_5)_2$$

$$(C_6H_5)_2CClCH(C_6H_5)_2 \xrightarrow[\text{of HCl}]{\beta\text{-Elimination}} (C_6H_5)_2C{=}C(C_6H_5)_2$$

The alkylation of disodioacetylacetone with dihaloalkanes has been used to synthesize *bis*-β-diketones. Tri-, tetra-, nona-, and deca-methylene bromides have been employed successfully.[28]

$$2\ CH_3COCHNaCOCH_2Na + Br(CH_2)_nBr \xrightarrow[n\,=\,3,\,4,\,9,\text{ or }10]{} CH_3COCHNaCOCH_2(CH_2)_nCH_2COCHNaCOCH_3$$

Failures in the preparation of *bis*-β-diketones have been reported with methylene chloride, ethylene chloride, and ethylene bromide.[28] With ethylene bromide a mixture of two unidentified products was obtained.[28] Attempted alkylation by *o*-, *m*-, and *p*-xylylene bromide has been reported.[29] The desired alkylation probably occurred, but the products

[28] K. G. Hampton, R. J. Light, and C. R. Hauser, *J. Org. Chem.*, **30**, 1413 (1965).

[29] M. M. Coombs and R. P. Houghton, *J. Chem. Soc.*, 5015 (1961).

were not characterized. The reaction of disodioacetylacetone with dichlorodiphenylmethane afforded a 70% yield of tetraphenylethylene.[28] The mode of formation of this hydrocarbon has been suggested to be that shown in the accompanying scheme.[28] However, other mechanisms involving free radicals or carbenes are also possible.

$$CH_3COCHNaCOCH_2Na + (C_6H_5)_2CCl_2 \rightarrow CH_3COCHNaCOCH_2Cl + (C_6H_5)_2CNaCl$$

$$(C_6H_5)_2CNaCl \xrightarrow{(C_6H_5)_2CCl_2} (C_6H_5)_2CCl{-}CCl(C_6H_5)_2$$

$$(C_6H_5)_2CCl{-}CCl(C_6H_5)_2 \xrightarrow[\text{salt}]{\text{Disodium}} (C_6H_5)_2C{=}C(C_6H_5)_2 + CH_3COCHNaCOCH_2Cl$$

One rather unusual alkylation has been employed for preparation of a *bis*-β-diketone. The alkylation of dipotassioacetylacetone by 3-chloro-2,4-pentanedione has been reported.[30, 30a] The yield was low.

$$CH_3COCHKCOCH_2K + CH_3COCHClCOCH_3 \rightarrow CH_3COCH_2COCH_2CH(COCH_3)_2$$

Alkylations of Unsymmetrical β-Diketones (Table IV)

A problem of practical importance in the alkylation of the dianions of β-diketones is that many diketones are at least theoretically capable of forming two different dianions and therefore of producing two isomeric products upon alkylation. This situation arises in unsymmetrical diketones whenever both carbonyl groups are flanked by carbon atoms bearing one or more hydrogen atoms.

$$RCH_2COCH_2COCH_2R'$$

$$RCH_2CO\overset{\ominus}{C}HCO\overset{\ominus}{C}HR' \xrightarrow{1.\ R''X;\ 2.\ H^{\oplus}} RCH_2COCH_2COCHR'R''$$

$$R\overset{\ominus}{C}HCO\overset{\ominus}{C}HCOCH_2R' \xrightarrow{1.\ R''X;\ 2.\ H^{\oplus}} RR''CHCOCH_2COCH_2R'$$

Results of studies of the effect of diketone structure on the formation of dianions under these circumstances appear to have considerable generality for prediction of the site of secondary ionization and subsequent alkylation in unsymmetrical systems. Because a high degree of specificity has been observed in most of the systems investigated, the alkylation of unsymmetrical diketones through dianions has usually led to only one product.

[30] R. J. Gritter and E. L. Patmore, *Proc. Chem. Soc.*, 328 (1962).

[30a] E. L. Patmore, Doctoral Dissertation, University of Connecticut, 1963 (*Dissertation Abstr.*,64–3553).

It appears well established that methyl-methylene β-diketones having one aliphatic γ substituent **(7)** undergo alkylation through dicarbanions to

$$\underset{\mathbf{7}}{RCH_2COCH_2COCH_3} \qquad \underset{\mathbf{8}}{RCH_2COCH_2COCH_2R'}$$

form mainly or entirely the methyl alkylation product **8**. The most rigorously investigated substrate, 2,4-hexanedione **(9)**, was treated with 2 equivalents of sodium amide.[31] This might have led to dianion **10** or **11** or a mixture of the two. Treatment of the reaction mixture with methyl iodide gave an 89:11 mixture, as determined by gas chromatography, of the methyl and the methylene alkylation products **12** and **13**, respectively. Thus alkylation at the methyl group greatly predominated over that at the methylene group.

$$\underset{\mathbf{9}}{CH_3CH_2COCH_2COCH_3}$$

$$\underset{\mathbf{10}}{CH_3CH_2COCHNaCOCH_2Na} \xleftarrow{\quad NaNH_2 \quad} \cdot \xrightarrow{\quad} \underset{\mathbf{11}}{CH_3CHNaCOCHNaCOCH_3}$$

$$\mathbf{10} \xrightarrow{1.\ CH_3I;\ 2.\ H^{\oplus}} \underset{\mathbf{12}}{CH_3CH_2COCH_2COCH_2CH_3} \qquad \mathbf{11} \xrightarrow{1.\ CH_3I;\ 2.\ H^{\oplus}} \underset{\mathbf{13}}{(CH_3)_2CHCOCH_2COCH_3}$$

It has been concluded that the formation of the salts **10** and **11** is essentially an irreversible process (see p. 164) so that they are probably not interconvertible under the reaction conditions.[31] Therefore the formation of the dianions and the alkylation products is under kinetic rather than thermodynamic control, and the dianion that predominates as an intermediate is not necessarily the less basic of the two.*

Additional evidence that the ratio of alkylation products reflects reasonably well the ratio of dianions formed lies in the fact that acetyl- and propionyl-type dianions do not differ greatly in their nucleophilicities and, actually, propionyl-type dianions are more reactive than the acetyl type.[20] Thus, even if the dianions had been interconvertible, the effect on the interpretation of the experiment would not be great, because the alkylation product **13** of the propionyl-type dianion was formed in only minor amounts from 2,4-hexanedione **(9)**.

Alkylation of dipotassiopropionylacetone gave similar results.[31] It is quite likely that the isomeric dipotassium salts of this β-diketone equilibrate with one another, particularly if the monoanion or another proton

[31] K. G. Hampton, T. M. Harris, and C. R. Hauser, *J. Org. Chem.*, **31,** 663 (1966).

* The term basicity is used in the thermodynamic sense. The less basic anion is the more stable and would be the predominant species if the anions could be equilibrated with one another.

donor of comparable pK_a is present. Moreover, the dipotassium salt(s) is largely precipitated in the reaction medium. Thus the extent of equilibration and the relative reactivity of the two dianions probably reflect solubility factors as well as anion stability. Nevertheless alkylation at the methyl position predominates.

A study has also been made of the competitive formation of disodium salts in which the competing sites were on two different diketone molecules.[31] A sufficient amount of sodium amide was added to an equimolar mixture of acetylacetone and dipropionylmethane to convert one-half of the mixture to dianions and the remainder to monoanions. The mixture was treated with excess n-butyl bromide. Analysis of the products by gas chromatography indicated that the dianion of acetylacetone had predominated over that of dipropionylmethane in a ratio of 61:39.

$$\begin{array}{c} CH_3COCH_2COCH_3 \\ + \\ C_2H_5COCH_2COC_2H_5 \end{array} \xrightarrow{2\ NaNH_2} \begin{array}{c} CH_3COCHNaCOCH_3 \\ + \\ CH_3CH_2COCHNaCOCH_2CH_3 \end{array} \xrightarrow[(50\%\ \text{conversion})]{1\ NaNH_2}$$

$$\begin{array}{c} CH_3COCHNaCOCH_2Na \\ + \\ C_2H_5COCHNaCOCHNaCH_3 \end{array} \xrightarrow[2.\ H^{\oplus}]{1.\ n\text{-}C_4H_9Br} \begin{array}{cr} CH_3COCH_2COCH_2C_4H_9\text{-}n & (61\%) \\ + & \\ C_2H_5COCH_2COCH(CH_3)C_4H_9\text{-}n & (39\%) \end{array}$$

The two disodium salts were not in equilibrium with one another. When a mixture of disodiodipropionylmethane and monosodioacetylacetone in liquid ammonia was stirred for 30 minutes before treatment with n-butyl bromide, only the alkylation product of dipropionylmethane was obtained.[31]

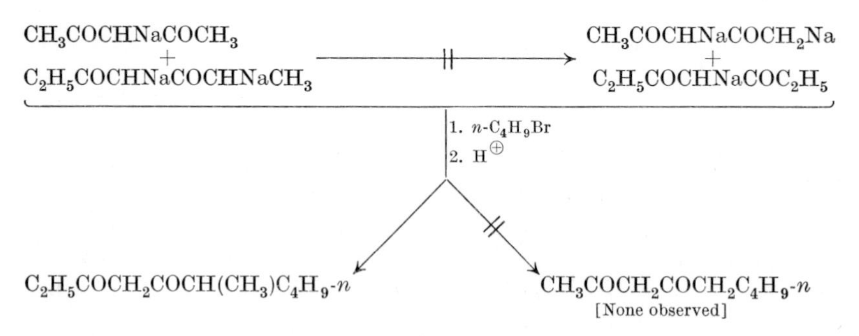

The competitive ionization of a mixture of acetylacetone and dipropionylmethane also has been conducted with potassium amide.[31] The same preference for alkylation of acetylacetone was observed. In fact, the reaction was somewhat more selective, and an 83:17 ratio of the products was obtained.

In addition to the alkylation of simple aliphatic methyl-methylene diketones two cyclic systems have been studied.[32] α-Acetylcyclohexanone and α-acetylcyclopentanone both undergo alkylation by means of dipotassium salts exclusively at methyl positions.

$$\text{2-}COCH_3\text{-cycloalkanone } [(CH_2)_n],\ n = 0 \text{ or } 1 \xrightarrow[\text{2. } C_6H_5CH_2Cl \quad \text{3. } H^{\oplus}]{\text{1. } 2\,KNH_2} \text{2-}COCH_2CH_2C_6H_5\text{-cycloalkanone } [(CH_2)_n],\ n = 0\ (62\%);\ n = 1\ (58\%)$$

The specificity of reaction at methyl groups has been observed in the dialkylation of two *bis*-β-diketones by means of tetra-anions.[33] One of the products was also synthesized independently by a dianion alkylation with a dihalide which required alkylation to occur preferentially at a methyl position.[28]

$$CH_3COCH_2CO(CH_2)_nCOCH_2COCH_3 \xrightarrow[n = 5 \text{ or } 6]{4\,NaNH_2}$$

$$NaCH_2COCHNaCO(CH_2)_nCOCHNaCOCH_2Na \xrightarrow[\text{2. } H^{\oplus}]{\text{1. } 2\,C_6H_5CH_2Cl}$$

$$C_6H_5CH_2CH_2COCH_2CO(CH_2)_nCOCH_2COCH_2CH_2C_6H_5$$

$$2\,C_6H_5CH_2CH_2COCH_2COCH_3 \xrightarrow{4\,NaNH_2}$$

$$2\,C_6H_5CH_2CH_2COCHNaCOCH_2Na \xrightarrow[H^{\oplus}]{Br(CH_2)_4Br}$$

$$C_6H_5CH_2CH_2COCH_2CO(CH_2)_6COCH_2COCH_2CH_2C_6H_5$$

Treatment of one tetrasodio-*bis*-β-diketone with a dihalide afforded a polymeric β-diketone.[33]

$$NaCH_2COCHNaCO(CH_2)_5COCHNaCOCH_2Na \xrightarrow[\text{2. } H^{\oplus}]{\text{1. } BrCH_2CH_2CH_2Br}$$

$$(\text{—}CH_2COCH_2COCH_2CH_2CH_2CH_2\text{—})_n \quad n = 12\text{–}13$$

Methyl-methylene β-diketones having aryl groups at the γ position can be expected to undergo alkylation and other condensations at the γ-methylene group rather than at the methyl group. Alkylations of the disodium salt of 1-phenyl-2,4-pentanedione with four halides have all afforded the 1-alkylation products.[18a, 31]

$$CH_3COCH_2COCH_2C_6H_5 \xrightarrow[\text{Liq. } NH_3]{2\,NaNH_2}$$

$$CH_3COCHNaCOCHNaC_6H_5 \xrightarrow[\text{2. } H^{\oplus}]{\text{1. RX}} CH_3COCH_2COCH(R)C_6H_5$$

[32] T. M. Harris and C. R. Hauser, *J. Am. Chem. Soc.*, **81**, 1160 (1959).

[33] K. G. Hampton and C. R. Hauser, *J. Org. Chem.*, **30**, 2934 (1965).

It is possible that an equilibrium exists between the 1,3- and 3,5-disodium salts of 1-phenyl-2,4-pentanedione because a slow intermolecular proton transfer has been observed between the monosodium salt of this diketone and disodioacetylacetone.[11]

$$\begin{array}{c} CH_3COCHNaCOCH_2Na \\ + \\ CH_3COCHNaCOCH_2C_6H_5 \end{array} \underset{\not\leftarrow}{\xrightarrow{\text{Slow}}} \begin{array}{c} CH_3COCHNaCOCH_3 \\ + \\ CH_3COCHNaCOCHNaC_6H_5 \end{array}$$

However, the reverse reaction is not observed with these diketones, indicating that the phenylacetyl-type dianion is substantially more stable, i.e., less basic, than that of acetylacetone.[11] As a consequence the 1,3-dianion of 1-phenyl-2,4-pentanedione can be expected to be more stable than the 3,5-dianion. This raises the question whether the 3,5-dianion might not be initially formed and then gradually converted to the 1,3-dianion.

$$\overset{\ominus}{C}H_2CO\overset{\ominus}{C}HCOCH_2C_6H_5 \longrightarrow CH_3CO\overset{\ominus}{C}HCO\overset{\ominus}{C}HC_6H_5$$

The possibility receives some support from studies of phenylacetone; although the methylene anion of this ketone is more stable,[34] kinetically controlled formylation occurs at the methyl position.[35] However, there is at present no direct evidence bearing on this problem with 1-phenyl-2,4-pentanedione, and the slowness of proton interchange between the sodium salt of 1-phenyl-2,4-pentanedione and disodioacetylacetone indicates that initial ionization of the aryl β-diketone at the 1-methylene position is more likely than at the 5-methyl position.

A study has been made of the alkylation of the disodium salts of two other types of unsymmetrical diketones; in one, the competition was between an isobutyryl group and an acetyl group, and in the other between an isobutyryl group and a propionyl group.[31] In both cases alkylation was observed almost exclusively at the position of less substitution, as indicated by arrows in the accompanying formulas.

$$\underset{\text{(CH}_3)_2\text{CHCOCH}_2\text{COCH}_3}{\overset{\text{Alkylation}}{\downarrow}} \qquad \underset{\text{(CH}_3)_2\text{CHCOCH}_2\text{COCH}_2\text{CH}_3}{\overset{\text{Alkylation}}{\downarrow}}$$

The alkylation of two diketones, 2-propionylcyclopentanone and 2-propionylcyclohexanone, in which the substituents are both methylene groups has been studied. Alkylation of 2-propionylcyclopentanone

[34] H. O. House and V. Kramar, *J. Org. Chem.*, **28**, 3362 (1963).

[35] L.-M. Roch, *Ann. Chim. (Paris)*, [3] **6**, 105 (1961).

occurred preferentially on the propionyl group, whereas with 2-propionylcyclohexanone little selectivity between ring and chain alkylation was observed.[36, 37]

$$\text{2-propionylcyclopentanone (ring C=O, }COC_2H_5\text{)} \xrightarrow[\text{2. } n\text{-}C_4H_9Br \;\; \text{3. } H^{\oplus}]{\text{1. } 2\,NaNH_2} \text{cyclopentanone-2-}COCH(CH_3)C_4H_9\text{-}n$$

$$\text{2-propionylcyclohexanone (}COC_2H_5\text{)} \xrightarrow[\text{2. } n\text{-}C_4H_9Br \;\; \text{3. } H^{\oplus}]{\text{1. } 2\,NaNH_2} n\text{-}C_4H_9\text{-cyclohexanone-}COC_2H_5 + \text{cyclohexanone-2-}COCH(CH_3)C_4H_9\text{-}n$$

The specificity of γ-alkylation and other γ-condensations of unsymmetrical β-diketones can be summarized as follows: When disodium salts are employed, the preferred site of reaction will be phenylacetyl > acetyl > propionyl > isobutyryl. The same sequence may in many cases be observed with potassium salts, but strict adherence is not assured.

It would be of considerable theoretical interest to determine the relative basicities of the various types of dianions, particularly with a closely related group of β-diketones. However, the inability of sodium mono- and di-anions to equilibrate in liquid ammonia is a complicating factor, as is the insolubility of many of the dipotassium salts in liquid ammonia.

Recent studies have shown that the kinetically preferred ionization of unsymmetrical monoketones often does not lead to the more stable anion.[34, 35, 38–41] However, it has been suggested that the order of ease of formation of disodium salts of diketones qualitatively represents the order of acidity of the γ-hydrogen atoms.[20] No experimental verification of this has been made.

Alkylations of Other β-Diketones (Table V)

The other β-diketones that have been alkylated through their dianions may be considered in two groups: the benzoylacylmethanes and symmetrical diketones other than acetylacetone.

[36] K. G. Hampton and C. R. Hauser, unpublished results.
[37] D. R. Bryant and C. R. Hauser, unpublished results.
[38] H. O. House, *Modern Synthetic Reactions*, Chapter 7, Benjamin, New York, 1965.
[39] H. O. House and B. M. Trost, *J. Org. Chem.*, **30**, 1341, 4395 (1965).
[40] D. Caine, *J. Org. Chem.*, **29**, 1868 (1964).
[41] H. M. E. Caldwell, *J. Chem. Soc.*, 2442 (1951).

Benzoylacetone and the higher benzoylacylmethanes are incapable of alkylation through more than one dianion. Benzoylacetone has been alkylated with benzyl chloride and ethyl bromide.[2, 20] The ethylation of

$$C_6H_5CO\overset{\ominus}{C}HCO\overset{\ominus}{C}HR \xrightarrow[2.\ H^{\oplus}]{1.\ R'X} C_6H_5COCH_2COCHRR'$$

$R = H, R'X = C_6H_5CH_2Cl$ (77%)
$R = H, R'X = C_2H_5Br$ (85%)
$R = CH_3, R'X = C_2H_5Br$ (84%)

1-phenyl-1,3-pentanedione has also been effected.[20] The dianion of 1-phenyl-1,3-hexanedione has been prepared. Alkylation has not been reported, although an aldol condensation proceeded satisfactorily.[33]

The alkylation of dianions of benzoylacetone with dihaloalkanes has been reported. The reactions give rise to *bis*-β-diketones. Dihalides that have been employed successfully include tri- and tetra-methylene bromide and the *o*-, *m*-, and *p*-xylylene halides.[28, 29]

$$2\ C_6H_5CO\overset{\ominus}{C}HCO\overset{\ominus}{C}H_2 \xrightarrow[2.\ H^{\oplus}]{1.\ RX_2} C_6H_5COCH_2COCH_2RCH_2COCH_2COC_6H_5$$

$RX_2 = BrCH_2CH_2CH_2Br, BrCH_2CH_2CH_2CH_2Br,$
o-, *m*- and p-$C_6H_4(CH_2Br)_2$, p-$C_6H_4(CH_2Cl)_2$

In addition, the two *bis*-dianions shown by the accompanying formula have been prepared and twofold condensations have been effected with them.[33] Although alkylation has not been reported, these anions would presumably yield *bis*-alkylation products.

$$C_6H_5CO\overset{\ominus}{C}HCO\overset{\ominus}{C}H(CH_2)_n\overset{\ominus}{C}HCO\overset{\ominus}{C}HCOC_6H_5$$

$n = 3$ or 4

The dipotassium salt of 2-acetyl-6-methoxycoumaran-3-one has been prepared. Treatment of this dianion with a stoichiometric amount of benzyl chloride afforded the dibenzylation product in 21% yield. No monobenzylation product was detected, but much of the coumaran-3-one was recovered.[42]

CH₃O-substituted coumaran-3-one with 2-COCH₃ $\xrightarrow[\begin{subarray}{l}2.\ C_6H_5CH_2Cl\\3.\ H^{\oplus}\end{subarray}]{1.\ 2\ KNH_2}$ CH₃O-substituted coumaran-3-one with 2-$COCH(CH_2C_6H_5)_2$

Alkylation of 1,3-cyclohexanedione with three alkyl halides has been reported. The dipotassium salt was employed and the 4-alkyl derivatives were isolated in 17–45% yield.[24] The disodium salt did not benzylate as smoothly as the dipotassium salt.

[42] W. I. O'Sullivan and C. R. Hauser, *J. Org. Chem.*, **25**, 839 (1960).

$$\text{1,3-cyclohexanedione} \xrightarrow{2\,KNH_2} \text{K,K dianion} \xrightarrow[2.\ H^{\oplus}]{1.\ RX} \text{R-substituted 1,3-cyclohexanedione}$$

$RX = CH_3I,\ CH_2{=}CHCH_2Br,\ C_6H_5CH_2Cl$

Other symmetrical diketones that have been alkylated include dipropionylmethane and diisobutyrylmethane.[20, 31]

$$CH_3CH_2COCH_2COCH_2CH_3 \qquad (CH_3)_2CHCOCH_2COCH(CH_3)_2$$

Only one alkylation of the dianion of an acyclic diketone having a single substituent at the reactive methylene position has been reported, and it proceeded with difficulty.[43] The dipotassium salt of 3-phenyl-2,4-pentanedione is readily formed but is very insoluble in ammonia. No benzylation was observed after a 2-hour reaction period in that solvent. When the alkylation was conducted in diethyl ether, reaction took place to only a slight extent. However, in a mixture of ammonia and pyridine alkylation took place rapidly and the benzylation product was obtained in 62% yield.

$$CH_3COCH(C_6H_5)COCH_3 \xrightarrow{2\,KNH_2} CH_3COCK(C_6H_5)COCH_2K \xrightarrow[2.\ H^{\oplus}]{1.\ C_6H_5CH_2Cl} CH_3COCH(C_6H_5)COCH_2CH_2C_6H_5$$

Further study is necessary to determine whether substituents on the active methylene group of diketones invariably lead to difficulty in effecting subsequent condensations of the dianions.

Relative Reactivities of Diketone Dianions

In most synthetic applications the need to know the relative nucleophilicity of dianions does not arise; however, such information might be useful in situations in which two different dianions are present in a molecule or a solution and the alkylation of only one is desired. One report has appeared describing studies of the reactivity of different types of disodio-β-diketones toward alkyl halides.[20] In that investigation two dianions were formed in equal concentration in liquid ammonia and the mixture was treated with an alkyl halide. The amount of alkyl halide employed was equal to no more than one-half the total quantity of the dianions. The compositions of the product mixtures were determined by gas chromatography, and the nucleophilicity ratios were computed on the

[43] W. I. O'Sullivan and C. R. Hauser, *J. Org. Chem.*, **25**, 1110 (1960).

TABLE I. Competitive Alkylation of Disodio-β-diketones with Alkyl Halides in Liquid Ammonia

Disodium Salt A	Disodium Salt B	Alkyl Halide	Nucleophilicity Ratio, A/B
$CH_3COCHNaCOCH_2Na$	$CH_3CH_2COCHNaCOCHNaCH_3$	n-C_4H_9Br	1/6.4
,,	,,	n-C_4H_9Cl	1/5.1
,,	,,	n-C_4H_9I	1/4.3
,,	,,	*sec*-C_4H_9Br	1/3.8
,,	,,	i-C_4H_9Br	1/7.5
$CH_3COCHNaCOCH_2Na$	$CH_3COCHNaCOCHNaC_6H_5$	n-C_4H_9Br	1/0.52
$CH_3CH_2COCHNaCOCHNaCH_3$	$(CH_3)_2CHCOCHNaCOCNa(CH_3)_2$	n-C_4H_9Br	1/1.0
$C_6H_5COCHNaCOCH_2Na$	$C_6H_5COCHNaCOCHNaCH_3$	C_2H_5Br	1/2.7

assumption that only bimolecular alkylation reactions were occurring and that there were no side reactions. The results are shown in Table I.

From the results in Table I it can be concluded that, over the range of compounds studied, the nucleophilicities do not vary much more than one order of magnitude. The order of reactivity, isobutyryl $\simeq$ propionyl $>$ acetyl $>$ phenylacetyl, is reasonably independent of both alkyl halide and diketone. If meaningful data are to be obtained, studies of this type must be limited to cases in which both dianions are soluble.

It should be noted that a relationship exists between the nucleophilicity of a dianion and its ease of formation; that is, the most readily formed is the least reactive. One exception is that an isobutyryl-type dianion, although more difficult to form than a propionyl type, is of approximately the same reactivity. It has been suggested that the isobutyryl-type dianion is more basic than the propionyl type (see p. 171).[20] Thus it seems probable that the reactivity of the isobutyryl dianion is reduced by steric hindrance to the approach of the alkyl halide.

Alkylations of β-Keto Aldehydes (Table VI)

A variety of β-keto aldehydes has been alkylated through dianions. They include cases in which the anion undergoing alkylation is primary, secondary, or tertiary.

Primary anions are involved in the alkylations of acetoacetaldehyde and its α-substituted derivatives. Acetoacetaldehyde itself is unstable* and polymerizes to 1,3,5-triacetylbenzene,[45] but its monosodium derivative is stable. Treatment of monosodioacetoacetaldehyde with 1 equivalent of potassium amide affords the dianion which is relatively soluble in

* Free acetoacetaldehyde has been isolated under carefully controlled conditions and characterized.[44] β-Keto aldehydes with an α-substituent cannot undergo this self-condensation.

[44] D. Dahm, R. Johnson, and F. H. Rathmann, *Proc. N. Dakota Acad. Sci.*, **12,** 19 (1958) *C.A.*, **53,** 2084a (1959)].

[45] R. L. Frank and R. H. Varland, *Org. Syntheses, Coll. Vol.* **3,** 829 (1955).

liquid ammonia. Alkylation at the γ position has been effected with methyl, *n*-butyl, *n*-octyl, and benzyl halides in good yields.[21] The products are unstable and, if isolated as the keto aldehyde, condense gradually to give triacylbenzenes. The alkylated keto aldehydes have been isolated as their copper chelates by treatment with cupric acetate of the salts remaining after evaporation of the ammonia from the alkylation reaction. In addition, three of the reaction products have been treated with cyanoacetamide to give the 6-alkyl-3-cyano-2-pyridones in high yields.[21] These reactions are illustrated in the accompanying scheme.

$$CH_3COCHNaCHO \xrightarrow{KNH_2} \overset{\ominus}{C}H_2CO\overset{\ominus}{C}HCHO \xrightarrow{RX} RCH_2CO\overset{\ominus}{C}HCHO \xrightarrow{Cu^{2\oplus}} \text{Copper chelate}$$

$$RCH_2CO\overset{\ominus}{C}HCHO \xrightarrow{NCCH_2CONH_2} \text{6-}RCH_2\text{-3-cyano-2-pyridone}$$

$$RCH_2CO\overset{\ominus}{C}HCHO \xrightarrow{H^{\oplus}} (RCH_2COCH_2CHO) \longrightarrow 1,3,5\text{-}(RCH_2CO)_3C_6H_3$$

The dianions of α-phenyl- and α-benzyl-acetoacetaldehyde have been prepared.[8, 21] The keto aldehydes having an α-substituent are sufficiently stable to permit handling. Dianion formation has been effected by treatment with 2 equivalents of potassium amide. Alkylation of the resulting dipotassium salts afforded keto aldehydes that were also sufficiently stable to be isolated.[8, 21]

$$CH_3COCHRCHO \xrightarrow{2\,KNH_2} (CH_2COCRCHO)K_2 \xrightarrow[2.\ H^{\oplus}]{1.\ R'X} R'CH_2COCHRCHO$$

$R = C_6H_5$, $R' = C_6H_5CH_2$ (42%)
$R = C_6H_5CH_2$, $R' = C_6H_5CH_2$ (51%)
$R = C_6H_5CH_2$, $R' = n\text{-}C_4H_9$ (27%)

Alkylations of the formyl derivatives of several cyclic ketones have been effected. The α-formylcycloalkanones are sufficiently stable that they can be handled directly. However, most reactions have employed the monosodium salts as starting materials, and the alkylation products usually have been isolated as derivatives.

α-Formylcyclopentanone and α-formylcyclohexanone have both been converted to secondary dianions by treatment of their monosodium salts with 1 equivalent of potassium amide in liquid ammonia.[23] In addition, the dianion of α-formylcyclohexanone has been prepared from the free keto aldehyde by treatment with 2 equivalents of potassium amide.[24] Alkylation of these dianions has been effected with several alkyl halides.[23, 24] Most products have been isolated as copper chelates, 3-cyano-2-pyridones, or monosubstituted cycloalkanones.[23, 24] The over-all yields in the alkylations followed by hydrolyses ranged from 32% to

41%. The procedure for preparation of monoalkylcyclopentanones is illustrated; cyclohexanone derivatives were prepared similarly.

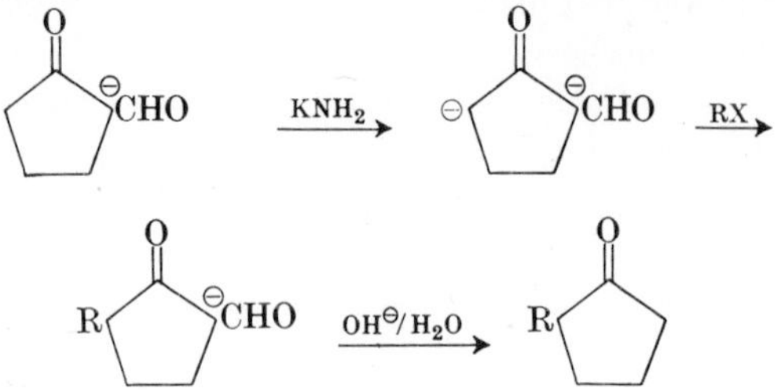

$RX = n\text{-}C_4H_9Br$ (38%)
$RX = C_6H_5CH_2Cl$ (32%)

Two dianions have been studied in which alkylation took place at a tertiary site yielding a product with quarternary substitution. The alkylation products of 2-formyl-6-methylcyclohexanone were not isolated directly, although they presumably could be.[23] Instead they were converted to copper chelates, cyanopyridones, or α-alkyl-α-methylcyclohexanones. The last of these reactions is illustrated in the accompanying scheme.

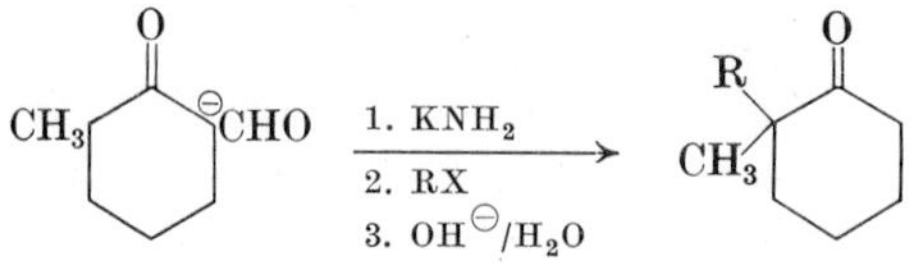

1. KNH_2
2. RX
3. $OH^{\ominus}/H_2O$

$RX = CH_3I$ (60%)
$RX = n\text{-}C_4H_9Br$ (72%)
$RX = C_6H_5CH_2Cl$ (55%)

2-Formyl-1-decalone has been alkylated at the 9-position.[3, 23] The initial reaction products were hydrolyzed to the 9-alkyl-1-decalones or converted to other derivatives. The alkylation with methyl iodide afforded an approximately equal mixture of *cis*- and *trans*-9-methyl-1-decalone. Alkylation with *n*-butyl bromide yielded only 9-butyl-*cis*-1-decalone, and benzylation afforded only one stereoisomer which was presumably 9-benzyl-*cis*-1-decalone.

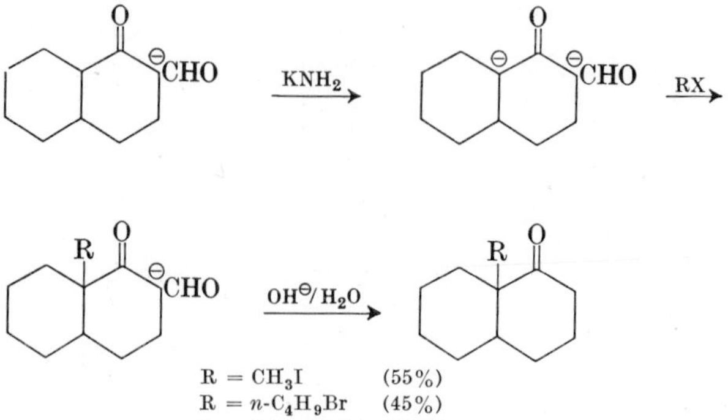

$R = CH_3I$ (55%)
$R = n\text{-}C_4H_9Br$ (45%)
$R = C_6H_5CH_2Cl$ (58%)

However, methylation of the dianion of a formyldecalone having *trans* alkyl groups at positions 7 and 10 failed.[46] Apparently, the alkyl substituents shielded both sides of the enolate anion from nucleophilic attack on the alkyl halide.

$$(CH_3)_2CH\text{-decalone dianion (CHO, }CH_3) \xrightarrow{CH_3I} \text{No reaction}$$

Alkylations of β-Keto Esters (Table VII)

Four β-keto esters have been studied with regard to their ability to form dianions and subsequently to undergo alkylation at the γ position. Additional study is needed to define fully the optimum conditions and the limitations of these reactions.

The dipotassium salt of ethyl acetoacetate has been prepared by addition of the ester to 2 equivalents of potassium amide in liquid ammonia.[4] Alkylation with methyl iodide and ethyl bromide gave the γ-alkylated products in yields of 37% and 29%, respectively.[4] These yields did not change significantly when the period during which the dianion was allowed to form was varied from a few minutes to an hour. Alkylation failed with *n*-butyl bromide and benzyl chloride. However, benzylation of ethyl disodioacetoacetate has been effected in low yield.[25]

$$CH_3COCH_2CO_2C_2H_5 \xrightarrow{2\ NH_2^{\ominus}} \overset{\ominus}{C}H_2CO\overset{\ominus}{C}HCO_2C_2H_5 \xrightarrow[2.\ H^{\oplus}]{1.\ RX} RCH_2COCH_2CO_2C_2H_5$$

One possible explanation for the poor results obtained in these alkylations is that ethyl acetoacetate may be only partially converted to the dianion by amide ion. As a result, amide ion would be still present in the reaction mixtures to compete with dianion for the alkyl halide.

Alkylation of ethyl α,γ-diphenylacetoacetate has also been reported.[4] The dipotassium salt with benzyl chloride gave, after decarboxylation, a 44% yield of 1,3,4-triphenyl-2-butanone. Alkylation is presumed to have occurred at the γ position, not at the α.

$$C_6H_5CH_2COCH(C_6H_5)CO_2C_2H_5 \xrightarrow{2\ KNH_2} C_6H_5CHKCOCK(C_6H_5)CO_2C_2H_5 \xrightarrow[\text{2. Hydrolysis and decarboxylation}]{1.\ C_6H_5CH_2Cl} C_6H_5CH_2CH(C_6H_5)COCH_2C_6H_5$$

[46] J. A. Marshall, W. I. Fanta, and G. L. Bundy, *Tetrahedron Letters*, 4807 (1965).

Two substituted tetronic acids have been alkylated at the 5 position by means of their dipotassium salts. Three alkylations have been reported. The yields varied widely.[24]

$$\xrightarrow[\text{2. R'X} \quad \text{3. H}^{\oplus}]{\text{1. 2 KNH}_2}$$

$R = CH_3$, $R' = C_6H_5CH_2$ (90%)
$R = CH_3$, $R' = CH_3$ (39%)
$R = C_2H_5$, $R' = C_6H_5CH_2$ (18%)

Arylations of β-Dicarbonyl Compounds (Table VIII)

The arylation of dianions of β-dicarbonyl compounds has been investigated principally with β-diketones. The preferred method of phenylation and of most other arylations involves treatment of the disodium salt with the appropriate diaryliodonium chloride.[11] The reaction is mechanistically complex and appears to involve a single electron transfer (see p. 160). The stoichiometry is difficult to determine because the product-forming arylation reaction is followed by a slower proton transfer from the initial product to the disodiodiketone leading to the eventual consumption of 2 equivalents of disodium salt. Although it has not been observed, the resulting anion could also conceivably undergo phenylation.

$$RCOCHNaCOCH_2Na \xrightarrow{(C_6H_5)_2ICl} RCOCHNaCOCH_2C_6H_5 + C_6H_5I$$

$$\xrightarrow[\text{Slow}]{RCOCHNaCOCH_2Na} RCOCHNaCOCHNaC_6H_5 + RCOCHNaCOCH_3$$

An additional problem in determining the proper stoichiometry of the reaction is that, under the reaction conditions, iodobenzene acts as a phenylating agent (see p. 161).[11] Thus between 1 and 2 equivalents of disodio-β-diketone are required for each equivalent of product produced, but possibly more than 1 equivalent of product is formed from each equivalent of diaryliodonium salt. This problem of stoichiometry leads to difficulty in the evaluation of yields. The yields reported in the next paragraph and in Table VIII are calculated on the basis of an assumed stoichiometry of 2:1 of disodio-β-diketone:diaryliodonium salt. This neglects the contribution of phenylation by iodobenzene, a contribution which can be estimated as no more than 10% of the total observed arylation.

Phenylation of disodio-β-diketones by this method has been achieved in yields of 50–98%. *p*-Tolylation and *p*-chlorophenylation have been accomplished with the corresponding 4,4′-dimethyldiphenyliodonium chloride and 4,4′-dichlorodiphenyliodonium chloride. The yields (21–44%) were somewhat less than those obtained in phenylation. With unsymmetrical aliphatic diketones of the methyl-methylene type, arylation at the methyl position was observed exclusively, as illustrated with 2,4-nonanedione.[11]

$$n\text{-}C_5H_{11}COCHNaCOCH_2Na \xrightarrow[2.\ H^{\oplus}]{1.\ (C_6H_5)_2ICl} n\text{-}C_5H_{11}COCH_2COCH_2C_6H_5 \quad (98\%)$$

No arylations of the dianions of β-keto aldehydes or β-keto esters with diaryliodonium salts have been reported.

Phenylation of disodio- and dipotassio-acetylacetone with benzyne has been reported. Bromobenzene was added to a mixture of the dianion and amide ion in liquid ammonia. However, reactions other than phenylation of the dianion by benzyne apparently predominated inasmuch as the yields were only 13–18%.[2, 11]

$$CH_3CO\overset{\ominus}{C}HCO\overset{\ominus}{C}H_2 \xrightarrow[2.\ H^{\oplus}]{1.\ C_6H_5Br,\ NH_2^{\ominus}} CH_3COCH_2COCH_2C_6H_5 \quad \text{Low yield}$$

Arylations by most substituted bromobenzenes would be expected to be even poorer because mixtures of isomeric aromatic substitution products could arise. However, aryne phenylation is satisfactory when the arylation reaction is intramolecular and a ring closure is effected. One example of this is the ring closure of 3-(o-chlorophenyl)-2,4-pentanedione.[47]

$$o\text{-}ClC_6H_4CH(COCH_3)_2 \xrightarrow{2KNH_2} o\text{-}ClC_6H_4C(K)(COCH_3)COCH_2K \xrightarrow{KNH_2} \text{(benzyne)}\,C(K)(COCH_3)COCH_2K \xrightarrow[2.\ H^{\oplus}]{1.\ \text{Closure}} \text{1-acetyl-2-indanone} \quad (98\%)$$

Another example of this reaction is the cyclization of 3-(o-chlorobenzyl)-2,4-pentanedione.[47]

[47] T. M. Harris and C. R. Hauser, *J. Org. Chem.*, **29**, 1391 (1964).

$$o\text{-ClC}_6\text{H}_4\text{CH}_2\text{CH(COCH}_3)_2 \xrightarrow[\text{2. H}^{\oplus}]{\text{1. Excess KNH}_2} \text{3-acetyl-2-tetralone}$$

(65%)

Interestingly, similar treatment of 6-(*o*-chlorophenyl)-2,4-hexanedione gave 1-acetyl-2-tetralone.[48] It is not known whether the cyclization

$$o\text{-ClC}_6\text{H}_4\text{CH}_2\text{CH}_2\text{COCH}_2\text{COCH}_3 \xrightarrow[\text{2. H}^{\oplus}]{\text{1. Excess KNH}_2} \text{1-acetyl-2-tetralone}$$

involved the monoanion or the dianion of the diketone. If the dianion is involved, this is an unusual situation in which steric conditions lead to preferential reaction of the normally unreactive α-methylene anion.

One cyclization involving a β-keto aldehyde has been reported. Although the initial cyclization probably proceeded normally, the resulting 3-formyl-2-tetralone underwent subsequent transformations leading to a binaphthol derivative under the conditions of the reaction and isolation.[47]

$$o\text{-ClC}_6\text{H}_4\text{CH}_2\text{CH(COCH}_3)\text{CHO} \xrightarrow[\text{KNH}_2]{\text{Excess}} [\text{3-formyl-2-tetralone}] \rightarrow [\text{3-(CH}_2\text{OH)-2-naphthol}]_2$$

COMPARISON WITH OTHER METHODS

β-Dicarbonyl compounds have been prepared by a number of methods. Most important among these is the Claisen condensation. This, in its many modifications, has been used widely for the preparation of β-diketones, β-keto aldehydes, and β-keto esters. The scope of this reaction is summarized in two earlier volumes of *Organic Reactions*.[49, 50] Examples are shown in the accompanying scheme.

[48] J. F. Bunnett and J. A. Skorcz, *J. Org. Chem.*, **27**, 3836 (1962).

[49] C. R. Hauser and B. E. Hudson, *Org. Reactions*, **1**, Chapter 9 (1942).

[50] C. R. Hauser, F. W. Swamer, and J. T. Adams, *Org. Reactions*, **8**, Chapter 3 (1954).

$$CH_3CH_2CO_2C_2H_5 + CH_3COCH_2CH_3 \xrightarrow{NaNH_2} CH_3CH_2COCH_2COCH_2CH_3 \text{ (57\%)} \quad \text{(Ref. 51)}$$

$$\text{cyclohexanone} \xrightarrow[Na]{HCO_2CH_2CH_2CH(CH_3)_2} \text{2-formylcyclohexanone (CHO) (83\%)} \quad \text{(Ref. 52)}$$

$$2\ CH_3CH_2CO_2C_2H_5 \xrightarrow{NaOC_2H_5} CH_3CH_2COCH(CH_3)CO_2C_2H_5 \text{ (81\%)} \quad \text{(Ref. 53)}$$

An important consideration in the comparison of the dianion alkylation reaction with the Claisen condensation and other synthetic methods is that the Claisen procedure and most other methods assemble dicarbonyl compounds by carbon-carbon condensations between a carbonyl group and the α-methylene carbon (points *a*). The dianion method, on the other hand, couples alkyl, aryl, and other groups to the γ position (points *b*) of existing dicarbonyl compounds. Thus the choice of synthetic routes is often dictated by the relative availability of starting materials. If the alkyl halide is more readily available than the corresponding alkylated acid or ketone, the dianion method will often be the preferred synthetic method.

$$R \overset{b}{—} CH_2CO \overset{a}{—} CHR' \overset{a}{—} COR'' \qquad R \overset{b}{—} CH_2COCHR' \overset{a}{—} CHO$$

$$R \overset{b}{—} CH_2CO \overset{a}{—} CHR' \overset{a}{—} CO_2CH_3$$

γ-Alkylated Acetylacetones

For the preparation of β-diketones of the 1-alkylated 2,4-pentanedione type, alkylation of disodioacetylacetone is often preferable to Claisen condensations. The dianion reaction has to commend it the high yields that have been obtained and the absence of failures except with tertiary alkyl halides and certain very sensitive halides. In addition, acetylacetone and a multitude of the alkyl halides are readily available commercially.

Several basic reagents have been employed for effecting Claisen condensations. The most common is sodium, although sodium amide is probably of equal merit. The acetylation of an aliphatic methylmethylene ketone occurs principally at the methyl group; however, self-condensation of the ester or of the ketone often competes. The product

[51] J. T. Adams and C. R. Hauser, *J. Am. Chem. Soc.*, **66**, 1220 (1944).
[52] H. Rupe and O. Klemm, *Helv. Chim. Acta*, **21**, 1539 (1938).
[53] S. M. McElvain, *J. Am. Chem. Soc.*, **51**, 3124 (1929).

commonly is isolated by use of the copper chelate as an intermediate. A comparison of the yields obtained in the preparation of diketones of this type by the dianion method and by the base-catalyzed Claisen condensation method involving acylations of acetone with esters and of ketones with ethyl acetate is shown in Table II.

TABLE II. COMPARISON OF METHODS FOR PREPARATION OF 1-ALKYLATED 2,4-PENTANEDIONES

Diketone	Dianion Method		Ethyl Acetate, Ketone, and Na		Acetone, Ester, and Base		
	Yield (%)	Ref.	Yield (%)	Ref.	Yield (%)	(Base)	Refs.
$C_2H_5COCH_2COCH_3$	59, 65	16	60	54	60	($NaNH_2$)	56
n-$C_3H_7COCH_2COCH_3$	70	16	59	54	25	(Na)	59
$(CH_3)_2CHCH_2$-$COCH_2COCH_3$	66	16	64	55	—	—	—
n-$C_5H_{11}COCH_2COCH_3$	67, 73	16	61*	56	65–80	(NaH)	60
n-$C_8H_{17}COCH_2COCH_3$	77	26	69	57	—	—	—
n-$C_9H_{19}COCH_2COCH_3$	66, 79	16	69	54	92	(NaH)	61, 62
$CH_2{=}CHCH_2CH_2$-$COCH_2COCH_3$	65	26	Good	58	—	—	—
$C_6H_5CH_2CH_2$-$COCH_2COCH_3$	73	16	—	—	35	(Na)	63

* $NaNH_2$ was used in this reaction.

The boron fluoride method for acylation of ketones has not been widely used for preparation of 1-alkyl-2,4-pentanediones. Acetylation of methyl-methylene type ketones occurs mainly on the methylene side to give 3-alkyl-2,4-pentanediones as shown in the accompanying scheme. There has been relatively limited experience with boron fluoride-catalyzed acylation of acetone.[64–66]

$$(CH_3CO)_2O + n\text{-}C_3H_7COCH_3 \xrightarrow{BF_3} \underset{(31\%)}{CH_3COCH(C_2H_5)COCH_3} + \underset{(3\%)}{n\text{-}C_3H_7COCH_2COCH_3} \quad \text{(Ref. 65)}$$

$$(n\text{-}C_3H_7CO)_2O + CH_3COCH_3 \xrightarrow{BF_3} \underset{(46\%)}{n\text{-}C_3H_7COCH_2COCH_3} \quad \text{(Ref. 66)}$$

[54] C. Weygand and H. Baumgärtel, *Ber.*, **62**, 574 (1929).
[55] J. M. Sprague, L. J. Beckham, and H. Adkins, *J. Am. Chem. Soc.*, **56**, 2665 (1934).
[56] R. Levine, J. A. Conroy, J. T. Adams, and C. R. Hauser, *J. Am. Chem. Soc.*, **67**, 1510 (1945).
[57] G. T. Morgan and E. Holmes, *J. Chem. Soc.*, **127**, 2891 (1925).
[58] G. Leser, *Bull. Soc. Chim. France*, [3] **27**, 64 (1902).
[59] G. T. Morgan, H. D. K. Drew, and C. R. Porter, *J. Chem. Soc.*, **125**, 737 (1924).
[60] F. W. Swamer and C. R. Hauser, *J. Am. Chem. Soc.*, **72**, 1352 (1950).
[61] V. L. Hansley, U.S. pat. 2,158,071 [*C.A.*, **33**, 6342 (1939)].
[62] V. L. Hansley, U.S. pat. 2,218,023 [*C.A.*, **35**, 1066 (1941)].
[63] G. T. Morgan and C. R. Porter, *J. Chem. Soc.*, **125**, 1273 (1924).
[64] C. R. Hauser, F. W. Swamer, and J. T. Adams, *Org. Reactions*, **8**, 186 (1954).
[65] C. R. Hauser and J. T. Adams, *J. Am. Chem. Soc.*, **66**, 345 (1944).
[66] J. T. Adams and C. R. Hauser, *J. Am. Chem. Soc.*, **67**, 284 (1945).

Other methods for the preparation of these diketones include acylation of ethyl or *t*-butyl acetoacetate to form diketo esters. The ester linkage is cleaved and decarboxylation is effected to give the corresponding diketone. Acylating agents include acid anhydrides, acid chlorides, and tetra-acyloxysilanes.[67–69]

$$\underset{\substack{X\,=\,OCOR,\ Cl,\\ \text{or } OSi(OCOR)_3}}{RCOX} + CH_3COCH_2CO_2R' \xrightarrow{\text{Base}} CH_3COCH(COR)CO_2R' \longrightarrow CH_3COCH_2COR$$

Other Aliphatic Diacylmethanes

2,4-Pentanediones with one or more aliphatic substituents at both the 1 and the 5 position can be prepared by alkylation of the appropriate diketones through their dianions; the specificity of the alkylation of unsymmetrical diketones has been discussed on p. 166. The other major method for preparation of these compounds is by base-catalyzed Claisen condensations.[50] The Claisen procedure is often the one of choice when the appropriate esters and ketones are available, since the diketones can be prepared in one step. The dianion procedure requires the use of β-diketones as starting materials and hence often entails an additional step since relatively few β-diketones are commercially available. Nevertheless, because the yields that have been observed in the alkylation of the dianions of aliphatic diacylmethanes have been very good, this is an attractive method.

Special methods for the preparation of the symmetrical diacylmethanes include the aluminum chloride-catalyzed acylation of vinyl acetate with 2 equivalents of an acid chloride and the diacylation of diethyl malonate with tetra-acyloxysilanes.[69, 70]

$$2\ RCOCl + CH_2{=}CHOCOCH_3 \xrightarrow{AlCl_3} RCOCH_2COR + CO + CH_3COCl$$

$$(RCO_2)_4Si + CH_2(CO_2C_2H_5)_2 \xrightarrow[Cu(OCOCH_3)_2]{MgO} RCOCH_2COR$$

1-Phenyl-1,3-alkanediones

The γ alkylation products of benzoylacetone have been successfully prepared by the dianion method. Two examples of the advantageous

[67] W. H. Reeder and G. A. Lescisin, U.S. pat. 2,369,250 [*C.A.*, **40**, 901 (1946)].

[68] A. Treibs and K. Hintermeier, *Chem. Ber.*, **87**, 1163 (1954).

[69] Y. K. Yur'ev and Z. V. Belyakova, *J. Gen. Chem. USSR* (*Eng. Trans.*), **29**, 1432 (1959).

[70] A. Sieglitz and O. Horn, *Chem. Ber.*, **84**, 607 (1951).

use of the dianion procedure involve synthesis of the *bis*-β-diketones **14** and **15** in yields of 58% and 89%, respectively.[28]

$$C_6H_5COCH_2CO(CH_2)_nCOCH_2COC_6H_5$$

14, $n = 5$; **15**, $n = 6$

The alternative syntheses of these compounds by condensation of sodioacetophenone with pimelyl chloride and diethyl suberate have been effected in yields of 25% and 5%, respectively.[71] For the preparation of most simple diketones the dianion method and the Claisen method are of equal merit, since the base- or boron fluoride-catalyzed acylation of acetophenone and the base-catalyzed acylation of ketones with methyl benzoate also usually occur in high yield.

In addition, hydration of phenyl acetylenic ketones[72] and treatment of benzoylacetonitrile with Grignard reagents[73, 74] have been employed in the synthesis of 1-phenyl-1,3-alkanediones. These routes are shown in the accompanying scheme.

$$C_6H_5COC{\equiv}C(CH_2)_4CH_3 \xrightarrow[H_2O]{H_2SO_4} C_6H_5COCH_2CO(CH_2)_4CH_3$$

$$C_6H_5COCH_2CN \xrightarrow[2.\ H^{\oplus}]{1.\ \text{Excess RMgX}} C_6H_5COCH_2COR$$

α-Acyl Derivatives of Cyclic Ketones

α-Acetylcyclohexanone, α-acetylcyclopentanone, and presumably the α-acetyl derivatives of most other cyclic ketones undergo alkylation at the methyl position through the use of dianions (see p. 169). Higher acyl derivatives of cyclopentanone, but not of cyclohexanone, undergo condensations largely at the side chain methylene position (see p. 171).

Other methods for synthesizing diketones of this type include intermolecular Claisen condensations effected by strong bases or by boron fluoride and Dieckmann condensations.[75] Acylations of cyclopentanone under basic conditions are generally unsatisfactory because of rapid self-condensation of the cyclic ketone. (See reactions at top of p. 185.)

The acylation of enamines of cyclic ketones has been employed for the

[71] D. F. Martin, M. Shamma, and W. C. Fernelius, *J. Am. Chem. Soc.*, **80**, 4891 (1958).
[72] C. Moureu and R. Delange, *Compt. Rend.*, **131**, 710 (1900).
[73] C. E. Rehberg and H. R. Henze, *J. Am. Chem. Soc.*, **63**, 2785 (1941).
[74] A. Mavrodin, *Bull. Soc. Chim. Romania*, **15**, 99 (1933) [*C.A.*, **28**, 3396 (1934)].
[75] J. P. Schaefer and J. J. Bloomfield, *Org. Reactions*, **15**, 1–203 (1967).

$$\text{cyclohexanone} \xrightarrow[\text{NaNH}_2]{C_6H_5CH_2CH_2CO_2CH_3} \text{2-}(COCH_2CH_2C_6H_5)\text{cyclohexanone} \ (24\%) \quad \text{(Ref. 32)}$$

$$\text{cyclopentanone} \xrightarrow[\text{BF}_3]{(n\text{-}C_3H_7CO)_2O} \text{2-}(COCH_2CH_2CH_3)\text{cyclopentanone} \ (88\%) \quad \text{(Ref. 76)}$$

$$C_6H_5CH_2CH_2COCH_2CH_2CH_2CH_2CO_2CH_3 \xrightarrow{\text{NaNH}_2} \text{2-}(COCH_2CH_2C_6H_5)\text{cyclopentanone} \ (57\%) \quad \text{(Ref. 32)}$$

synthesis of α-acyl derivatives of cyclic ketones, although its practicality has not been established for acyclic systems.[77, 78]

$$\text{1-morpholinocyclohexene} \xrightarrow[\text{2. Hydrolysis}]{\text{1. } n\text{-}C_7H_{15}COCl} \text{2-}(COC_7H_{15}\text{-}n)\text{cyclohexanone} \ (75\%) \quad \text{(Ref. 78)}$$

γ-Aryl Diketones

The use of diaryliodonium chlorides as arylating agents for β-diketones appears to be an excellent method for the preparation of γ-aryl diketones.[11] Certain of these compounds have also been prepared by the acylation of the appropriate ketone with ethyl phenylacetate in approximately 50% yield.

$$C_6H_5CH_2CO_2C_2H_5 + CH_3COR \xrightarrow{\text{Base}} C_6H_5CH_2COCH_2COR$$

$R = CH_3$ (38%) (Ref. 79)
$R = C_2H_5$ (54%) (Ref. 56)
$R = C_6H_5$ (50%) (Ref. 80)

An attempted boron fluoride-catalyzed acylation of acetone with phenylacetic anhydride was unsuccessful.[66]

[76] R. M. Manyik, F. C. Frostick, J. J. Sanderson, and C. R. Hauser, *J. Am. Chem. Soc.*, **75**, 5030 (1953).

[77] J. Szmuszkovicz, *Advan. Org. Chem.* **4**, 1–113 (1963).

[78] G. Stork, A. Brizzolara, H. Landesman, J. Szmuszkovicz, and R. Terrell, *J. Am. Chem. Soc.*, **85**, 207 (1963).

[79] G. T. Morgan and C. R. Porter, *J. Chem. Soc.*, **125**, 1271 (1924).

[80] C. Bülow and H. Grotowsky, *Ber.*, **34**, 1483 (1901).

Phenylacetone is not a suitable starting material for preparation of 1-phenyl-2,4-alkanediones since boron fluoride-catalyzed acylation occurs principally at the 1 position.[81] Basic methods give a mixture of 1- and 3-acylation with a low combined yield.[82]

Acyclic β-Keto Aldehydes

The dianion procedure appears to be a highly effective method for preparation of acyclic β-keto aldehydes. One alternative method that has been widely used is the formylation of ketones with a formate ester and sodium or a sodium alkoxide.[22] This is an excellent procedure with acetone and other symmetrical ketones. The point of formylation of unsymmetrical methyl-methylene ketones is reported to be dependent upon the acylation conditions.[35] In alcoholic solvents the point of formylation is predominantly the 3-methylene group, whereas in aprotic systems formylation at the 1 position usually predominates. However,

$$C_6H_5CH_2CH_2COCH_3 \xrightarrow[\text{Benzene}]{HCO_2C_2H_5,\ Na} C_6H_5CH_2CH_2COCH_2CHO$$

$$C_6H_5CH_2CH_2COCH_3 \xrightarrow[C_2H_5OH]{HCO_2C_2H_5,\ NaOC_2H_5} C_6H_5CH_2CH(CHO)COCH_3$$

there are numerous reports that formylation in aprotic solvents does not give exclusive formation of the 1-formyl derivatives, and that significant amounts of the 3-formyl derivatives are also formed.[21, 83–87]

For the synthesis of acyclic β-keto aldehydes with a single substituent at the α position, the dianion procedure provides a highly specific route. In the alternative procedure, formylation of an unsymmetrical ketone would in most cases be expected to occur on both sides of the carbonyl group with little specificity. Nevertheless formylation of one such ketone has been reported in which only one of the isomeric products was observed.[88]

$$n\text{-}C_4H_9COC_2H_5 \xrightarrow[NaOC_2H_5]{HCO_2C_2H_5} n\text{-}C_4H_9COCH(CH_3)CHO$$

[81] C. R. Hauser and R. M. Manyik, *J. Org. Chem.*, **18**, 588 (1953).
[82] G. T. Morgan, H. D. K. Drew, and C. R. Porter, *Ber.*, **58**, 341 (1925).
[83] R. P. Mariella and E. Godar, *J. Org. Chem.*, **22**, 566 (1957).
[84] E. E. Royals and K. C. Brannock, *J. Am. Chem. Soc.*, **75**, 2050 (1953).
[85] E. E. Royals and K. C. Brannock, *J. Am. Chem. Soc.*, **76**, 1180 (1954).
[86] R. P. Mariella and R. Stansfield, *J. Am. Chem. Soc.*, **73**, 1368 (1951).
[87] E. E. Royals and E. R. Covington, *J. Am. Chem. Soc.*, **77**, 3155 (1955).
[88] R. P. Mariella and V. Kvinge, *J. Am. Chem. Soc.*, **70**, 3126 (1948).

Formyl Derivatives of Alkylated Cyclic Ketones

The principal alternative to the alkylation of α-formyl derivatives of cyclic ketones through their dianions is the acylation of the corresponding alkylated cyclic ketones with formate esters. The latter reaction usually can be effected in high yield and is free of side reactions since base-catalyzed formylation is not observed at methinyl positions. For example, 2-methylcyclohexanone affords only one formylation product.[89]

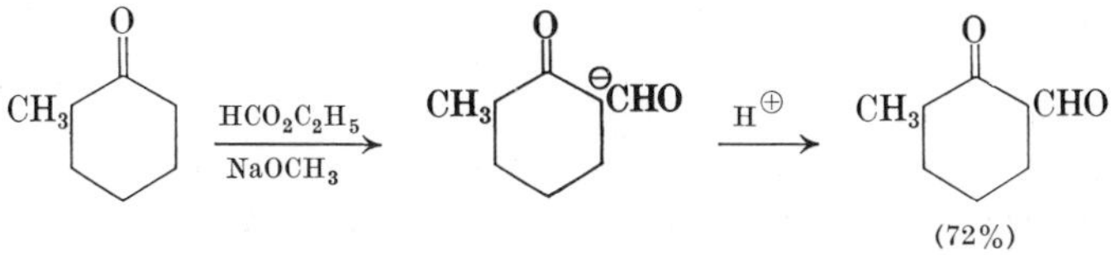

$HCO_2C_2H_5$, $NaOCH_3$; $H^{\oplus}$ (72%)

The shortcoming of the acylation reaction is that the α-alkylated cyclic ketones are often not readily available because direct alkylation of cyclic ketones usually affords mixtures of mono-, di-, and poly-alkylation products.[34, 38–40]

As a consequence of the difficulty associated with the direct synthesis of α-alkylated cyclic ketones, alkylation of the dianions of their formyl derivatives followed by removal of the formyl group provides a useful method for the preparation of these ketones. One example of the utility of the dianion procedure is in the preparation of 9-methyldecalone. Direct alkylation of decalone affords mainly the 2-methyl derivative.[90]

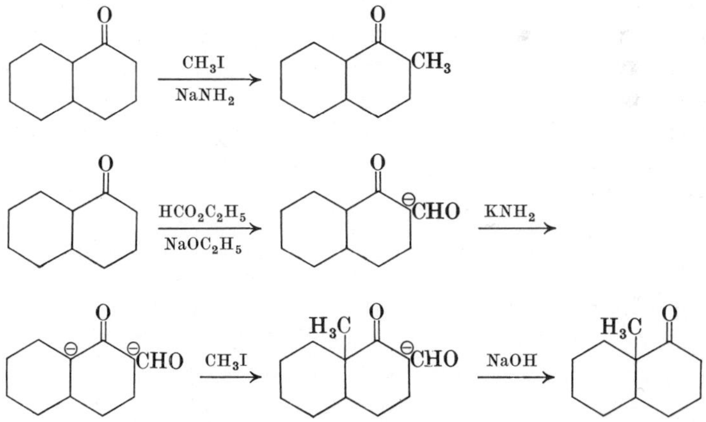

CH_3I, $NaNH_2$

$HCO_2C_2H_5$, $NaOC_2H_5$; KNH_2; CH_3I; $NaOH$

[89] W. S. Johnson and H. Posvic, *J. Am. Chem. Soc.*, **69**, 1361 (1947).

[90] J. W. Cook and C. A. Lawrence, *J. Chem. Soc.*, 823 (1937).

Alternative procedures involve alkylating compounds in which the formyl group has been converted to an enol ether,[89] enol thioether,[91] or enamine.[92]

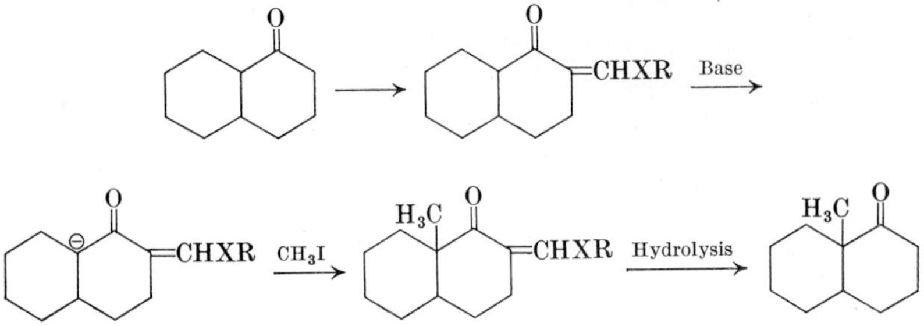

=CHXR Base

=CHXR CH_3I H_3C =CHXR Hydrolysis H_3C

XR = $OCH(CH_3)_2$, SC_4H_9-*n*, or $N(CH_3)C_6H_5$

In addition the 9-anion of 1-decalone has been prepared by cleavage of 1-acetoxy-$\Delta^{1,9}$-octalin with methyllithium.[93] Because under aprotic conditions the 9-anion is not in equilibrium with the 2-anion, exclusive 9-alkylation can be effected.

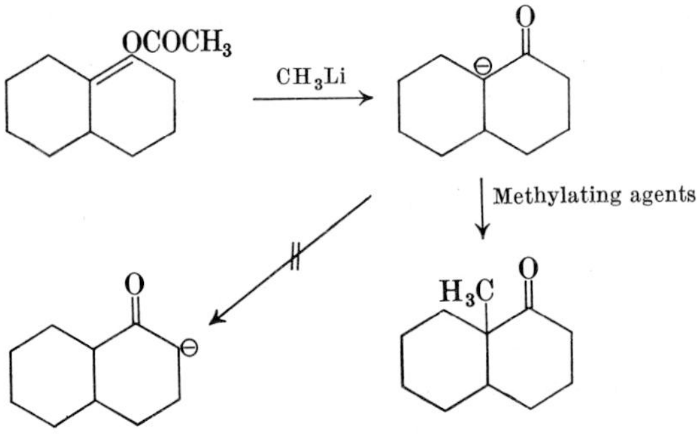

$OCOCH_3$ CH_3Li

Methylating agents

H_3C

β-Keto Esters

The alkylation of dianions derived from simple β-keto esters has not yet been widely explored, and only limited success has been achieved in the alkylation of ethyl acetoacetate, the simplest member of the series.[4] The preparation of β-keto esters by the acylation of esters was reviewed in Volume 1 of *Organic Reactions*.[49] The use of the dianion reaction for the preparation of 5-alkyltetronic acids appears to be practical.[24] Some alternative syntheses of these compounds are by the following routes.

[91] R. E. Ireland and J. A. Marshall, *J. Org. Chem.*, **27**, 1615 (1962).

[92] A. J. Birch and R. Robinson, *J. Chem. Soc.*, 501 (1944).

[93] H. O. House and B. M. Trost, *J. Org. Chem.*, **30**, 2502 (1965).

$$CH_3CHBrCOBr + CH_2(CO_2C_2H_5)_2 \xrightarrow[2.\ H^{\oplus}]{1.\ Na}$$

(lactone with $CO_2C_2H_5$, CH_3, =O) $\xrightarrow[H_2O]{HCl}$ (lactone with CH_3, =O) (Ref. 94)

$$CH_3CH_2COCH(CH_3)CO_2C_2H_5 \xrightarrow{Br_2} CH_3CHBrCOCBr(CH_3)CO_2C_2H_5 \xrightarrow{Base}$$

(lactone with Br, CH_3, CH_3, =O) $\xrightarrow{Pd/C}$ (lactone with CH_3, CH_3, =O) (Ref. 95)

RELATED MULTIPLE ANIONS OF CARBONYL COMPOUNDS

Dianion formation from several compounds closely related to β-diketones has been investigated. Certain of these dianions have been alkylated.

The amino analogs of β-diketones have been studied briefly. The product arising from condensation of aniline and benzoylacetone at the acetyl carbonyl group was alkylated at the methyl position in 63% yield via the dipotassium salt.[96]

$$C_6H_5COCH{=}C(NHC_6H_5)CH_3 \xrightarrow{2\ KNH_2} C_6H_6COCH{=}C(KNC_6H_5)CH_2K \xrightarrow[2.\ H^{\oplus}]{1.\ C_6H_5CH_2Cl} C_6H_5COCH{=}C(NHC_6H_5)CH_2CH_2C_6H_5$$

The reaction of the anilino derivative of acetylacetone was also investigated. Secondary ionization of this compound apparently occurred at the methyl group adjacent to the carbonyl group in preference to that next to the anilino group, since treatment of the dipotassium salt with benzyl chloride gave only the corresponding benzylation product **16**.[96]

$$CH_3COCH{=}C(NHC_6H_5)CH_3 \xrightarrow{KNH_2} KCH_2COCH{=}C(KNC_6H_5)CH_3 \xrightarrow[2.\ H^{\oplus}]{1.\ C_6H_5CH_2Cl} C_6H_5CH_2CH_2COCH{=}C(NHC_6H_5)CH_3$$

16 (40%)

$$CH_3COCH{=}C(NHC_6H_5)CH_3 \not\longrightarrow CH_3COCH{=}C(KNC_6H_5)CH_2K$$

[94] E. Benary, *Ber.*, **44,** 1759 (1911).

[95] N. M. Chopra, W. Cocker, B. E. Cross, J. T. Edward, D. H. Hayes, and H. P. Hutchison, *J. Chem. Soc.*, 594 (1955).

[96] S. Boatman and C. R. Hauser, *J. Org. Chem.*, **31,** 1785 (1966).

The condensation products of methylamine with acetylacetone and with benzoylacetone were not successfully converted to dianions under comparable conditions.[96] Presumably the phenyl group of the anilino compounds participated in charge delocalization of the dipotassium salts.

A sulfone analog of benzoylacetone, phenylsulfonylacetone, has been alkylated in fair yield at the terminal position.[97] The isomeric 2-methylsulfonylacetophenone has been alkylated similarly. The reaction failed in liquid ammonia, but a low yield was obtained in a mixture of ammonia and pyridine.[98]

$$C_6H_5SO_2CH_2COCH_3 \xrightarrow[\text{2. } C_6H_5CH_2Cl \text{; 3. } H^{\oplus}]{\text{1. 2 } KNH_2} C_6H_5SO_2CH_2COCH_2CH_2C_6H_5 \ (39\%)$$

$$C_6H_5COCH_2SO_2CH_3 \xrightarrow[\text{2. } C_6H_5CH_2Cl \text{; 3. } H^{\oplus}]{\text{1. 2 } KNH_2} C_6H_5COCH_2SO_2CH_2CH_2C_6H_5 \ (10\%)$$

N-Benzoylacetamide has been benzylated at the methyl position. Other imides have been alkylated similarly.[99]

$$C_6H_5CONHCOCH_3 \xrightarrow[\text{2. } C_6H_5CH_2Cl \text{; 3. } H^{\oplus}]{\text{1. } KNH_2} C_6H_5CONHCOCH_2CH_2C_6H_5 \ (64\%)$$

Some 1,3,5-triketones are capable of conversion to trianions by treatment with 3 equivalents of alkali amide. Again, the least stabilized anion is the site of alkylation.[100, 101]

$$C_6H_5COCH_2COCH_2COCH_3 \xrightarrow[\text{2. } C_6H_5CH_2Cl \text{; 3. } H^{\oplus}]{\text{1. 3 } NaHH_2} C_6H_5COCH_2COCH_2COCH_2CH_2C_6H_5 \ (63\%)$$

A similar result has been obtained with dehydracetic acid. Aldol condensations occur at the acetyl position by means of piperidine catalysis.[102] When dehydracetic acid is treated with 3 equivalents of sodium amide, ionization of the hydroxyl hydrogen, an acetyl hydrogen, and a 6-methyl hydrogen occur to form a tri-anion. The 6-methyl anion is most reactive, and condensations with *n*-propyl bromide, benzophenone, and methyl benzoate occur at that position.[103]

[97] W. I. O'Sullivan, D. F. Tavares, and C. R. Hauser, *J. Am. Chem. Soc.*, **83**, 3453 (1961).

[98] N. M. Carroll and W. I. O'Sullivan, *J. Org. Chem.*, **30**, 2830 (1965).

[99] S. D. Work, D. R. Bryant, and C. R. Hauser, *J. Am. Chem. Soc.*, **86**, 872 (1964).

[100] K. G. Hampton, T. M. Harris, C. M. Harris, and C. R. Hauser, *J. Org. Chem.*, **30**, 4263 (1965).

[101] T. M. Harris and R. L. Carney, *J. Am. Chem. Soc.*, **89**, 6734 (1967).

[102] R. H. Wiley, C. H. Jarboe, and H. G. Ellert, *J. Am. Chem. Soc.*, **77**, 5102 (1955).

[103] T. M. Harris and C. M. Harris, *Chem. Commun.*, 699 (1966).

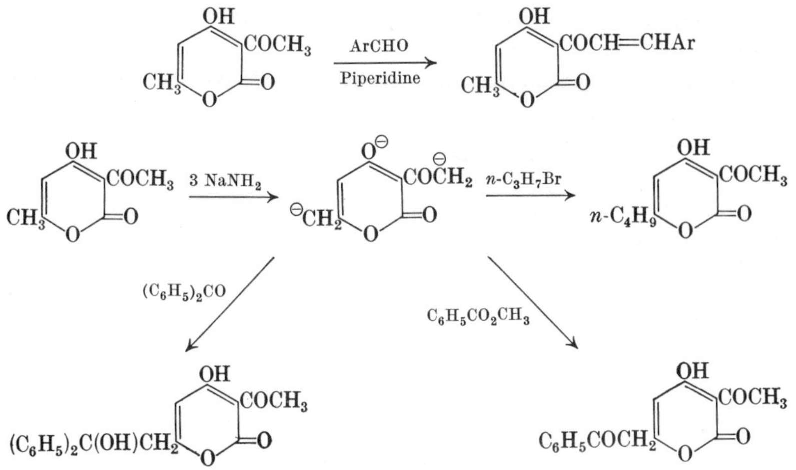

EXPERIMENTAL CONDITIONS

The amides of the three common alkali metals, lithium, sodium, and potassium, have all been employed for the preparation of the dianions of diketones, but only the last two have been used with keto aldehydes and keto esters. Dilithioacetylacetone has been alkylated, but the reaction proceeds too slowly to be generally practical.[16] Although sodium amide and potassium amide are of equal usefulness for the alkylation of many dicarbonyl compounds, sodium amide is preferred for alkylation of acetylacetone for reasons discussed on p. 163. In general the disodium salt of the dicarbonyl compound is preferred over the dipotassium salt whenever proton transfer to the dianion from other carbonyl compounds or from ammonia competes substantially with alkylation.

The effective stoichiometry of the alkylation reaction should be considered. The ratio of alkali amide to dicarbonyl compound must be carefully regulated. Insufficient alkali amide leads to incomplete conversion of monoanion to dianion. Excess alkali amide reacts with alkyl halide, probably as fast as or faster than does the dianion. Excess alkyl halide can be employed without causing dialkylation to occur since the monoanion is not sufficiently reactive for secondary alkylation at the α-methylene group. However, because most dianions are highly reactive, excess alkyl halide usually serves no useful purpose because the reactions are complete within a few minutes.

Moisture and carbon dioxide contamination of the reaction mixture reduces the amount of base present, thereby reducing the yield. In most reactions it is probably desirable to use about 10% excess of alkali amide, followed by a corresponding excess of alkyl halide to minimize losses from these sources.

With few exceptions, alkylations have been conducted in liquid ammonia or liquid ammonia containing some diethyl ether or tetrahydrofuran at the normal boiling point of the solution. Diethyl ether alone has been reported to be a somewhat inferior solvent for alkylation, and low yields were obtained despite the higher reaction temperature.[26] The use of pyridine as a solvent has been reported to give good results with 3-phenyl-2,4-pentanedione; the dianion was too insoluble in ammonia for alkylation to proceed at a reasonable rate.[43]

In some diketone alkylations the reaction mixtures have been neutralized with ammonium chloride before evaporation of the ammonia. In others this step has been omitted. It is not clear whether neutralization in this manner serves any definite purpose.

In the alkylation of diketones the products have usually been isolated by addition of diethyl ether and dilute hydrochloric acid to the residue that remained after evaporation of the liquid ammonia. The ethereal solution was collected and the aqueous solution re-extracted with ether to remove dissolved diketone. The combined ethereal extracts were dried and evaporated to afford in most cases relatively pure product. Further purification has been effected by distillation or recrystallization.

Initial isolation as a copper chelate has generally not been necessary. This is in contrast to the base-catalyzed Claisen condensations where preparation of the copper chelates has often been required for adequate purification of the reaction products.

In the alkylation of keto aldehydes the reaction mixtures have not been treated with ammonium chloride because it is advantageous to keep the keto aldehydes as monoanions to avoid self-condensation or reaction with ammonia. The alkylation products of α-substituted keto aldehydes have been isolated by acidification of the salts that remained after evaporation of the ammonia. When an α-substituent was not present, acidification yielded the unstable keto aldehydes which gradually underwent self-condensation to form 1,3,5-triacylbenzenes. However, in most cases the products have been converted directly to derivatives. Treatment of the residual salts with copper acetate, cyanoacetamide, or alkali afforded the copper chelates, 3-cyano-2-pyridones, or de-formylated ketones, respectively.

The optimum conditions for arylations of β-diketones with diaryliodonium salts are not so well defined as are the conditions for alkylation reactions. The molar ratio of reactants that has been employed is 4:2:1 of sodium amide, diketone, and diaryliodonium salt.[11] This ratio is based on the theoretical reaction sequence in the accompanying scheme.

$$\mathrm{RCO\overset{\ominus}{C}HCO\overset{\ominus}{C}H_2} \xrightarrow{(C_6H_5)_2ICl}$$

$$\mathrm{RCO\overset{\ominus}{C}HCOCH_2C_6H_5 + C_6H_5I} \xrightarrow{\mathrm{RCO\overset{\ominus}{C}HCO\overset{\ominus}{C}H_2}} \mathrm{RCO\overset{\ominus}{C}HCO\overset{\ominus}{C}HC_6H_5 + RCO\overset{\ominus}{C}HCOCH_3}$$

This reaction sequence is not totally accurate in that some phenylation probably occurs by means of the iodobenzene and the second step, conversion of the phenylation product to dianion, is relatively slow. Evidence for this has been discussed on p. 161. These complications are believed to be of minor importance, and the yields in Table VIII have been calculated on the basis of this stoichiometry.

The arylated diketones have been isolated by procedures similar to those employed with alkylated diketones. It has sometimes been found useful to wash the ethereal solution of the product with sodium thiosulfate solution if the product is contaminated with iodine.

EXPERIMENTAL PROCEDURES

2,4-Nonanedione.* A suspension of sodium amide (1.1 moles) in liquid ammonia is prepared in a 1-l. three-necked flask equipped with an air condenser, a ball-sealed mechanical stirrer, and a glass stopper. In the preparation of this reagent, commercial, anhydrous, liquid ammonia (800 ml.) is introduced from a cylinder through an inlet tube. To the stirred ammonia is added a small piece of sodium. After the appearance of a blue color, a few crystals of ferric nitrate hexahydrate (about 0.25 g.) are added, followed by small pieces of freshly cut sodium until 25.3 g. (1.1 g. atoms) has been added.

After the sodium amide formation is complete, as indicated by the disappearance of the blue color, the glass plug is replaced by a pressure-compensated addition funnel containing 60.0 g. (0.60 mole) of acetylacetone in 40 ml. of anhydrous diethyl ether. The top of the addition funnel is fitted with a nitrogen inlet tube. The reaction flask is immersed at least 3 in. into a dry ice-acetone bath, and simultaneously the slow introduction of dry nitrogen through the inlet tube is begun. After the reaction mixture is thoroughly cooled (about 20 minutes), acetylacetone is added intermittently in small portions over 10 minutes. The acetylacetone should be added in spurts which fall on the surface of the reaction mixture rather than on the wall of the flask. The cooling bath is removed and,

* This procedure is adapted from that of Hampton, Harris, and Hauser[16] which appears in *Organic Syntheses*.[104]

[104] K. G. Hampton, T. M. Harris, and C. R. Hauser, *Org. Syntheses*, **47**, 92 (1967).

after 20 minutes, the nitrogen purge is stopped and 68.5 g. (0.50 mole) of *n*-butyl bromide in 40 ml. of anhydrous diethyl ether is introduced dropwise during 10–20 minutes. The addition funnel is rinsed with a small volume of anhydrous ether, which is added to the reaction mixture.

After 30 minutes, 400 ml. of anhydrous ether is added, and the ammonia is removed by cautious heating on the steam bath. Addition of 200 g. of crushed ice causes a thick slurry to form. Next a mixture of 60 ml. of concentrated hydrochloric acid and 10 g. of crushed ice is added. The reaction mixture is stirred until all solids are dissolved, then transferred to a separatory funnel, the flask being washed with a little ether and dilute hydrochloric acid. The ether layer is separated, and the water layer, which should be acidic, is further extracted three times with 100-ml. portions of ether. The combined ethereal extracts are dried over anhydrous magnesium sulfate. After filtration and removal of the solvent, the residue is distilled through a 12-in. Vigreux column to give initially a small forerun, b.p. 32–110°/19 mm., principally acetylacetone, and then 63.0–63.6 g. (81–82%) of 2,4-nonanedione, b.p. 100–103°/19 mm., as a colorless liquid.

The addition of acetylacetone to liquid ammonia is highly exothermic. Moreover ammonia vapor reacts with the β-diketone to produce an insoluble ammonium salt, which tends to clog the tip of the addition funnel. Cooling the reaction mixture to dry ice temperature reduces the vigor of the reaction and minimizes the clogging of the addition funnel.

Other methods have also been employed to add acetylacetone to sodium amide in liquid ammonia. They include addition of the ammonium salt of the diketone[2, 26] and the direct addition of the diketone.[6] The latter procedure has been employed satisfactorily when the diketone is added in a fine stream from a syringe.[15] Other diketones do not offer this problem.

1,12-Diphenyl-1,3,10,12-dodecanetetrone (Alkylation with a Dihalide).[28] To a suspension of 0.2 mole of sodium amide in 400–700 ml. of liquid ammonia, prepared as described on p. 193, 16.2 g. (0.10 mole) of benzoylacetone is added in solid form from an Erlenmeyer flask through Gooch tubing. After 30 minutes a solution of 10.8 g. (0.05 mole) of tetramethylene bromide in 20 ml. of diethyl ether is added during 10 minutes. After 1 hour an additional 250 ml. of diethyl ether is added, and the reaction mixture is placed on a steam bath. When the liquid ammonia has evaporated, the resulting suspension is cooled to about 10° and a mixture of 100 g. of crushed ice and 60–80 ml. of cold, concentrated hydrochloric acid is added. A portion of the product crystallizes and is collected by filtration. The two layers of the filtrate are separated. The ethereal layer is combined with three ethereal extracts of the aqueous layer and the solvent is evaporated. The residue is combined with the previously collected solid and recrystallized from a mixture of benzene and methanol to give 16.8 g. (89%) of 1,12-diphenyl-1,3,10,12-dodecanetetrone, m.p. 108–109°.

Copper Chelate of 5-Phenyl-1,3-pentanedione.[21] Sodioacetoacetaldehyde is prepared in the following manner. In a 2-l. three-necked flask equipped with a calcium chloride drying tube, a nitrogen inlet tube, and a ball-sealed mechanical stirrer are placed 43 g. (0.8 mole) of sodium methoxide and 1 l. of anhydrous diethyl ether. The flask is purged with dry nitrogen and cooled with an ice bath. The inlet tube is replaced by a pressure-compensated addition funnel containing 49.3 g. (0.85 mole) of acetone and 74.0 g. (1.0 mole) of ethyl formate. The solution is added over 15 minutes and the funnel is then replaced by the nitrogen inlet tube. The reaction mixture is allowed to warm to room temperature and then to stand for 10 hours. The thick suspension is filtered with suction and the filter cake washed with ether. A rubber dam is secured tightly over the top of the funnel. It is pulled down onto the surface of the filter cake by the suction to protect the salt from atmospheric moisture. After most of the solvent has been removed, the salt is dried further in a vacuum oven at about 70°, powdered, and stored in a tightly capped bottle. The solid must be finely powdered to allow complete reaction in the following procedure; this is most easily accomplished if the powdering is done before the solid is completely dry. The fine powder is returned to the vacuum oven for completion of drying. Approximately 75 g. (87%) of sodioacetoacetaldehyde is obtained. It can be stored without apparent decomposition if protected from moisture.

In a 500-ml. three-necked flask equipped with a vapor-tight stirrer, an air condenser, and a glass stopper is placed 250 ml. of commercial, anhydrous, liquid ammonia. A minimum amount of potassium is added to the stirred solution to produce a permanent blue color. To the liquid is added approximately 0.05 g. of hydrated ferric chloride followed by 4.3 g. (0.11 g. atom) of potassium. The resulting dark blue solution of potassium is converted to a colorless solution of potassium amide in 10–30 minutes.

To the potassium amide solution is added cautiously 10.8 g. (0.10 mole) of sodioacetoacetaldehyde by means of a powder funnel. An olive green solution is formed. After 15 minutes, 13.9 g. (0.11 mole) of benzyl chloride is added. During the early part of the addition of benzyl chloride, the reaction develops a dark purple coloration, indicating formation of stilbene. This color disappears before more than about 15% of the halide has been added. The mixture is stirred for 30 minutes, and then the ammonia is evaporated as 250 ml. of anhydrous diethyl ether is added. The resulting suspension is shaken with 125 ml. of cold water and, after most of the suspended material has dissolved, the two layers are separated. The ether layer is extracted twice with water, and the aqueous solutions are combined and washed once with ether.

The aqueous solution, which contains the monoanion of 5-phenyl-1,3-pentanedione, is added to 200 ml. of saturated aqueous cupric acetate to which 6.6 g. (0.11 mole) of acetic acid has been added. The resulting heavy grey-blue precipitate which immediately forms is separated by extractions with chloroform (4 × 100 ml.). The chloroform solutions are combined and dried over magnesium sulfate. The solvent is evaporated and the residue is recrystallized from a mixture of ethanol and chloroform to give 16.5 g. (80%) of fairly pure copper chelate of 5-phenyl-1,3-pentanedione, m.p. 171–174°. Recrystallization from a mixture of chloroform and ethanol gives with good recovery dark blue plates, m.p. 176–178°.

The monoanion of 5-phenyl-1,3-pentanedione can be converted to a pyridone. Cyanoacetamide (6 g., 0.072 mole) and piperidinium acetate (0.5 g.) are added to an aqueous solution of the monoanion (derived from 0.056 mole of the dianion of acetoacetaldehyde and 7.08 g. (0.056 mole) of benzyl chloride). The solution is heated under reflux for 2 hours, cooled, and acidified with acetic acid to give 9.6 g. (73%) of 6-phenethyl-3-cyano-2(1)-pyridone, m.p. 186–190°. The melting point is raised to 204–205° by recrystallization from a mixture of acetone and ethanol.

2-*n*-Butyl-2-methylcyclohexanone (Formylation, Alkylation, and Removal of the Formyl Group).* A mixture of 123.2 g. (1.1 moles) of 2-methylcyclohexanone and 81.4 g. (1.1 moles) of ethyl formate is added dropwise (rapidly) to a stirred suspension of 54 g. (1.0 mole) of commercial, methanol-free sodium methoxide in 2 l. of anhydrous diethyl ether at 0–10°. After the addition is complete, the mixture is stirred for 15 minutes at this temperature and then for 12 hours at room temperature. Preferably the formylation is conducted under a dry nitrogen atmosphere. The thick suspension is filtered by suction, and the filter cake is washed with anhydrous ether, care being taken to protect the product from atmospheric moisture. The salt is partially dried in a vacuum oven at 70°, powdered, dried completely, and stored in a tightly capped bottle. Sodio-2-formyl-6-methylcyclohexanone, a cream-colored powder, is obtained in 80–85% yield (130–138 g.).

In a 1-l. three-necked flask equipped with a dry ice condenser and a ball-sealed mechanical stirrer is placed 700 ml. of commercial, anhydrous,

* The procedure was furnished by Dr. S. Boatman. It is a modification of the procedure described by Boatman, Harris, and Hauser.[23] It has been submitted to *Organic Syntheses*.[105]

[105] S. Boatman, T. M. Harris, and C. R. Hauser, submitted to *Org. Syntheses*.

liquid ammonia. A solution of 0.18 mole of potassium amide is prepared from 7.0 g. of potassium by the method described on p. 195. To the solution is added through a powder funnel 24.9 g. (0.15 mole) of sodio-2-formyl-6-methylcyclohexanone. The escaping ammonia blows away some of the finely powdered salt if the addition is not made carefully.

After 1 hour, a solution of 27.4 g. (0.2 mole) of *n*-butyl bromide in 50 ml. of anhydrous ether is added dropwise through a pressure-compensated addition funnel. The mixture is stirred for 3 hours, the dry ice condenser is replaced with a water-cooled condenser, and the ammonia is then evaporated carefully on a steam bath as 400 ml. of anhydrous ether is added. After the ether has been heated under reflux for 5 minutes, 100 g. of ice is added cautiously, followed by 300 ml. of cold water. When the solid has dissolved, the layers are separated, and the ether layer is extracted twice with 50-ml. portions of cold water.

The combined water layers are placed in a round-bottomed flask equipped with an efficient water condenser, and 6.4 g. of sodium hydroxide is added. The mixture is warmed to about 60° and the condenser is removed briefly to allow ether vapor to escape. The condenser is replaced and the mixture is heated under reflux until an acidified aliquot no longer gives a red color with ethanolic ferric chloride. The mixture is cooled and extracted with three 200-ml. portions of ether. The combined ethereal extracts are washed with dilute hydrochloric acid and dried over anhydrous magnesium sulfate. The ether is evaporated and the residue is distilled at reduced pressure to give 17–19 g. (67–75%) of 2-*n*-butyl-2-methylcyclohexanone, b.p. 116–118°/20 mm. A higher-boiling fraction of 2-formyl-6-*n*-butyl-6-methylcyclohexanone, b.p. 201–203°/20 mm., will be obtained if sufficient time is not allowed for hydrolysis.

An alternative isolation procedure involves steam distillation of the basic aqueous solution of the alkylation product until no further organic material distils. This may be done either after or instead of boiling the basic solution. The steam distillate is extracted with ether; the ethereal solution is dried, evaporated, and the residue distilled.

1-Phenyl-2,4-pentanedione (Arylation with Diphenyliodonium Chloride).[11] Diphenyliodonium chloride is prepared from benzene, potassium iodate, sulfuric acid, and acetic anhydride by the method of Beringer and co-workers.[106] The salt is recrystallized from methanol to give colorless crystals, m.p. 225–226°.

A solution of 0.1 mole of disodioacetylacetone is prepared from 10.0 g. of acetylacetone and 0.2 mole of sodium amide in 600 ml. of liquid ammonia as described on p. 193. During 5–10 minutes 15.8 g. (0.05 mole) of diphenyliodonium chloride is added to the solution through Gooch

[106] F. M. Beringer, E. J. Geering, I. Kuntz, and M. Mausner, *J. Phys. Chem.*, **60**, 141 (1956).

tubing from an Erlenmeyer flask. After 1 hour, 500 ml. of diethyl ether is added to the suspension while the ammonia is carefully evaporated on the steam bath. The ether suspension is heated under reflux for 1 hour, cooled in an ice bath, and 100 g. of ice and 20 ml. of concentrated hydrochloric acid are added. The ethereal layer is separated and the aqueous layer is extracted several times with diethyl ether. If the combined ethereal solution contains iodine and appears very dark, it should be extracted with aqueous sodium thiosulfate until no further diminution of color occurs. The ethereal solution is dried with magnesium sulfate, filtered, and the ether removed. The residual liquid is distilled under reduced pressure to give 8.1 g. (92%) of 1-phenyl-2,4-pentanedione, b.p. 138–141°/13 mm. The crude reaction mixture contains traces of benzene, biphenyl, and 3-phenyl-2,4-pentanedione, as indicated by gas chromatography.

TABULAR SURVEY

Tables III through VIII list examples of alkylation and arylation of the dianions of dicarbonyl compounds which were reported before October 1966. Certain results reported after this date are also included.

In each table the entries are in order of increasing number of carbon atoms in the dianion being alkylated, and, for a given dianion, in order of increasing number of carbon atoms in the halide.

TABLE III. ALKYLATIONS OF ACETYLACETONE DIANION

Alkyl Halide	Alkali Amide	Product(s)	Yield, (%)	Refs.
CH_3I	$NaNH_2$	$CH_3COCH_2COCH_2CH_3$	59, 65	16
	KNH_2	$CH_3COCH_2COCH_2CH_3$ +	35, 46	16
		$CH_3CH_2COCH_2COCH_2CH_3$	26, 18	
CH_2Cl_2	$NaNH_2$	Alkylation failed	—	28
C_2H_5Br	$NaNH_2$	$CH_3COCH_2COCH_2C_2H_5$	70	16
$ClCH_2CH_2Cl$	$NaNH_2$	Alkylation failed	—	28
$BrCH_2CH_2Br$	$NaNH_2$	Alkylation failed	—	28
n-C_3H_7Br	$NaNH_2$	$CH_3COCH_2COCH_2C_3H_7$-n	77	15
i-C_3H_7Br	$NaNH_2$	$CH_3COCH_2COCH_2C_3H_7$-i	66	16
	KNH_2	$CH_3COCH_2COCH_2C_3H_7$-i	27	26
$CH_2{=}CHCH_2Br$	KNH_2	$CH_3COCH_2COCH_2CH_2CH{=}CH_2$	65	26
$Br(CH_2)_3Br$	$NaNH_2$	$CH_3COCH_2COCH_2(CH_2)_3CH_2COCH_2COCH_3$	69–78	28
	KNH_2	$CH_3COCH_2COCH_2(CH_2)_3CH_2COCH_2COCH_3$	40	28
n-C_4H_9Cl	$NaNH_2$	$CH_3COCH_2COCH_2C_4H_9$-n	—	20
n-C_4H_9Br	$LiNH_2$	$CH_3COCH_2COCH_2C_4H_9$-n	25	16
	$NaNH_2$	$CH_3COCH_2COCH_2C_4H_9$-n	67–82	11, 16, 20, 31, 104
	KNH_2	$CH_3COCH_2COCH_2C_4H_9$-n	68	26
	KNH_2	$CH_3COCH_2COCH_2C_4H_9$-n +	43, 53	16
		n-$C_4H_9CH_2COCH_2COCH_2C_4H_9$-$n$	16, 14	
n-C_4H_9I	$NaNH_2$	$CH_3COCH_2COCH_2C_4H_9$-n	—	20
i-C_4H_9Br	$NaNH_2$	$CH_3COCH_2COCH_2C_4H_9$-i	78	16, 20
sec-C_4H_9Br	$NaNH_2$	$CH_3COCH_2COCH_2C_4H_9$-sec	78	16, 20
t-C_4H_9Cl	KNH_2	Alkylation failed	—	26
$Br(CH_2)_4Br$	$NaNH_2$	$CH_3COCH_2COCH_2(CH_2)_4CH_2COCH_2COCH_3$	80	28
$CH_3COCHClCOCH_3$	KNH_2	$CH_3COCH_2COCH_2CH(COCH_3)_2$	Low	30, 30a

TABLE III. ALKYLATIONS OF ACETYLACETONE DIANION (*Continued*)

Alkyl Halide	Alkali Amide	Product(s)	Yield, (%)	Refs.
$CH_3COCH_2COCH_2Br$	KNH_2	Alkylation failed	—	30*a*
n-$C_6H_{13}Br$	$NaNH_2$	$CH_3COCH_2COCH_2C_6H_{13}$-$n$	69	101
n-$C_7H_{15}Br$	KNH_2	$CH_3COCH_2COCH_2C_7H_{15}$-$n$	77	26
$C_6H_5CH_2Cl$	$LiNH_2$	Alkylation failed	—	26
	$NaNH_2$	$CH_3COCH_2COCH_2CH_2C_6H_5$	69–73	16, 26
	KNH_2	$CH_3COCH_2COCH_2CH_2C_6H_5$	60	2, 26
	KNH_2	$CH_3COCH_2COCH_2CH_2C_6H_5$ +	41, 39	16
		$C_6H_5CH_2CH_2COCH_2COCH_2CH_2C_6H_5$	22, 16	
n-$C_8H_{17}Br$	$NaNH_2$	$CH_3COCH_2COCH_2C_8H_{17}$-$n$	66, 79	16
	KNH_2	$CH_3COCH_2COCH_2C_8H_{17}$-$n$ +	31–67	26, 16
		n-$C_8H_{17}CH_2COCH_2COCH_2C_8H_{17}$-$n$	14–42	
$C_6H_5CH_2CH_2Cl$	KNH_2	Alkylation failed	—	26
o-$C_6H_4(CH_2Br)_2$	KNH_2	Alkylation failed	—	29
m-$C_6H_4(CH_2Br)_2$	KNH_2	Alkylation failed	—	29
p-$C_6H_4(CH_2Br)_2$	KNH_2	Alkylation failed	—	29
$Br(CH_2)_9Br$	$NaNH_2$	$CH_3COCH_2COCH_2(CH_2)_9CH_2COCH_2COCH_3$	81	28
$Br(CH_2)_{10}Br$	$NaNH_2$	$CH_3COCH_2COCH_2(CH_2)_{10}CH_2COCH_2COCH_3$	73	28
$(C_6H_5)_2CHCl$	KNH_2	Alkylation failed	—	26
$(C_6H_5)_2CCl_2$	$NaNH_2$	Alkylation failed	—	28

TABLE IV. ALKYLATIONS OF THE DIANIONS OF UNSYMMETRICAL ALIPHATIC AND ALICYCLIC DIKETONES

Diketone	Alkali Amide	Alkyl Halide	Product(s)	Yield (%)	Refs.
$C_2H_5COCH_2COCH_3$	$NaNH_2$	CH_3I	$C_2H_5COCH_2COCH_2CH_3$ + $(CH_3)_2CHCOCH_2COCH_3$ 89:11	56–66	31
	KNH_2	CH_3I	$C_2H_5COCH_2COCH_2CH_3$ + $(CH_3)_2CHCOCH_2COCH_3$ 87:13	— —	31
i-$C_3H_7COCH_2COCH_3$	$NaNH_2$	CH_3I	i-$C_3H_7COCH_2COCH_2CH_3$ + $(CH_3)_3CCOCH_2COCH_3$ 99:1	56	31
2-($COCH_3$)cyclopentanone	KNH_2	$C_6H_5CH_2Cl$	2-($COCH_2CH_2C_6H_5$)cyclopentanone	62	32
i-$C_3H_7COCH_2COC_2H_5$	$NaNH_2$	CH_3I	i-$C_3H_7COCH_2COCH(CH_3)_2$ + $(CH_3)_3CCOCH_2COC_2H_5$ 99:1	58	31
2-(COC_2H_5)cyclopentanone	$NaNH_2$	n-C_4H_9Br	2-($COCH(C_4H_9$-$n)CH_3$)cyclopentanone	60	36
	KNH_2	$C_6H_5CH_2Cl$	2-($COCH(CH_2C_6H_5)CH_3$)cyclopentanone + Principal isomer 2-($C_6H_5CH_2$)-5-(COC_2H_5)cyclopentanone	63	37

TABLE IV. ALKYLATIONS OF THE DIANIONS OF UNSYMMETRICAL ALIPHATIC AND ALICYCLIC DIKETONES (*Continued*)

Diketone	Alkali Amide	Alkyl Halide	Product(s)	Yield (%)	Refs.
2-($COCH_3$)cyclohexanone	KNH_2	$C_6H_5CH_2Cl$	2-($COCH_2CH_2C_6H_5$)cyclohexanone	58	32
n-$C_5H_{11}COCH_2COCH_3$	$NaNH_2$	C_2H_5Br	n-$C_5H_{11}COCH_2COCH_2C_2H_5$	70	16
	$NaNH_2$	n-C_4H_9Br	n-$C_5H_{11}COCH_2COCH_2C_4H_9$-$n$	68	16
	KNH_2	$C_6H_5CH_2Cl$	n-$C_5H_{11}COCH_2COCH_2CH_2C_6H_5$	52	26
2-(COC_2H_5)cyclohexanone	$NaNH_2$	n-C_4H_9Br	2-[$COCH(C_4H_9$-$n)CH_3$]cyclohexanone + 2-n-C_4H_9-6-COC_2H_5-cyclohexanone 55:45	82	36
	KNH_2	$C_6H_5CH_2Cl$	2-[$COCH(CH_2C_6H_5)CH_3$]cyclohexanone + 2-$C_6H_5CH_2$-6-COC_2H_5-cyclohexanone	69	37

$C_6H_5CH_2COCH_2COCH_3$	$NaNH_2$	CH_3I	$C_6H_5CH(CH_3)COCH_2COCH_3$	—	31
	$NaNH_2$	n-C_4H_9Br	$C_6H_5CH(C_4H_9$-$n)COCH_2COCH_3$	77	11, 20, 31
	$NaNH_2$	$C_6H_5CH_2Cl$	$C_6H_5CH(CH_2C_6H_5)COCH_2COCH_3$	77	31
	$NaNH_2$	$C_6H_5CHClCH_3$	$C_6H_5CH(CH_3)CH(C_6H_5)COCH_2COCH_3$ (erythro)	44	18a
$C_6H_5CH_2CH_2COCH_2COCH_3$	KNH_2	n-C_4H_9Br	$C_6H_5CH_2CH_2COCH_2COCH_2C_4H_9$-$n$	67	26
	$NaNH_2$	$Br(CH_2)_4Br$	$C_6H_5CH_2CH_2COCH_2COCH_2(CH_2)_4$-$CH_2COCH_2COCH_2CH_2C_6H_5$	38	28
	KNH_2	$C_6H_5CH_2Cl$	$C_6H_5CH_2CH_2COCH_2COCH_2CH_2C_6H_5$	59	32
n-$C_9H_{19}COCH_2COCH_3$	KNH_2	$C_6H_5CH_2Cl$	n-$C_9H_{19}COCH_2COCH_2CH_2C_6H_5$	52	26
	KNH_2	n-$C_8H_{17}Br$	n-$C_9H_{19}COCH_2COCH_2C_8H_{17}$-$n$	71	26
$CH_3COCH_2CO(CH_2)_5$-$COCH_2COCH_3$	$NaNH_2$	$Br(CH_2)_3Br$	$(CH_2COCH_2COCH_2CH_2CH_2CH_2)_n$ (n = 12–13)	77	33
	$NaNH_2$	$C_6H_5CH_2Cl$	$C_6H_5CH_2CH_2COCH_2CO(CH_2)_5$-$COCH_2COCH_2CH_2C_6H_5$	67	33
$CH_3COCH_2CO(CH_2)_6$-$COCH_2COCH_3$	$NaNH_2$	$C_6H_5CH_2Cl$	$C_6H_5CH_2CH_2COCH_2CO(CH_2)_6$-$COCH_2COCH_2CH_2C_6H_5$	62	33

TABLE V. ALKYLATIONS OF THE DIANIONS OF DIKETONES NOT INCLUDED IN TABLES III AND IV

Diketone	Alkali Amide	Alkyl Halide	Product	Yield (%)	Refs.
1,3-cyclohexanedione	KNH_2	CH_3I	CH_3 (2-methyl-1,3-cyclohexanedione, ring)	17	24
	KNH_2	$CH_2{=}CHCH_2Br$	$CH_2{=}CHCH_2$ (ring)	21	24
	$NaNH_2$	$C_6H_5CH_2Cl$	$C_6H_5CH_2$ (ring)	—	24
	KNH_2	$C_6H_5CH_2Cl$	$C_6H_5CH_2$ (ring)	45	24
$C_2H_5COCH_2COC_2H_5$	$NaNH_2$	n-C_4H_9Cl	$C_2H_5COCH_2COCH(C_4H_9\text{-}n)CH_3$	—	20
	$NaNH_2$	n-C_4H_9Br	$C_2H_5COCH_2COCH(C_4H_9\text{-}n)CH_3$	73	20, 31
	$NaNH_2$	n-C_4H_9I	$C_2H_5COCH_2COCH(C_4H_9\text{-}n)CH_3$	—	20
	$NaNH_2$	*sec*-C_4H_9Br	$C_2H_5COCH_2COCH(C_4H_9\text{-}sec)CH_3$	83	20
	$NaNH_2$	i-C_4H_9Br	$C_2H_5COCH_2COCH(C_4H_9\text{-}i)CH_3$	78	20

$i\text{-}C_3H_7COCH_2COC_3H_7\text{-}i$	$NaNH_2$	$n\text{-}C_4H_9Br$	$i\text{-}C_3H_7COCH_2COC(CH_3)_2C_4H_9\text{-}n$	26	20
$C_6H_5COCH_2COCH_3$	$NaNH_2$	C_2H_5Br	$C_6H_5COCH_2COCH_2C_2H_5$	85	20
	$NaNH_2$	$ClCH_2CH_2Cl$	Alkylation failed	—	28
	$NaNH_2$	$BrCH_2CH_2Br$	Alkylation failed	—	28
	$NaNH_2$	$Br(CH_2)_3Br$	$C_6H_5COCH_2COCH_2(CH_2)_3CH_2COCH_2COC_6H_5$	58	28
	$NaNH_2$	$Br(CH_2)_4Br$	$C_6H_5COCH_2COCH_2(CH_2)_4CH_2COCH_2COC_6H_5$	89	28
	KNH_2	$C_6H_5CH_2Cl$	$C_6H_5COCH_2COCH_2CH_2C_6H_5$	77	2
	$NaNH_2$	$o\text{-}C_6H_4(CH_2Br)_2$	$(C_6H_5COCH_2COCH_2CH_2)_2C_6H_4\text{-}o$	65	28
	KNH_2	$o\text{-}C_6H_4(CH_2Br)_2$	$(C_6H_5COCH_2COCH_2CH_2)_2C_6H_4\text{-}o$	14	29
	KNH_2	$m\text{-}C_6H_4(CH_2Br)_2$	$(C_6H_5COCH_2COCH_2CH_2)_2C_6H_4\text{-}m$	16	29
	$NaNH_2$	$p\text{-}C_6H_4(CH_2Cl)_2$	$(C_6H_5COCH_2COCH_2CH_2)_2C_6H_4\text{-}p$	67	28
	KNH_2	$p\text{-}C_6H_4(CH_2Br)_2$	$(C_6H_5COCH_2COCH_2CH_2)_2C_6H_4\text{-}p$	22	29
$C_6H_5COCH_2COC_2H_5$	$NaNH_2$	C_2H_5Br	$C_6H_5COCH_2COCH(C_2H_5)CH_3$	84	20
$CH_3COCH(C_6H_5)COCH_3$	KNH_2	$C_6H_5CH_2Cl$	$CH_3COCH(C_6H_5)COCH_2CH_2C_6H_5$	62	43
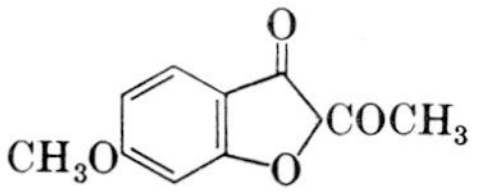 CH_3O (ring structure: O, O) $COCH_3$	KNH_2	$C_6H_5CH_2Cl$	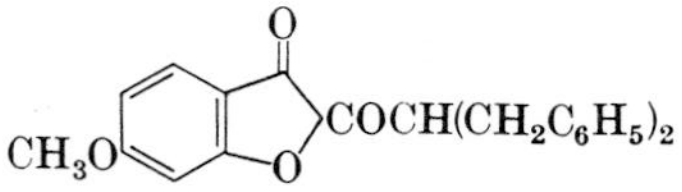 CH_3O (ring structure: O, O) $COCH(CH_2C_6H_5)_2$	21	42
$C_6H_5COCH_2COCH_2C_6H_5$	$NaNH_2$	$C_6H_5CHClCH_3$	$C_6H_5COCH_2COCH(C_6H_5)CH(CH_3)C_6H_5$ (erythro)	—	18*a*

TABLE VI. ALKYLATIONS OF THE DIANIONS OF KETO ALDEHYDES

Keto Aldehyde	Dialkali Salt	Alkyl Halide	Product	Yield (%)	Refs.
CH_3COCH_2CHO	Na, K	CH_3Br	$(CH_3CH_2COCHCHO)_2Cu$	66	21
	Na, K	CH_3I	CH_3CH_2–[pyridone ring: CN, =O, N–H]	61	21
	Na, K	CH_3I	$C_6H_3(COCH_2CH_3)_3$-1,3,5	27	21
	Na, K	n-C_4H_9Br	$(n\text{-}C_4H_9CH_2COCHCHO)_2Cu$	72	21
	Na, K	$C_6H_5CH_2Cl$	$(C_6H_5CH_2CH_2COCHCHO)_2Cu$	80	21
	Na, K	$C_6H_5CH_2Cl$	$C_6H_5CH_2CH_2$–[pyridone ring: CN, =O, N–H]	73	21
	Na, K	$C_6H_5CH_2Cl$	$C_6H_3(COCH_2CH_2C_6H_5)_3$-1,3,5	29	21
	Na, K	n-$C_8H_{17}Br$	$(n\text{-}C_8H_{17}CH_2COCHCHO)_2Cu$	55	21
[cyclopentanone ring: O, CHO]	Na, K	n-C_4H_9Br	n-C_4H_9–[cyclopentanone ring: O]	38	23
	Na, K	$C_6H_5CH_2Cl$	$C_6H_5CH_2$–[cyclopentanone ring: O]	32	23

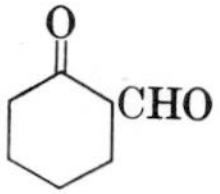 O, CHO	Na, K	CH_3I	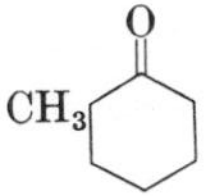 CH_3, O	36	23
	Na, K	n-C_4H_9Br	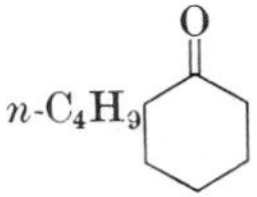 n-C_4H_9, O	41	23
	Na, K	$C_6H_5CH_2Cl$	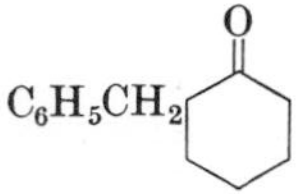 $C_6H_5CH_2$, O *	40	23
	Na, K	$C_6H_5CH_2Cl$	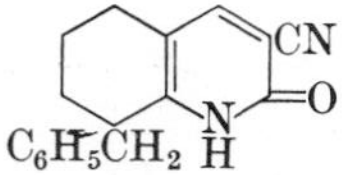 CN, O, N, H, $C_6H_5CH_2$	40	23
	K	$C_6H_5CH_2Cl$	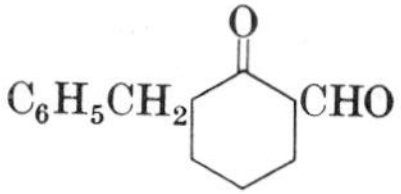 $C_6H_5CH_2$, O, CHO	20	24
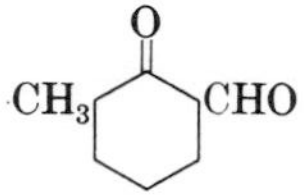 CH_3, O, CHO	Na, K	CH_3I	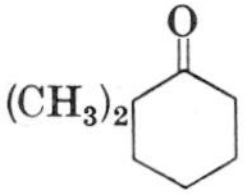 $(CH_3)_2$, O	60	23
	Na, K	n-C_4H_9Br	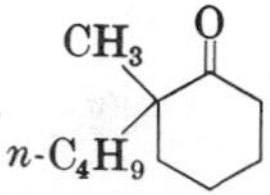 CH_3, O, n-C_4H_9	67–75	23, 105

* This product was also isolated as the copper chelate of the keto aldehyde.

TABLE VI. ALKYLATIONS OF THE DIANIONS OF KETO ALDEHYDES (*Continued*)

Keto Aldehyde	Dialkali Salt	Alkyl Halide	Product	Yield (%)	Refs.
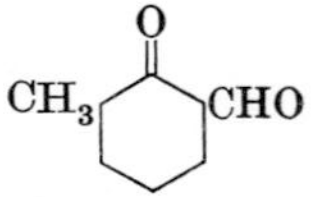 [structure: 2-methyl-6-formylcyclohexanone; O, CH_3, CHO] (*contd.*)	Na, K	$C_6H_5CH_2Cl$	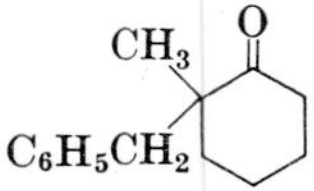 [structure: O, CH_3, $C_6H_5CH_2$ on cyclohexanone] *	55	23
	Na, K	$C_6H_5CH_2Cl$	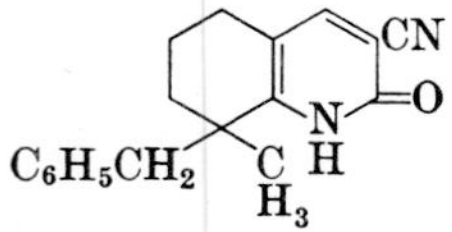 [structure: CN, N, H, =O, $C_6H_5CH_2$, CH_3 on tetrahydroquinolinone]	80	23
$CH_3COCH(C_6H_5)CHO$	K	$C_6H_5CH_2Cl$	$C_6H_5CH_2CH_2COCH(C_6H_5)CHO$	42	8
$CH_3COCH(CH_2C_6H_5)CHO$	K	n-C_4H_9Br	n-$C_4H_9CH_2COCH(CH_2C_6H_5)CHO$	27	21
	K	$C_6H_5CH_2Cl$	$C_6H_5CH_2CH_2COCH(CH_2C_6H_5)CHO$	51	21
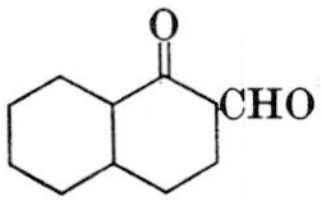 [structure: O, CHO on decalone]	Na, K	CH_3I	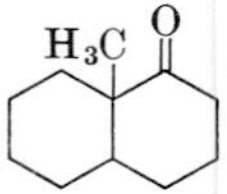 [structure: H_3C, O on decalone] * (*cis:trans*/46:54)	55	23

	Na, K	n-C_4H_9Br	C_4H_9-n, O (*cis*)	45	23
	Na, K	$C_6H_5CH_2Cl$	$CH_2C_6H_5$, O * (probably *cis*)	58	23
i-C_3H_7, O, CHO, CH_3	—	CH_3I	Alkylation failed	—	46

* This product was also isolated as the copper chelate of the keto aldehyde.

TABLE VII. ALKYLATIONS OF THE DIANIONS OF KETO ESTERS

Keto Ester	Alkali Amide	Alkyl Halide	Product	Yield (%)	Ref.
$CH_3COCH_2CO_2C_2H_5$	$NaNH_2$	CH_3I	$CH_3CH_2COCH_2CO_2C_2H_5$	Low	25
	KNH_2	CH_3I	$CH_3CH_2COCH_2CO_2C_2H_5$	36, 37	4
	KNH_2	C_2H_5Br	$C_2H_5CH_2COCH_2CO_2C_2H_5$	27, 29	4
	$LiNH_2$	n-C_4H_9Br	Alkylation failed	—	25
	$NaNH_2$	n-C_4H_9Br	Alkylation failed	—	4
	KNH_2	n-C_4H_9Br	Alkylation failed	—	25
	$NaNH_2$	$C_6H_5CH_2Cl$	$C_6H_5CH_2CH_2COCH_2CO_2H_5$	41	25
	KNH_2	$C_6H_5CH_2Cl$	Alkylation failed	—	4
O=, CH_3, =O, O (ring)	KNH_2	CH_3I	O=, CH_3, CH_3, =O, O (ring)	39	24
	KNH_2	$C_6H_5CH_2Cl$	O=, CH_3, $C_6H_5CH_2$, =O, O (ring)	90	24
O=, C_2H_5, =O, O (ring)	KNH_2	$C_6H_5CH_2Cl$	O=, C_2H_5, $C_6H_5CH_2$, =O, O (ring)	18	24
$C_6H_5CH_2COCH(C_6H_5)CO_2C_2H_5$	KNH_2	$C_6H_5CH_2Cl$	$C_6H_5CH_2CH(C_6H_5)COCH_2C_6H_5$	44	4

TABLE VIII. ARYLATIONS OF THE DIANIONS OF DICARBONYL COMPOUNDS

Dicarbonyl Compound	Alkali Amide	Arylating Agent	Product	Yield (%)	Ref.
$CH_3COCH_2COCH_3$	$NaNH_2$	C_6H_5Br	$CH_3COCH_2COCH_2C_6H_5$	18	11
	KNH_2	C_6H_5Br	$CH_3COCH_2COCH_2C_6H_5$	13	2
	$NaNH_2$	$(C_6H_5)_2ICl$	$CH_3COCH_2COCH_2C_6H_5$	92	11
	$NaNH_2$	$(p\text{-}ClC_6H_4)_2ICl$	$CH_3COCH_2COCH_2C_6H_4Cl\text{-}p$	44	11
$C_2H_5COCH_2COC_2H_5$	$NaNH_2$	$(C_6H_5)_2ICl$	$C_2H_5COCH_2COCH(C_6H_5)CH_3$	50	11
$n\text{-}C_3H_7COCH_2COCH_3$	$NaNH_2$	$(C_6H_5)_2ICl$	$n\text{-}C_3H_7COCH_2COCH_2C_6H_5$	98	11
$n\text{-}C_5H_{11}COCH_2COCH_3$	$NaNH_2$	$(C_6H_5)_2ICl$	$n\text{-}C_5H_{11}COCH_2COCH_2C_6H_5$	78	11
	$NaNH_2$	$(p\text{-}CH_3C_6H_4)_2ICl$	$n\text{-}C_5H_{11}COCH_2COCH_2C_6H_4CH_3\text{-}p$	21	11
$C_6H_5COCH_2COCH_3$	$NaNH_2$	$(C_6H_5)_2ICl$	$C_6H_5COCH_2COCH_2C_6H_5$	61	11
	$NaNH_2$	$(p\text{-}CH_3C_6H_4)_2ICl$	$C_6H_5COCH_2COCH_2C_6H_4CH_3\text{-}p$	44	11
$o\text{-}ClC_6H_4CH(COCH_3)_2$	KNH_2	—	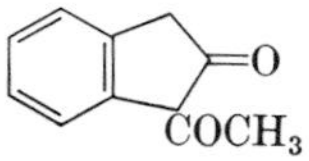 =O, $COCH_3$	98	47
$o\text{-}ClC_6H_4CH_2CH(CHO)COCH_3$	KNH_2	—	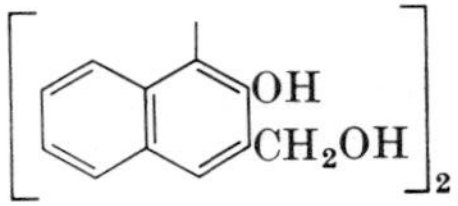 [OH, CH_2OH]$_2$	29	47
$o\text{-}ClC_6H_4CH_2CH(COCH_3)_2$	KNH_2	—	 =O, $COCH_3$	65	47
$n\text{-}C_9H_{19}COCH_2COCH_3$	$NaNH_2$	$(C_6H_5)_2ICl$	$n\text{-}C_9H_{19}COCH_2COCH_2C_6H_5$	53	11

CHAPTER 3

THE RITTER REACTION

L. I. KRIMEN AND DONALD J. COTA
Abbott Laboratories

CONTENTS

INTRODUCTION

The formation of N-substituted amides by the addition of nitriles to alkenes in the presence of concentrated sulfuric acid, described first by Ritter in 1948,[1] is one of the newer synthetic reactions available to the organic chemist. The reaction has since been extended to the addition of nitriles to a wide variety of compounds capable of forming a carbonium ion, and it constitutes the only really useful procedure for the preparation of amides of tertiary carbinamines,

$$RR'R''C\text{—}NHCOR'''$$

In its most general form the Ritter reaction involves the nucleophilic addition of a nitrile to a carbonium ion in the presence of sulfuric acid.

$$RC{\equiv}N: + R_3C^{\oplus} \xrightarrow{H_2SO_4} R_3C\text{—}\overset{\oplus}{N}{\equiv}CR \quad \mathbf{1}$$

[1] J. J. Ritter and P. P. Minieri, *J. Am. Chem. Soc.*, **70**, 4045 (1948).

Subsequent dilution with water yields the amide. When hydrogen cyanide is employed as the nitrile, the resulting N-alkylformamide can be

$$R_3C{:}\overset{\oplus}{N}{\equiv}CR + H_2O \xrightarrow{-H^{\oplus}} R_3C{-}N(H){-}C({=}O){-}R$$

readily hydrolyzed to the corresponding carbinamine; thus the tertiary carbinamines are readily available.

$$R_3C{-}N(H){-}C({=}O){-}H \xrightarrow{H_2O} R_3C{-}NH_2$$

If a nucleophilic center is suitably placed in the primary adduct **1**, a wide variety of heterocyclic compounds can be prepared.

Attention has been given primarily to reactions in which the carbonium ion is generated in sulfuric acid or, less commonly, in sulfonic acids, phosphoric acid, or boron trifluoride. For comparison analogous reactions which are initiated by a nitrilium ion, formed from a nitrile and a Friedel-Crafts catalyst, have been included in the discussion of related reactions.

Cursory surveys of the Ritter reaction have appeared in *Russian Chemical Reviews* where Zil'berman dealt with the formation of new nitrogen-carbon bonds[2] and in *Advances in Heterocyclic Chemistry* in which Johnson reviewed its utility in heterocyclic syntheses.[3]

MECHANISM

The mechanism suggested by Ritter and his associates in their initial papers is illustrated in the following scheme for the reaction between isobutene and acetonitrile.[1, 4, 5]

$$(CH_3)_2C{=}CH_2 + H_2SO_4 \rightleftharpoons (CH_3)_3C^{\oplus} + HSO_4^{\ominus} \rightleftharpoons (CH_3)_3COSO_3H \quad \text{(Eq. 1)}$$

$$(CH_3)_3C^{\oplus} + N{\equiv}CCH_3 \rightarrow (CH_3)_3CN{=}\overset{\oplus}{C}CH_3 \quad \text{(Eq. 2)}$$

$$(CH_3)_3CN{=}\overset{\oplus}{C}CH_3 + HSO_4^{\ominus} \rightarrow (CH_3)_3CN{=}C(OSO_3H)CH_3 \quad \text{(Eq. 3)}$$

$$(CH_3)_3CN{=}C(OSO_3H)CH_3 + H_2O \rightarrow (CH_3)_3CNHCOCH_3 + H_2SO_4 \quad \text{(Eq. 4)}$$

[2] E. N. Zil'bermann, *Usp. Khim.*, **29**, 709 (1960) [*Russ. Chem. Rev., English Transl.*, 311 (1960)].

[3] F. Johnson and R. Madroñero, in A. R. Katritzky and A. J. Boulton, eds., *Advances in Heterocyclic Chemistry*, Vol. 6, pp. 95–146, Academic Press, New York, 1966.

[4] J. J. Ritter and J. Kalish, *J. Am. Chem. Soc.*, **70**, 4048 (1948).

[5] J. J. Ritter, U.S. pat. 2,573,673 [*C.A.*, **46**, 9584h (1952)].

The possibility that the reaction could proceed by the addition of acetamide under acid conditions was discounted when substitution of acetamide for acetonitrile gave no reaction product.[1]

Ritter's proposal that the reaction proceeds via a carbonium ion intermediate has been substantially supported by the studies of Jacquier and Christol.[6–17] They described the synthesis of amines by the reaction of cycloalkanols, secondary aliphatic alcohols, cyclosubstituted alcohols, and alicyclic alcohols containing a spiro structure with potassium cyanide in sulfuric acid. In many reactions there is an intramolecular rearrangement of the Meerwein type,[18] including ring expansion or contraction, leading to the amine resulting from the reaction of hydrogen cyanide with the most stable carbonium ion. For example, 1-isopropylcyclopentanol, 1-methylcyclopentylmethylcarbinol, cyclopentyldimethylcarbinol, 1,2-dimethylcyclohexanol, and 2,2-dimethylcyclohexanol all yield 1,2-dimethylcyclohexylamine.[16]

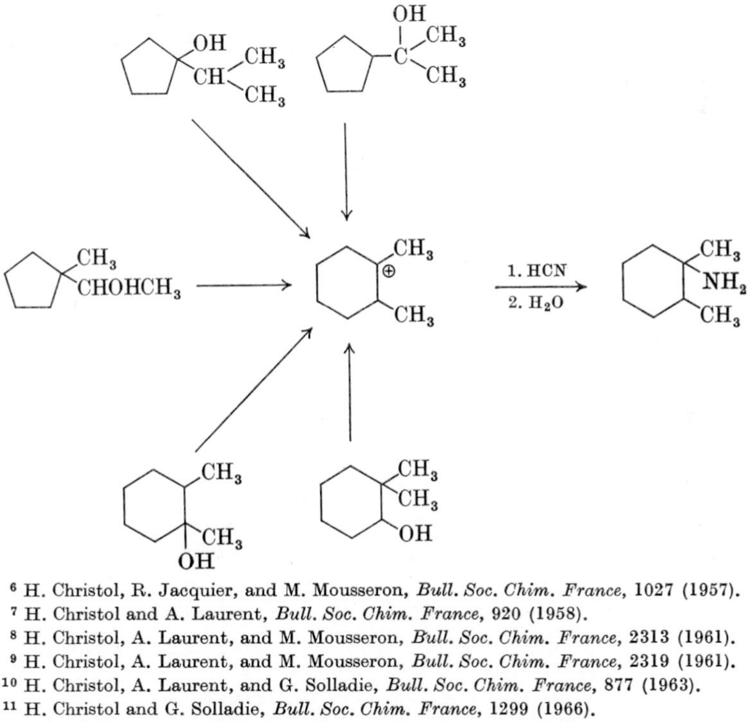

[6] H. Christol, R. Jacquier, and M. Mousseron, *Bull. Soc. Chim. France*, 1027 (1957).
[7] H. Christol and A. Laurent, *Bull. Soc. Chim. France*, 920 (1958).
[8] H. Christol, A. Laurent, and M. Mousseron, *Bull. Soc. Chim. France*, 2313 (1961).
[9] H. Christol, A. Laurent, and M. Mousseron, *Bull. Soc. Chim. France*, 2319 (1961).
[10] H. Christol, A. Laurent, and G. Solladie, *Bull. Soc. Chim. France*, 877 (1963).
[11] H. Christol and G. Solladie, *Bull. Soc. Chim. France*, 1299 (1966).
[12] H Christol and G Solladie, *Bull Soc. Chim. France*, 1307 (1966).
[13] H. Christol and G. Solladie, *Bull. Soc. Chim. France*, 3193 (1966).
[14] R. Jacquier and H. Christol, *Bull. Soc. Chim. France*, 556 (1954).
[15] R. Jacquier and H. Christol, *Bull. Soc. Chim. France*, 596 (1957).
[16] R. Jacquier and H. Christol, *Bull. Soc. Chim. France*, 600 (1957).
[17] R. Jacquier and H. Christol, *Bull. Soc. Chim. France*, 917 (1953).
[18] H. Meerwein, *Ann.*, **417**, 255 (1918).

Further support for a carbonium ion mechanism was furnished by Roe and Swern, who employed methanesulfonic acid in place of sulfuric acid.[19]

Similar investigations showed that C^{14}-labeled 1,2,2-triphenylethanol, when treated with acetonitrile and sulfuric acid, led to a mixture of substituted amides resulting from a carbonium ion equilibrium.[20, 21]

$$(C_6H_5)_2CH\overset{14}{C}H(OH)C_6H_5 \rightarrow (C_6H_5)_2CH\overset{14}{C}^{\oplus}HC_6H_5 \rightleftharpoons C_6H_5\overset{\oplus}{C}H\overset{14}{C}H(C_6H_5)_2$$

$$\downarrow \qquad\qquad \downarrow$$

$$(C_6H_5)_2CH\overset{14}{C}H(NHCOCH_3)C_6H_5 \qquad C_6H_5CH(NHCOCH_3)\overset{14}{C}H(C_6H_5)_2$$

From data obtained in a kinetic study of the reaction between *t*-butyl alcohol and acrylonitrile in 20–69% sulfuric acid solutions Deno concluded that the transition state was composed of acrylonitrile and protonated *t*-butyl alcohol.[22] This concept of a loosely solvated carbonium ion was also advanced by other investigators.[23] In a subsequent study using the α-olefin, 1-hexadecene, Weil and co-workers postulated a cyclic transition state which then proceeded to the "Ritter intermediate";[24] however, their

$$n\text{-}C_{14}H_{29}CH(CH_3)\text{—}O\text{—}SO_2OH \cdots N{\equiv}C\text{—}CH_3 \longrightarrow n\text{-}C_{14}H_{29}CH(CH_3)\text{—}N{=}C(CH_3)\text{—}O\text{—}SO_2(OH)$$

work was in agreement with a carbonium ion process. They formulated an irreversible reaction with water, leading to a stable intermediate, **2**, whose precise nature they were unable to determine. (See p. 219.)

Extension of the Ritter reaction to *t*-carboxylic acids furnished additional evidence for a carbonium ion process. Haaf showed that acids having this structure could be reversibly converted to *t*-alkylamido

[19] E. T. Roe and D. Swern, *J. Am. Chem. Soc.*, **75**, 5479 (1953).

[20] A. Laurent, E. Laurent-Dieuzeide, and P. Mison, *Bull. Soc. Chim. France*, 945 (1965).

[21] A. Laurent and P. Mison, *Bull. Soc. Chim. France*, 956 (1962).

[22] N. C. Deno, T. Edwards, and C. Perizzolo, *J. Am. Chem. Soc.*, **79**, 2108 (1957).

[23] C. W. Roberts and N. F. Nuenke, *J. Org. Chem.*, **24**, 1907 (1959).

[24] I. Weil, R. G. Goebel, E. R. Tulp, and A. Cahn, *Am. Chem. Soc., Div. Petrol. Chem., Preprints*, **8**, (2) B95 (1963).

$$\underset{\underset{OSO_3H}{|}}{RCH(CH_3)N{=}CCH_3} + H_2O \rightarrow$$

$$\left[RCH(CH_3)NHC(OH)(OSO_3H)CH_3 \equiv RCH(CH_3)\overset{\oplus}{N}H_2C(OH)(OSO_3^{\ominus})CH_3 \equiv (RCH(CH_3)NHC(=OH)CH_3)^{\oplus}HSO_4^{\ominus} \right]$$

2

compounds since treatment of the *t*-alkylamides with carbon monoxide in concentrated sulfuric acid yielded *t*-carboxylic acids.[25]

$$R_3CCO_2H \underset{}{\overset{H^{\oplus}}{\rightleftharpoons}} R_3CCO_2H_2^{\oplus} \overset{-H_2O}{\rightleftharpoons} R_3C\overset{\oplus}{C}O$$

$$R_3CNHCHO \overset{H_2O,\ -H^{\oplus}}{\rightleftharpoons} R_3CN{=}\overset{\oplus}{C}H \overset{HCN}{\rightleftharpoons} R_3C^{\oplus} + CO$$

After unsuccessful attempts by various workers[26, 27] to isolate and characterize the alkyl iminosulfate intermediate suggested by Ritter,[4] Glikmans and co-workers were able to verify its existence by showing that it is susceptible to hydrolysis in anhydrous acetic acid, obtaining the necessary water from the solvent with concomitant generation of acetic anhydride.[28, 29]

$$(CH_3)_2C{=}CH_2 + N{\equiv}CCH{=}CH_2 \xrightarrow{H_2SO_4} (CH_3)_3CN{=}C(OSO_3H)CH{=}CH_2 \xrightarrow{H_2O} (CH_3)_3CNHCOCH{=}CH_2$$

$$(CH_3)_2C{=}CH_2 + N{\equiv}CCH{=}CH_2 \xrightarrow[2CH_3CO_2H]{H_2SO_4} (CH_3)_3CNHCOCH{=}CH_2 + (CH_3CO)_2O$$

An earlier attempt to produce N-cumylacrylamide by this route was unsuccessful.[23]

Glikmans also showed that the alkene could be replaced by the corresponding alcohol or acetate and suggested the presence of an ion pair such as **3**, which would assist in the electronic transfer as indicated in formula **4**.

$$(CH_3)_2C{=}CH_2 + H_2SO_4 \rightleftharpoons [(CH_3)_3C^{\oplus}, SO_4H^{\ominus}]$$

$$(CH_3)_2C{=}CH_2 + H_2SO_4 \overset{CH_3CO_2H}{\rightleftharpoons} (CH_3)_3COCOCH_3 + H_2SO_4$$

$$(CH_3)_3COCOCH_3 + H_2SO_4 \overset{CH_3CO_2H}{\rightleftharpoons} [(CH_3)_3C^{\oplus}, SO_4H^{\ominus}]$$

3

[25] W. Haaf, *Chem. Ber.*, **96**, 3359 (1963).

[26] T. Clarke, J. Devine, and D. W. Dicker, *J. Am. Oil Chemists Soc.*, **41**, 78 [*C.A.*, **60**, 6733h (1964)].

[27] R. L. Holmes, J. P. Moreau, and G. Sumrell, *J. Am. Oil Chemists Soc.*, **42**, 922 (1965) [*C.A.*, **63**, 17892c (1965)].

[28] G. Glikmans, B. Torck, M. Hellin, and F. Coussemant, *Bull. Soc. Chim. France*, 1376 (1966).

[29] G. Glikmans, B. Torck, M. Hellin, and F. Coussemant, *Bull. Soc. Chim. France*, 1383 (1966).

Addition to the nitrile stabilizes the *t*-butyl cation in an extremely selective manner by the presence of the bisulfate group.[29]

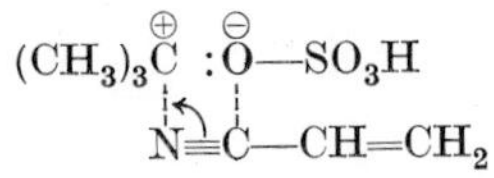

$$\begin{array}{l} (CH_3)_3\overset{\oplus}{C}\ :\overset{\ominus}{O}{-}SO_3H \\ \quad\quad\ \ N{\equiv}C{-}CH{=}CH_2 \end{array}$$

4

Thus, after eighteen years of investigation, the mechanism has been shown to be essentially the one originally proposed by Ritter.[1, 4, 5]

SCOPE AND LIMITATIONS

Although Ritter and his associates initially conducted their studies on rather simple alkenes and alcohols as carbonium ion sources, the reaction was rapidly extended. Examples now include alkanes, alkadienes, alicyclic and spiro alcohols, alkyl chlorides, glycols, aldehydes, chlorohydrins, N-methylolamides, ethers, carboxylic acids, esters, ketones, and ketoximes.

The nitrile source, too, has seen extensive investigation which has further broadened the synthetic value of this reaction. Not only hydrocyanic acid and aliphatic nitriles, but also cyanohydrins, cyano acids and their esters, and substituted nitriles have been used. Other compounds containing the nucleophilic nitrile group such as biuret, cyanogen, 1-cyanoformamide, cyanamide, dicyandiamide, and cyano complexes of inorganic acids have also been successfully employed.

Other significant extensions of the reactions lead to polyamides or to heterocyclic compounds. Polyamides are formed from alkadienes or glycols and dinitriles. Heterocyclic compounds result when the starting compound contains both a cyano substituent and a group easily converted to a carbon cation in strong acid.

Alkanes (Table I)

The extension of the Ritter reaction to the participation of isoparaffins as a carbonium ion source was made by Haaf. When *t*-butyl alcohol was employed as a hydride acceptor, methylcyclohexane reacted with hydrogen cyanide to give 1-methyl- and 2-methyl-cyclohexylamine in yields of 23% and 4%, respectively; the major product was *t*-butylamine (52% yield).[30] Using similar conditions, adamantane and acetonitrile gave 1-acetamidoadamantane (36%).[31]

[30] W. Haaf, *Chem. Ber.*, **97**, 3234 (1964).

[31] W. Haaf, *Angew. Chem.*, **73**, 144 (1961).

Alkenes (Tables II and III)

The addition of hydrogen cyanide and nitriles to alkenes has been investigated by numerous workers (Refs. 24, 26, 28, 29, 32, 33). At an early stage the reaction was elucidated for a variety of nitriles using diisobutylene.[1] As described for isobutene (p. 216), the reaction usually gives the alkylamido addition product corresponding to the carbonium ion initially formed by Markownikoff addition of a proton to the alkene. There are, however, several reports of products resulting from rearrangement of the incipient carbonium ion. Thus 1-phenylpropylamine is obtained from allylbenzene when operating in di-*n*-butyl ether.[8] On the other hand, in the absence of solvent the expected 2-phenylisopropylamine is obtained.[4, 34]

$$C_6H_5CH_2CH{=}CH_2 \rightarrow [C_6H_5CH_2\overset{\oplus}{C}HCH_3 \rightleftharpoons C_6H_5\overset{\oplus}{C}HCH_2CH_3]$$

$$\downarrow \qquad\qquad \downarrow$$

$$C_6H_5CH_2CH(NH_2)CH_3 \qquad C_6H_5CH(NH_2)CH_2CH_3$$

It is noteworthy that the carbonium ion from 3-methyl-1-phenylbutene did not undergo a similar rearrangement.[8] The initial secondary carbonium ion is stabilized to a greater extent by the adjacent phenyl group than is the tertiary carbonium ion that would result from a prototropic shift.

Conflicting results with camphene were reported by several groups of workers. Ritter and Minieri isolated the N-acylisobornylamines, a result of the Wagner rearrangement, when employing hydrogen cyanide or simple nitriles.[1] Other investigators also obtained N-acylisobornylamines from simple nitriles but found that hydrogen cyanide gave the unrearranged norcamphane derivative.[35] This anomaly was resolved by Kochetkov and co-workers who made a detailed study of the behavior of camphene. They showed that, when the reaction is conducted at temperatures in the $-15°$ to $-20°$ range, the formation of the isobornyl derivative with hydrogen cyanide is reduced to a minimum.[36] Similar results were obtained when 3-formamido-2,2,3-trimethylnorcamphane was formed by treating racemic camphene with hydrogen cyanide at 0–3°.[37]

[32] C. Malen and J. R. Boissier, *Bull. Soc. Chim. France*, 923 (1956).

[33] R. Maugé, C. Malen, and J. R. Boissier, *Bull. Soc. Chim. France*, 926 (1956).

[34] J. J. Ritter and F. X. Murphy, *J. Am. Chem. Soc.*, **74**, 763 (1952).

[35] G. A. Stein, M. Sletzinger, H. Arnold, D. Reinhold, and K. Pfister, III, *J. Am. Chem. Soc.*, **78**, 1514 (1956).

[36] N. K. Kochetkov, A. Ya. Khorlin, and K. I. Lopatina, *J. Gen. Chem. USSR* (*Eng. Transl.*), **29**, 77 (1959) [*C.A.*, 53, 22058i (1959)].

[37] C. A. Stone, M. L. Torchiana, K. L. Meckelnburg, J. Stavorski, M. Sletzinger, G. A. Stein, W. V. Ruyle, D. F. Reinhold, W. A. Gaines, H. Arnold, and K. Pfister, III, *J. Med. Pharm. Chem.*, **5**, 665 (1962) [*C.A.*, **57** 12341c (1962)].

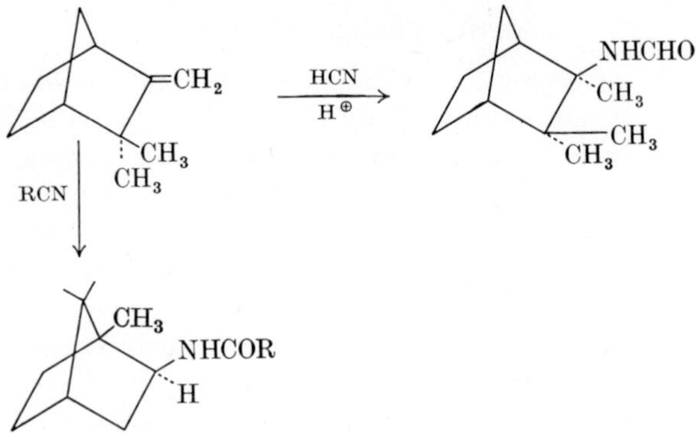

Kochetkov also showed that the highly active trichloro- and dichloroacetonitrile reacted with camphene at a reduced temperature without rearrangement, but even at −50° the rearranged products were obtained with aceto- and benzo-nitrile.[36] Rearrangement also occurs with α-pinene when it is treated with hydrogen cyanide at slightly elevated temperatures, giving 1,8-diformamido-*p*-menthane.[38,39]

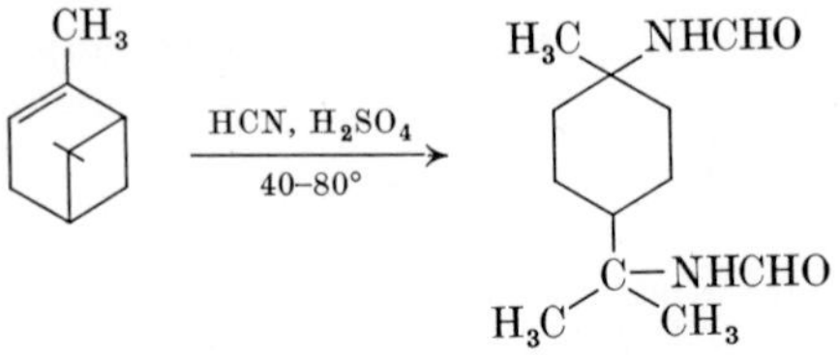

The failure of the reaction with alkenes has been noted in relatively few instances. The only structural similarity in these cases is the presence of one or two phenyl groups on the double-bonded carbon atom. The failure of 1,1-diphenyl-2,2-dimethylethylene to undergo the reaction was explained on the basis of the high stability of the derived carbonium ion and hence its lack of reactivity toward nitriles.[10] Similar reasoning could rationalize the failure of 1-(*p*-chlorophenyl)-1-phenylethylene and 1,1-di(*p*-chlorophenyl)-2,2-dichloroethylene.[40]

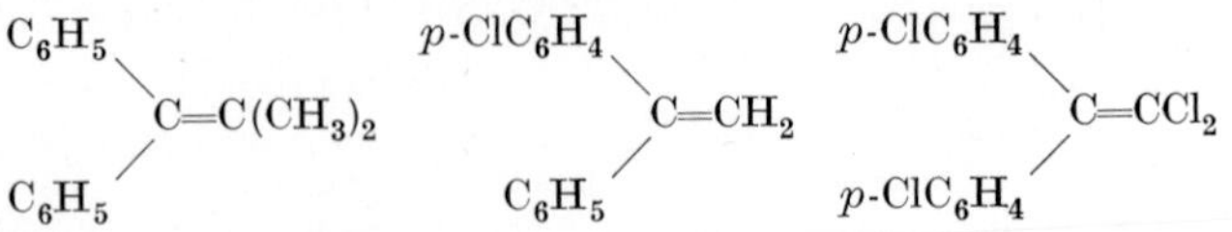

$(C_6H_5)_2C{=}C(CH_3)_2$ $\quad$ $p\text{-}ClC_6H_4(C_6H_5)C{=}CH_2$ $\quad$ $(p\text{-}ClC_6H_4)_2C{=}CCl_2$

Although Ritter reported the successful addition of acetonitrile to α-methylstyrene,[5] other investigators reported that α-methylstyrene and

[38] N. M. Bortnick, Brit. pat. 681,688 [*C.A.*, **48,** 727f (1954)].

[39] N. M. Bortnick, U.S. pat. 2,632,022 [*C.A.*, **48,** 4003h (1954)].

[40] A. Kluszyner, S. Blum, and E. D. Bergmann, *J. Org. Chem.*, **28,** 3588 (1963).

its *p*-substituted homologs appear to be stronger nucleophiles than the nitrogen in acrylonitrile.[23] Only when a strong *meta*-directing group was placed in the *meta* position did the resulting alkene give the normal Ritter product.[23] A recent patent discloses, however, that 1-(3,4-dichlorophenyl)-2,2-dimethylethylene gave the formamido derivative in 95% yield.[41]

$$3,4\text{-}Cl_2C_6H_3CH{=}C(CH_3)_2 \xrightarrow[H_2SO_4]{KCN,\ (n\text{-}C_4H_9)_2O} 3,4\text{-}Cl_2C_6H_3CH_2C(CH_3)_2NHCHO$$

Utilization of the non-conjugated ethylenic linkages in carboxylic acids has also been reported.[19, 27, 42, 43] Swern obtained good yields of substituted amidostearic acids from oleic acid and a number of nitriles. With linoleic acid, however, polymerization predominated.[42] When dinitriles were employed, diamides bearing free carboxylic acid groups were produced. For example, the addition of adiponitrile to oleic acid gave the diamido dicarboxylic acid **5** in an 82% yield.[43]

$$\begin{array}{c} CH_3(CH_2)_xCHNHCO(CH_2)_4CONHC(CH_2)_xCH_3 \\ \quad | \qquad\qquad\qquad\qquad\qquad\qquad | \\ (CH_2)_yCO_2H \qquad\qquad\qquad (CH_2)_yCO_2H \end{array}$$

5, $x + y = 15$

Halo alkenes such as $R_2C{=}CHX$ react with nitriles to give N-(2-halo-1-ethyl)amides in good yields. The halo amides were easily dehydrohalogenated to give oxazolines.[44]

$$RR_1C(NHCOR){-}CH_2X \xrightarrow{OH^{\ominus}} \text{2-R-4-}RR_1\text{-oxazoline } (RR_1C{-}CH_2{-}O{-}CR{=}N) + H_2O$$

An unusual intramolecular Ritter reaction of alkenes was encountered in a study of the Schmidt reaction with certain olefinic nitriles in polyphosphoric acid. The products resulted from the addition of the nitrile carbon atom across the olefinic linkage instead of the normal addition to produce secondary amides.[45, 46] An example is the formation of $\Delta^{8,9}$-4-hydrindenone from γ-cyclopentylidenebutyronitrile on heating in polyphosphoric acid. The accompanying four-step mechanism was proposed.[45]

[41] R. Kopf, D. Lorenz, and K. H. Boltze, Ger. pat. (West) 1,216,881 [*C.A.*, **65**, 3789f (1966)].

[42] E. T. Roe and D. Swern, *J. Am. Chem. Soc.*, **77**, 5408 (1955).

[43] A. E. Kulikova, S. B. Meiman, E. N. Zil'berman, *Zh. Prikl. Khim.*, **36**, 1367 (1963) [*C.A.*, **59**, 11240e (1963)].

[44] R. M. Lusskin and J. J. Ritter, *J. Am. Chem. Soc.*, **72**, 5577 (1950).

[45] R. T. Conley and B. E. Nowak, *J. Org. Chem.*, **26**, 692 (1961).

[46] R. K. Hill and R. T. Conley, *J. Am. Chem. Soc.*, **82**, 645 (1960).

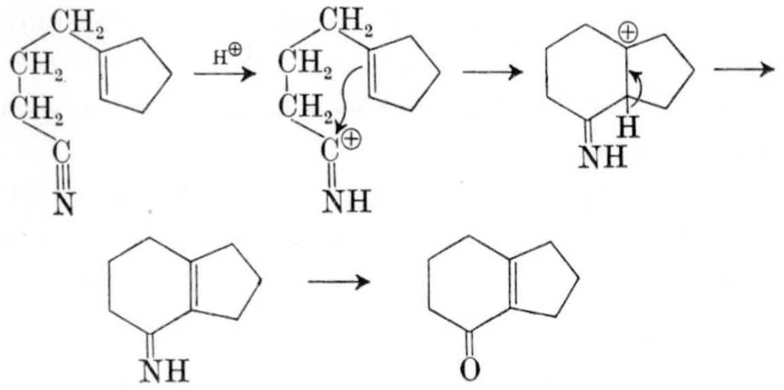

An analogous reaction had been reported earlier for γ-3-indenylbutyronitrile and described as a failure of the Ritter reaction.[47]

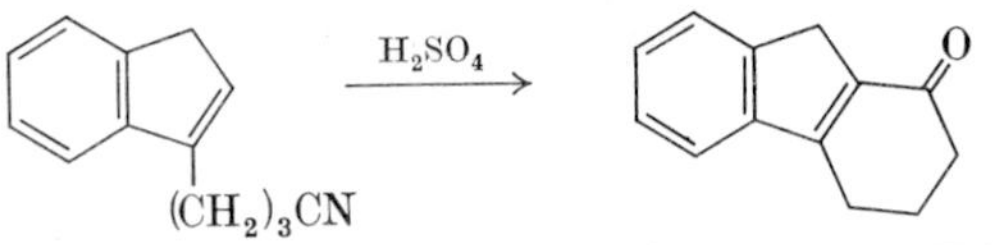

$(CH_2)_3CN$ $\xrightarrow{H_2SO_4}$

Conley suggested that these unsaturated nitriles do not react in the usual manner because the cyano group cannot approach the intermediate carbonium ion.[46,48–50] This proposal, however, was in disagreement with the results of Bobbitt and Doolittle who were able to cyclize 3-cyano-4-stilbazole to the lactam, 1-oxo-3-phenyl-1,2,3,4-tetrahydrocopyrine, in 90% yield.[51]

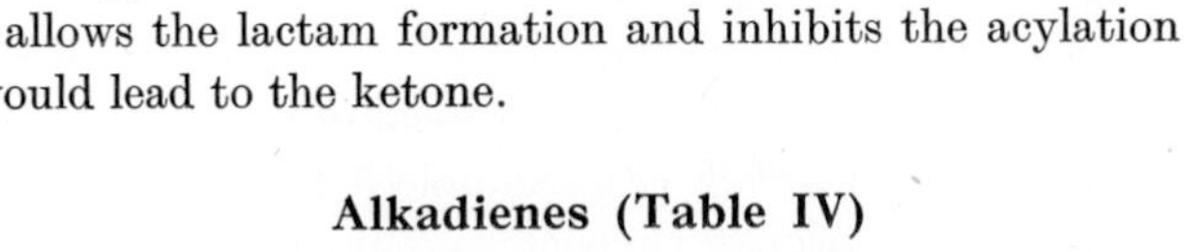

$CH{=}CHC_6H_5$, CN $\xrightarrow[\text{acid}]{\text{Polyphosphoric}}$ C_6H_5, NH, O

Bobbitt attributed the lactam formation to the electron-withdrawing effect of the pyridinium nucleus on the double bond. This electron deficiency allows the lactam formation and inhibits the acylation reaction which would lead to the ketone.

Alkadienes (Table IV)

The extension of the Ritter reaction to dienes to give dialkylamido derivatives has met with some success. Bortnick prepared 1,8-diamino-*p*-methane[38, 39] and 2,5-diamino-2,5-dimethylhexane[52] from limonene and

[47] F. H. Howell and D. A. H. Taylor, *J. Chem. Soc.*, 3011 (1957).
[48] R. T. Conley and M. C. Annis, *J. Org. Chem.*, **27**, 1961 (1962).
[49] R. T. Conley and B. E. Nowak, *J. Org. Chem.*, **27**, 1965 (1962).
[50] R. T. Conley and R. J. Lange, *J. Org. Chem.*, **28**, 210 (1963).
[51] J. M. Bobbitt and R. E. Doolittle, *J. Org. Chem.*, **29**, 2298 (1964).
[52] N. M. Bortnick, U.S. pat. 2,632,023 [*C.A.*, **49**, 1782a (1955)].

bimethallyl, respectively. Although the reactions attempted with dipentene and 2,5-dimethyl-1,5-hexadiene gave nitrogen-containing materials only in low yield along with substantial amounts of polymeric material, and butadiene reacted explosively,[26] Magat prepared a number of synthetic linear polyamides from ditertiary diolefins and dinitriles.[53] For example, the polymer, poly(α,α,α′,α′-tetramethyl)decamethyleneadipamide was prepared from adiponitrile and 2,11-dimethyl-1,11-dodecadiene.

$$n\mathrm{NC(CH_2)_4CN} + n\mathrm{CH_2{=}C(CH_3)(CH_2)_8C(CH_3){=}CH_2} \rightarrow \left[\mathrm{-NH-C(CH_3)_2-(CH_2)_8-C(CH_3)_2-NHCO(CH_2)_4CO-} \right]_n$$

The participation of a conjugated diene in combined Ritter and Diels-Alder reactions has been reported.[54] 2,3-Dimethyl-1,3-butadiene and acetonitrile furnish the substituted tetrahydropyridine. It was proposed that the diene reacted with 1 mole of acetonitrile to give the iminosulfate intermediate. The $\mathrm{-N{=}C(SO_3H)-}$ grouping then behaved as a dienophile, adding another mole of butadiene and acetonitrile, to give the product.

$$\mathrm{CH_2{=}C(CH_3)-C(CH_3){=}CH_2} \xrightarrow[\mathrm{H_2SO_4}]{\mathrm{CH_3CN}} \mathrm{CH_3C(CH_3){=}C(CH_3)CH_2N{=}C(OSO_3H)CH_3} \xrightarrow[\text{Diels-Alder}]{\mathrm{CH_2{=}C(CH_3)-C(CH_3){=}CH_2}}$$

$$\mathrm{CH_3C(CH_3){=}C(CH_3)CH_2N}\langle\text{ring: } \mathrm{N-CH_2-CCH_3{=}CCH_3-CH_2-C(CH_3)(OSO_3H)-N}\rangle \xrightarrow[\text{Ritter}]{\mathrm{CH_3CN,\ H_2O}}$$

$$\mathrm{CH_3C(CH_3){=}C(CH_3)CH_2N}\langle\text{ring: } \mathrm{N-CH_2-CCH_3{=}CCH_3-CH_2-C(CH_3)(NHCOCH_3)-N}\rangle$$

[53] E. E. Magat, U.S. pat. 2,628,219 [*C.A.*, **47,** 5130c (1953)].

[54] M. Lora-Tamayo, G. Gancía Muñoz, and R. Madroñero, *Bull. Soc. Chim. France*, 1334 (1958).

Alcohols (Tables V–VIII)

The use of primary, secondary, and tertiary alcohols, alicyclic and spiro alcohols, glycols, heterocyclic alcohols, and halohydrins in the Ritter reaction has been investigated.

The failure of primary aliphatic alcohols to react with nitriles even at elevated temperatures or on prolonged heating, or with fuming sulfuric acid was reported by Ritter.[55] Other investigators[56–58] observed, however, that primary aralkyl alcohols and glycols condensed smoothly with nitriles under mild conditions to give N-aralkylamides and N,N′-*bis*-aralkylamides in good yields. N-(2,4-Dimethylbenzyl)acetamide was obtained from 2,4-dimethylbenzyl alcohol and acetonitrile in 87% yield. An exception

$$2,4\text{-}(CH_3)_2C_6H_3CH_2OH + CH_3CN \longrightarrow 2,4\text{-}(CH_3)_2C_6H_3CH_2NHCOCH_3$$

to the reactivity of benzyl alcohols is shown by *p*-nitrobenzyl alcohol, which is inert to nitriles in concentrated sulfuric acid.[14]

Glikmans found that alcohols, in general, gave lower yields than the corresponding olefins and required the use of concentrated sulfuric acid.[28] On the other hand, there are reports that the reverse is true because polymerization of olefin competes with the Ritter reaction.[59]

From studies of a number of alcohols of varied structure Christol, Jacquier, and their associates concluded that tertiary alcohols react with hydrogen cyanide without rearrangement;[13, 15] for example, methylcyclohexanol gives the corresponding amine in 60% yield.[60] Although most secondary cycloalkanols react without rearrangement, the yields are somewhat lower; thus cyclohexanol gives only a 4% yield of cyclohexylamine.[14, 15]

It is an oversimplification to conclude that the reaction product always corresponds to the most stable incipient carbonium ion, yet this conclusion is valid for a majority of the cases.

In a study of the reaction of substituted cycloalkanols it was learned that the yields decrease with increasing separation of the alcohol function

[55] F. R. Benson and J. J. Ritter, *J. Am. Chem. Soc.*, **71**, 4128 (1949).
[56] C. L. Parris, *Org. Syntheses*, **42**, 16 (1962).
[57] C. L. Parris and R. M. Christenson, *J. Org. Chem.*, **25**, 331 (1960).
[58] J. A. Sanguigni and R. Levine, *J. Med. Chem.*, **7**, 573 (1964).
[59] A. I. Meyers, *J. Org. Chem.*, **25**, 1147 (1960).
[60] L. I. Krimen, unpublished results.

and the substituent.[16] 1- and 2-Methylcyclohexanol gave 1-methylcyclohexylamine (**6**) in 60% yield, whereas the yield from the 4-methyl derivative was only 5%.

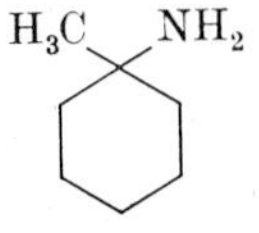

6

Similarly, when the ring is "farther away" from the alcohol as in the cyclohexylpropanols, the yields decrease. For example, ethylcyclohexylcarbinol and 1-cyclohexyl-2-propanol give 1-*n*-propylcyclohexylamine in yields of 40% and 5%, respectively.

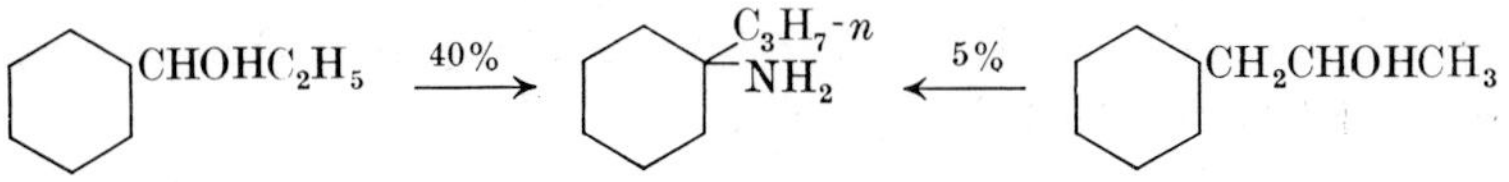

$$C_6H_{11}CHOHC_2H_5 \xrightarrow{40\%} 1\text{-}(C_3H_7\text{-}n)C_6H_{10}NH_2 \xleftarrow{5\%} C_6H_{11}CH_2CHOHCH_3$$

This effect of the separation of hydroxyl group and substituent is a characteristic specific to the Ritter reaction, and not necessarily a function of the number of successive prototropic rearrangements. 4-Methylcyclohexanol and 1-cyclopentyl-2-propanol rearrange, by means of three transitory carbonium ions, to methylcyclohexylamine and ethylcyclohexylamine in yields of 5% and 45%, respectively.

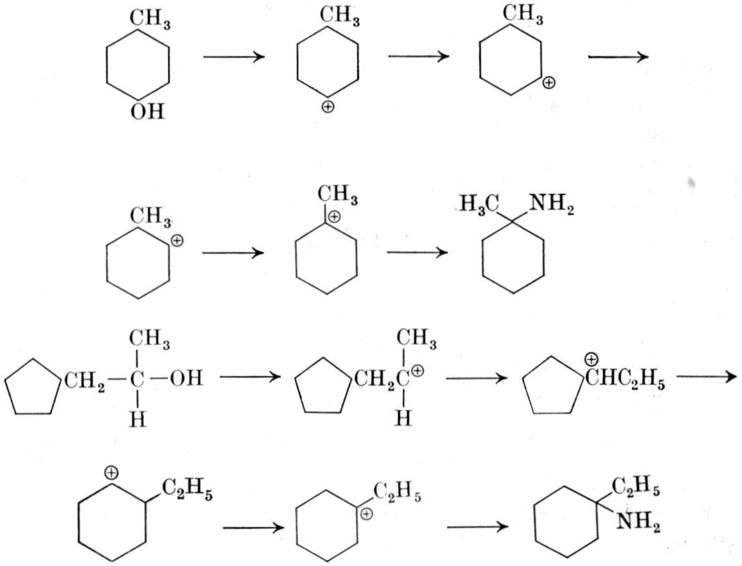

Christol and Jacquier also reported a number of alcohols which did not undergo a Ritter reaction, whereas their position or structural isomers did. Thus 2,2-dimethylcyclohexanol gave 1,2-dimethylcyclohexylamine in 20% yield, but 3,3-dimethylcyclohexanol failed to react. The absence of any reaction was also observed for 2-butanol. The authors attribute the failure of reaction to the difficulty of forming the corresponding ions rather than to the lack of reactivity of these ions.[14, 16]

Extending their investigations to alicyclic and spiro alcohols, Christol and co-workers found that most Ritter reactions were accompanied by a retropinacol rearrangement.[6, 14] When spiro[4.5]-6-decanol was subjected to the conditions of the Ritter reaction, ring expansion occurred and *trans*-9-aminodecalin was obtained. The stereospecific nature of the reaction was also noted with other bicyclic and spiro alcohols. For example, a mixture of *cis*- and *trans*-2-cyclopentylcyclopentanol, the diastereoisomers of β-decalol, and $\Delta^{9,10}$-octalin all gave *trans*-9-aminodecalin when subjected to the Ritter reaction.[6, 11, 17]

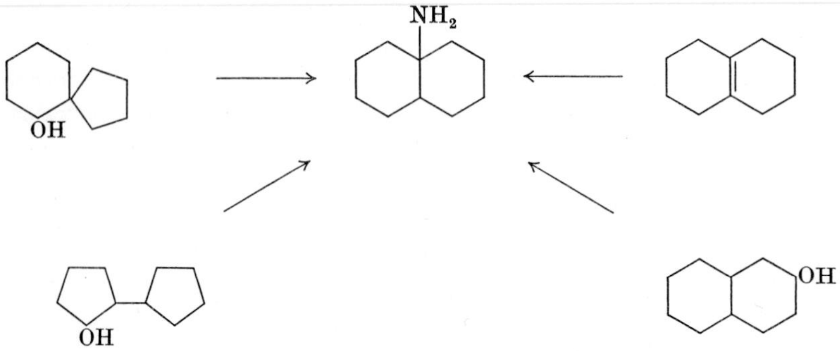

NH_2

OH

OH

OH

Similar stereospecificity was exhibited by spiro[4.4]-1-nonanol and *cis*- and *trans*-8-hydroxyhydrindane, which reacted to give only the *cis*-8-aminohydrindane. Only the thermodynamically more stable isomer of

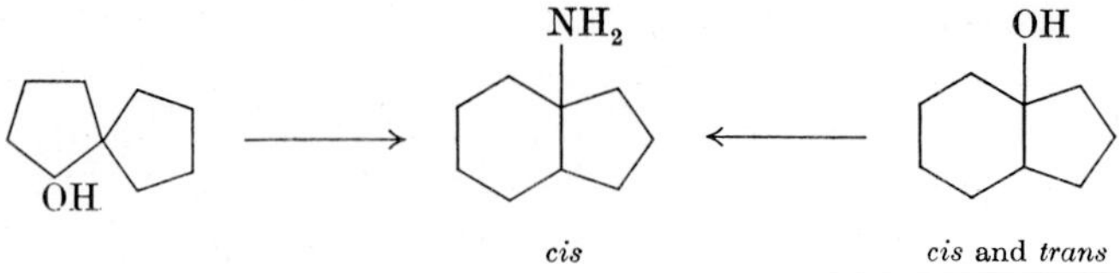

NH_2 OH

OH

cis *cis* and *trans*

1,2-dimethylcyclohexylamine, with the methyl groups *trans* in equatorial positions, was obtained from 1,2- and 2,2-dimethylcyclohexanol.[13]

Differences in behavior were also noted in the spiro and bicyclic alcohols.[6] For example, 4a-methyl-2-decahydronaphthol and bicyclo[5.4.0]-9-undecanol failed to react, but their isomers, 6-methylspiro[4.5]-6-decanol and bicyclo[5.4.0]-2-undecanol did react, to give in one case an angular amine, 1-aminobicyclo[5.4.0]undecane and in the other case a mixture of formamides **7** and **7a** that could not be hydrolyzed to the free amines.

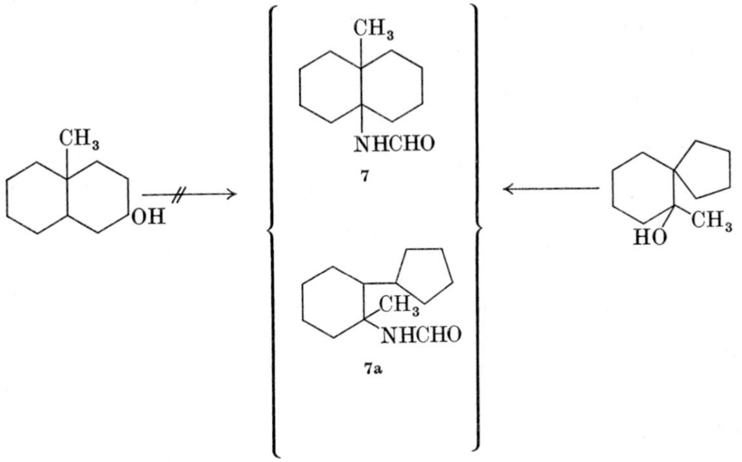

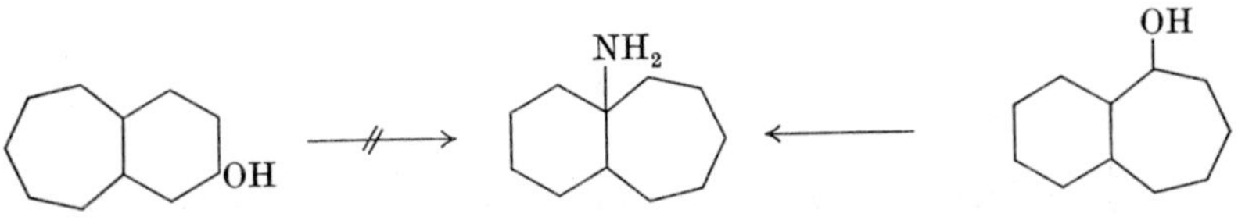

The extension of the Ritter reaction to secondary benzylic alcohols and benzylcarbinols,[8] tertiary benzylic alcohols,[9] diphenyl-carbinols and -propanols[10] enabled Christol and co-workers to establish the relative reactivities of the various derived carbonium ions and their susceptibility to prototropic rearrangements.

Further insight into the mode of these carbonium ion rearrangements was reported by Laurent,[20, 21] who employed C^{14}-labeled triphenylethanols and showed that the Ritter reaction of triphenylethanol **8** with alkyl or aryl nitriles proceeded via a rearrangement of **9** to **10** before the substitution occurred, thus ruling out the formation of the bridged phenonium ion **11**. (See p. 230.)

It is noteworthy that diphenylcarbinols such as **12**, when R is CH_3, C_2H_5, C_3H_7-*i*, CH_2Cl, or $CHCl_2$, yield only the corresponding 1,1-diphenylethylenes.[10, 40] A similar dehydration followed by dimerization was

$$(C_6H_5)_2CH\overset{14}{C}HC_6H_5(OH) \ (\mathbf{8}) \not\longrightarrow C_6H_5HC^{\oplus}\text{—}\overset{14}{C}HC_6H_5 \text{ (phenonium, } \mathbf{11}\text{)}$$

$$\mathbf{11} \not\longrightarrow (C_6H_5)_2CH\overset{14}{C}HC_6H_5(NHCOCH_3) + C_6H_5CH(NHCOCH_3)\overset{14}{C}H(C_6H_5)_2$$

$$\mathbf{8} \longrightarrow (C_6H_5)_2CH\overset{14}{C}^{\oplus}HC_6H_5 \ (\mathbf{9}) \rightleftharpoons C_6H_5C^{\oplus}H\overset{14}{C}H(C_6H_5)_2 \ (\mathbf{10})$$

described for 2-phenyl-2-propanol, indicating that the cumyl cation undergoes nucleophilic attack by α-methylstyrene more readily than by

$$(C_6H_5)_2C(CH_2R)OH \ (\mathbf{12}) \rightarrow (C_6H_5)_2C{=}CHR$$

the nitrile.[23] However, diphenylcarbinol itself and difluoromethyldiphenylcarbinol react with various nitriles to give the amides in yields ranging from 68% to 100%.[10, 40, 58] Magat was able to prepare linear

$$(C_6H_5)_2C(CHF_2)OH \xrightarrow{RCN} (C_6H_5)_2C(CHF_2)NHCOR$$

polyamides from disecondary alcohols and dinitriles. For example, adiponitrile and 1,10-dimethyl-1,10-decanediol yielded a polymer capable of being melt-spun to form fibers.[53]

The reaction has also been extended to acids and esters containing the hydroxy group. Both α- and β-acylamine derivatives have been prepared by this route.[61–63] The reaction of diphenylhydroxyacetic acid

[61] K. Hohenlohe-Oehringen, *Monatsh.*, **93**, 639 (1962).

[62] G. Jansen and W. Taub, *Acta. Chem. Scand.*, **19**, 1772 (1965).

[63] L. W. Hartzel and J. J. Ritter, *J. Am. Chem. Soc.*, **71**, 4130 (1949).

(benzilic acid) with chloroacetonitrile gave the chloroacetamidodiphenylacetic acid in 82% yield.[64]

$$(C_6H_5)_2C(OH)CO_2H + ClCH_2CN \rightarrow (C_6H_5)_2C(NHCOCH_2Cl)CO_2H$$

In the only study using heterocyclic alcohols Zagorevskii and Lopatina obtained 4-acetamido-4-alkyl-1-methylpiperidines from the corresponding 4-alkyl-1-methyl-4-piperidinols.[65]

$$\text{4-R-4-OH-1-}CH_3\text{-piperidine} \xrightarrow[H_2SO_4]{CH_3CN} \text{4-R-4-}NHCOCH_3\text{-1-}CH_3\text{-piperidine}$$

The Ritter reaction with halohydrins proceeds in a manner similar to that described for haloalkenes to give halogen-substituted amides.[44]

Alkyl Chlorides (Table IX)

The use of a tertiary alkyl halide in place of the corresponding alcohol was reported by Magat.[66] A carbonium ion initially is generated by the abstraction of the halogen atom, and then the reaction proceeds as with alcohols. Both formic and sulfuric acids were employed as reaction media,

$$(CH_3)_3CCl + H^{\oplus} \rightarrow (CH_3)_3C^{\oplus} + HCl$$

$$(CH_3)_3C^{\oplus} + C_6H_5CN \rightarrow (CH_3)_3CN{=}\overset{\oplus}{C}C_6H_5$$

and in both reactions the yields were lower than when the corresponding alkenes or alcohols were used.

Aldehydes and Ketones (Tables X, XI, XIII)

The reaction of aldehydes with nitriles to give methylene-*bis*-amides and with dinitriles to yield polyamides was studied by Mowry and Ringwald[67] and by Magat and co-workers.[68, 69] These investigators suggested

[64] K. Hohenlohe-Oehringen and H. Bretschneider, *Monatsh.*, **93,** 645 (1962).

[65] V. A. Zagorevskii and K. I. Lopatina, *J. Gen. Chem. USSR*, **33,** 2461 (1963) [*C.A.*, **60,** 495f (1964)].

[66] E. E. Magat, U.S. pat. 2,628,217 [*C.A.*, **47,** 5129g (1953)].

[67] D. T. Mowry and E. L. Ringwald, *J. Am. Chem. Soc.*, **72,** 4439 (1950).

[68] E. E. Magat, B. F. Faris, J. E. Reith, and L. F. Salisbury, *J. Am. Chem. Soc.*, **73,** 1028 (1951).

[69] E. E. Magat, L. B. Chandler, B. F. Faris, J. E. Reith, and L. F. Salisbury, *J. Am. Chem. Soc.*, **73,** 1031 (1951).

a mechanism in which the nitrile adds to the protonated aldehyde (equation 6) and formation of the iminosulfate follows (equation 7). Subsequent protonation produces a new carbonium ion which undergoes

$$\mathrm{RCHO} + \mathrm{H}^{\oplus} \rightleftharpoons \mathrm{R\overset{\oplus}{C}HOH} \qquad \text{(Eq. 5)}$$

$$\mathrm{R'CN} + \mathrm{R\overset{\oplus}{C}HOH} \rightleftharpoons \mathrm{R'\overset{\oplus}{C}{=}N\overset{R}{\overset{|}{C}}HOH} \qquad \text{(Eq. 6)}$$

$$\mathrm{R'\overset{\oplus}{C}{=}N\overset{R}{\overset{|}{C}}HOH} + \overset{\ominus}{\mathrm{O}}\mathrm{SO_3H} \rightleftharpoons \mathrm{R'\underset{\underset{OSO_3H}{|}}{C}{=}N\overset{R}{\overset{|}{C}}HOH} \qquad \text{(Eq. 7)}$$

reaction analogous to equations 6 and 7. Hydrolysis (equation 8) gives the methylene-*bis*-amides.

$$\mathrm{R'\underset{\underset{OSO_3H}{|}}{C}{=}N\overset{R}{\overset{|}{C}}HN{=}\underset{\underset{OSO_3H}{|}}{C}R'} + 2\mathrm{H_2O} \rightleftharpoons \mathrm{R'CONH\overset{R}{\overset{|}{C}}HNHCOR'} \qquad \text{(Eq. 8)}$$

Yields in excess of 90% are often obtained with formaldehyde, but other aldehydes such as acetaldehyde, butyraldehyde, and chloral can be used.[68] If a dinitrile is treated with formaldehyde, a polymer is formed which corresponds to the polyamide from monomethylenediamine and the parent acid of the dinitrile.[67, 69]

$$n\mathrm{NCRCN} + n\mathrm{CH_2O} \xrightarrow[\mathrm{H_2O}]{\mathrm{H_2SO_4}} \left[\mathrm{-CH_2NH\underset{\underset{O}{\|}}{C}R\underset{\underset{O}{\|}}{C}NH-}\right]_n$$

The reaction of nitriles with ketones provides another route to the difficultly accessible β-acylamino ketones (see α,β-unsaturated ketones, p. 234). In a detailed study of this reaction, Khorlin and co-workers showed that positive results were obtained with ketones that readily undergo acid-catalyzed condensations to ketols or α,β-unsaturated ketones.[70] They suggested that the reaction could proceed by a route similar to that described above for aldehydes or, more reasonably, that the ketol or the α,β-unsaturated ketone or both reacted with the nitrile.

$$2\mathrm{RCH_2COR'} \rightarrow \begin{cases} \mathrm{RCH_2C(R'){=}C(R)COR'} \\ \mathrm{RCH_2\underset{\underset{OH}{|}}{C}(R'){-}CH(R)COR'} \end{cases} \xrightarrow{\mathrm{R''CN}} \mathrm{RCH_2\underset{\underset{NHCOR''}{|}}{\overset{R'}{\overset{|}{C}}}{-}\overset{R}{\overset{|}{C}}HCOR'}$$

[70] A. Ya. Khorlin, O. S. Chizhov, and N. K. Kochetkov, *Zh. Obshch. Khim.*, **29**, 3411 (1959) [*C.A.*, **54**, 16418h (1960)].

The best results were given by methyl ketones (acetone, 62%; acetophenone, 74%). With an increase in the size of the alkyl group, the yields decrease (methyl ethyl ketone, 16%), and negative results are obtained with propiophenone and butyrone. Among cyclic ketones, only cyclohexanone gave the desired product.

In an unusual variation of this reaction Khorlin carried out a cross reaction in which the compound with the active methylene unit was not a ketone. Instead an aldehyde was the carbonyl component. Benzaldehyde and malonic ester condensed with acetonitrile to give the ethyl ester of 1-carbethoxy-2-phenyl-2-acetamidopropionic acid in 55% yield.[70]

$$C_6H_5CHO + CH_2(CO_2C_2H_5)_2 + CH_3CN \rightarrow C_6H_5\underset{\displaystyle NHCOCH_3}{\underset{|}{C}}HCH(CO_2C_2H_5)_2$$

Further reactions of ketones involving intramolecular Ritter reactions are described under Heterocyclic Syntheses (p. 236).

Ethers (Table XII)

The use of secondary and tertiary alkyl ethers as carbonium ion sources has received very limited attention. The only report to date describes the preparation of polyadipamides from the dimethyl ether of α,α'-tetramethyldecamethylene glycol and adiponitrile.[71]

$$n CH_3{-}O{-}\underset{CH_3}{\overset{CH_3}{C}}{-}(CH_2)_8{-}\underset{CH_3}{\overset{CH_3}{C}}{-}O{-}CH_3 + n NC(CH_2)_4CN \rightarrow$$

$$\left[{-}\underset{CH_3}{\overset{CH_3}{C}}{-}(CH_2)_8{-}\underset{CH_3}{\overset{CH_3}{C}}{-}NH\overset{O}{\overset{\|}{C}}{-}(CH_2)_4{-}\overset{O}{\overset{\|}{C}}{-}NH{-}\right]_n$$

α,β-Unsaturated Carbonyl Compounds (Table XIII)

The scope of the reaction was further broadened when Ritter found that α,β-unsaturated acids and esters having two alkyl groups or one aryl group on the β-carbon atom were very reactive. The reaction of benzonitrile

$$(CH_3)_2C{=}CHCO_2H + C_6H_5CN \rightarrow (CH_3)_2\underset{\displaystyle NHCOC_6H_5}{\underset{|}{C}}CH_2CO_2H$$

[71] E. E. Magat, U.S. pat. 2,518,156 [*C.A.*, **45**, 661a (1951)].

with these compounds enabled him to prepare a number of N-benzoylamino acids and esters.[63] β,β-Disubstituted-β-acylaminopropionamides were obtained from the corresponding α,β-unsaturated amides.[62]

The addition of benzonitrile to mesityl oxide led to the expected 4-methyl-4-benzamido-2-pentanone in 78% yield.[70]

$$(CH_3)_2C{=}CHCOCH_3 + C_6H_5CN \rightarrow (CH_3)_2\underset{\displaystyle NHCOC_6H_5}{\underset{|}{C}}CH_2COCH_3$$

However, in an analogous reaction, chalcone proved to be far less satisfactory as the olefin component and the corresponding benzamido ketone was obtained in very low yield.[72]

N-Methylolamides (Table XIV)

The reaction of nitriles with N-methylolphthalimide and N-methylolbenzamide, first reported by Buc,[73] was extended by Mowry[74] to a methylolsulfimide by employing N-methylolsaccharin. From N-methylolbenzamide and acrylonitrile, N-benzamidomethyl acrylamide was prepared in 83% yield.[74]

$$C_6H_5CONHCH_2OH + CH_2{=}CHCN \rightarrow C_6H_5CONHCH_2NHCOCH{=}CH_2$$

In a later study linear alkyl di-(N-methylolamides) were used to prepare methyl-*bis*-amides and polyamides.[75]

$$HOCH_2NHCORCONHCH_2OH + NC{-}R'{-}CN \rightarrow$$

$$HOCH_2NH{-}\left[{-}CORCONHCH_2NHCOR'CONHCH_2NH{-}\right]_n{-}CONHCH_2NHCOR'CN$$

These polymers are similar in character to polyamides prepared from dinitriles and formaldehyde,[69] and the reaction proceeds by the same mechanism.[68] (See Aldehydes and Ketones, p. 231.)

Carboxylic Acids and Esters (Tables II-C, II-I, VI-B, XIII–XVI)

The facile formation of carbonium ions from *t*-carboxylic acids and esters of *t*-alcohols in concentrated sulfuric acid suggested their utilization as carbonium ion sources in the Ritter reaction. And, indeed, treatment of trimethylacetic acid with hydrogen cyanide in 100% sulfuric acid gave *t*-butylamine in 68% yield.[25] In similar studies triphenylmethyl formate[76]

[72] P. J. Scheuer, H. C. Botelho, and C. Pauling, *J. Org. Chem.*, **22**, 674 (1957).

[73] S. R. Buc, *J. Am. Chem. Soc.*, **69**, 254 (1947).

[74] D. T. Mowry, U.S. pat. 2,529,455 [*C.A.*, **45**, 2980f (1951)].

[75] E. E. Magat and L. F. Salisbury, *J. Am. Chem. Soc.*, **73**, 1035 (1951).

[76] R. G. R. Bacon and J. Köchling, *J. Chem. Soc.*, **5609** (1964).

and *t*-butyl acetate[28] gave the corresponding N-trialkyl- or triaryl-acetamides. Ramp showed that *bis*(acetoxymethyl) alkylated benzenes could be used as carbonium ion sources.[77] *Bis*(acetoxymethyl)durene condenses

$$CH_3CO_2CH_2-C_6(CH_3)_4-CH_2OCOCH_3 + NC-R-CN \longrightarrow \left[-CH_2-C_6(CH_3)_4-CH_2NHC(=O)-R-C(=O)NH-\right]_n + 2CH_3CO_2H$$

with dinitriles to yield polyamides. Similar polyamides have been prepared from dinitriles and esters of disecondary alcohols.[53]

Oximes

While studying the course of the migrating group during the Beckmann rearrangement Hill[78] and Conley[49, 79, 80] discovered that α-trisubstituted and α,α′-tetrasubstituted oximes undergo an initial fragmentation to give an intermediate nitrile and a carbonium ion. The two fragments recombine in a Ritter reaction to form an amide. In the case of a cyclic

$$(CH_3)_3C-C(=NOH)-CH_3 \xrightarrow[\text{acid}]{\text{Polyphosphoric}} \left[\begin{matrix} CH_3CN \\ + \\ (CH_3)_3C^{\oplus} \end{matrix}\right] \rightarrow (CH_3)_3CNHCOCH_3$$

ketoxime, fragmentation yielded an unsaturated nitrile which on recombination produced a lactam.[49, 80–82] Treatment of 1,1-dimethyl-2-tetralone oxime with hot polyphosphoric acid resulted in the formation of 2-aza-1,1-dimethyl-3-benzosuberone in 24% yield.[50]

$$\text{1,1-dimethyl-2-tetralone oxime} \longrightarrow \left[o\text{-}(CH_2CH_2CN)C_6H_4C^{\oplus}(CH_3)_2\right] \longrightarrow \text{2-aza-1,1-dimethyl-3-benzosuberone}$$

[77] F. L. Ramp, *J. Polymer Sci.*, **3,** 1877 (1965).
[78] R. K. Hill and O. T. Chortyk, *J. Am. Chem. Soc.*, **84,** 1064 (1962).
[79] R. T. Conley, *J. Org. Chem.*, **28,** 278 (1963).
[80] R. T. Conley and W. N. Knopka, *Abstr.*, ACS Meeting, 57c (Jan. 1964).
[81] R. K. Hill and R. T. Conley, *Chem. Ind.* (*London*), 1314 (1956).
[82] K. Morita and Z. Suzuki, *J. Org. Chem.*, **31,** 233 (1966).

Heterocyclic Syntheses (Table XVII)

The discovery of Ritter and Tillmanns that the interaction of a suitably substituted diol with a nitrile resulted in the formation of a dihydro-1,3-oxazine, instead of the expected diamide, stimulated research in the synthesis of a variety of heterocyclic systems.[83]

In addition to dihydro-1,3-oxazines[83–88] the Ritter procedure has been successfully employed in the preparation of oxazolines,[89] pyrrolines,[90, 91] dihydropyridines,[86, 90, 92] Δ^2-thiazolines,[59, 90, 93, 94] thiazines,[59, 87, 95, 96] isoquinolines,[34] 2-quinolones,[97] 2-pyridones,[97–99] triazines,[100] azabicycloalkanes,[101–103] *bis*(heterocyclyl)alkanes,[104] and an oxazolone.[105]

Dihydro-1,3-oxazines. The synthesis of oxazines from a diol containing both a tertiary and a secondary hydroxyl group is illustrated in the reaction shown on p. 237; treatment of 2-methyl-2,4-pentanediol with acetonitrile gave the 1,3-oxazine **13** in 44% yield.[83]

Aryl,[83] aralkyl,[83] and unsaturated nitriles[88] and dinitriles[85] have also been successfully employed. Unsaturated tertiary alcohols permit the synthesis of spirooxazines.[84, 86] This unusual ring closure reaction was observed when either of the isomeric tertiary alcohols **13a** and **13b** was allowed to react with simple nitriles. A possible mechanism for the formation of 2,4,4-trimethyl-6,6-tetramethylene-5,6-dihydro-1,3-oxazine

[83] E.-J. Tillmanns and J. J. Ritter, *J. Org. Chem.*, **22**, 839 (1957).
[84] L. M. Trefonas, J. Schneller, and A. I. Meyers, *Tetrahedron Letters*, **22**, 785 (1961).
[85] A. I. Meyers, *J. Org. Chem.*, **25**, 145 (1960).
[86] A. I. Meyers, J. Schneller, and N. K. Ralhan, *J. Org. Chem.*, **28**, 2944 (1963).
[87] A. I. Meyers, *J. Org. Chem.*, **26**, 218 (1961).
[88] J. W. Lynn, *J. Org. Chem.*, **24**, 711 (1959).
[89] S. Julia and C. Papantoniou, *Compt. rend.*, **260**, 1440 (1965).
[90] A. I. Meyers and J. J. Ritter, *J. Org. Chem.*, **23**, 1918 (1958).
[91] A. I. Meyers *J. Org. Chem.*, **24**, 1233 (1959).
[92] A. I. Meyers, B. J. Betrus, N. K. Ralhan, and K. B. Rao, *J. Heterocyclic Chem.*, **1**, 13 (1964).
[93] J. R. Lowell, Jr., and G. K. Helmkamp., *J. Am. Chem. Soc.*, **88**, 768 (1966).
[94] G. K. Helmkamp, D. J. Pettit, J. R. Lowell, Jr., W. R. Mabey, and R. G. Wolcott, *J. Am. Chem. Soc.*, **88**, 1030 (1966).
[95] D. S. Tarbell, D. A. Buckley, P. P. Brownlee, R. Thomas, and J. S. Todd, *J. Org. Chem.*, **29**, 3314 (1964).
[96] A. I. Meyers and J. M. Greene, *J. Org. Chem.*, **31**, 556 (1966).
[97] A. I. Meyers and G. García-Muñoz, *J. Org. Chem.*, **29**, 1435 (1964).
[98] A. Vigier and J. Dreux, *Bull. Soc. Chim. France*, 2292 (1963).
[99] O. Yu. Magidson, *Zh. Obshch. Khim.*, **33**, 2173 (1963) [*C.A.*, **59**, 13942f (1963)].
[100] D. L. Trepanier and co-workers, *J. Med. Chem.*, **9**, 881 (1966).
[101] A. I. Meyers and W. Y. Libano, *J. Org. Chem.*, **26**, 4399 (1961).
[102] A. I. Meyers and N. K. Ralhan, *J. Org. Chem.*, **28**, 2950 (1963).
[103] A. I. Meyers and W. Y. Libano, *J. Org. Chem.*, **26**, 1682 (1961).
[104] A. I. Meyers, J. Org. Chem., **25**, 2231 (1960).
[105] C. W. Bird, *J. Org. Chem.*, **27**, 4091 (1962).

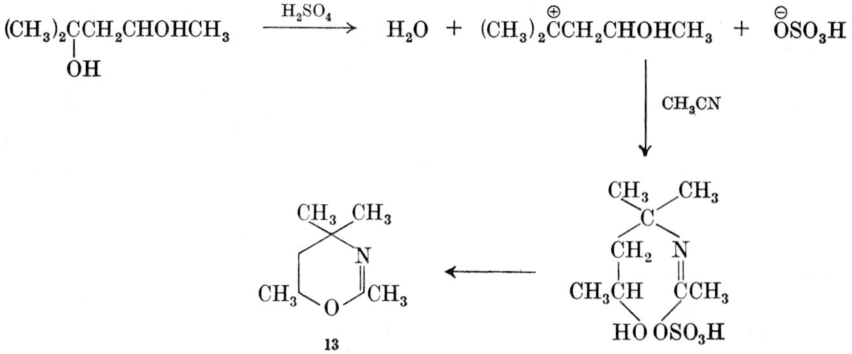

$$(CH_3)_2C(OH)CH_2CHOHCH_3 \xrightarrow{H_2SO_4} H_2O + (CH_3)_2\overset{\oplus}{C}CH_2CHOHCH_3 + \overset{\ominus}{O}SO_3H$$

$$\xrightarrow{CH_3CN} \text{cyclic intermediate (CH}_3\text{)}_2\text{C, CH}_2\text{, N, CH}_3\text{CH, CCH}_3\text{, HO OSO}_3\text{H} \longrightarrow \mathbf{13}$$

13

(**14**) follows. The yields from the 3-cyclopentenyl derivative, **13b**, were approximately 20–30% higher than those from the 1-cyclopentenyl compound, **13a**.[86]

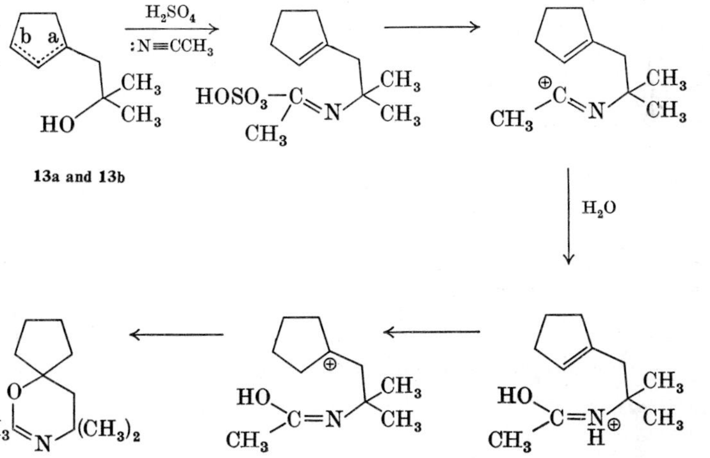

13a and 13b

$\xrightarrow[:N{\equiv}CCH_3]{H_2SO_4}$ HOSO$_3$–C(CH$_3$)=N–C(CH$_3$)$_2$– $\longrightarrow$ CH$_3$–$\overset{\oplus}{C}$=N–C(CH$_3$)$_2$– $\xrightarrow{H_2O}$ HO(CH$_3$)C=$\overset{\oplus}{N}$H–C(CH$_3$)$_2$– $\longrightarrow$ HO(CH$_3$)C=N–C(CH$_3$)$_2$– (cyclopentyl cation) $\longrightarrow$ **14**

14

Oxazolines. Julia and Papantoniou prepared a series of β-chloroamides from methallyl chloride and various nitriles.[89] The amides could be cyclized easily by either potassium ethoxide or silver fluoroborate to give oxazolines.

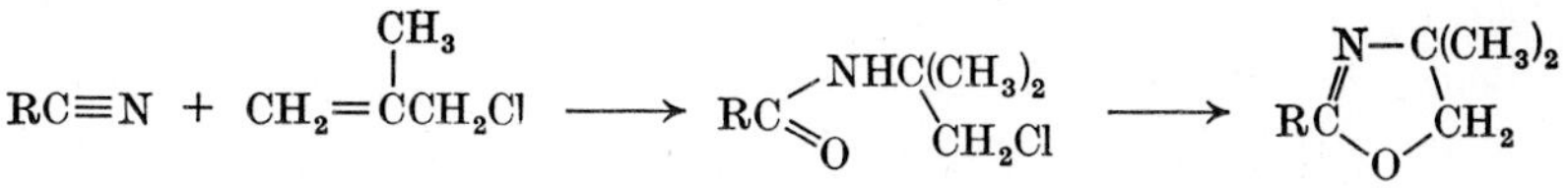

$$RC{\equiv}N + CH_2{=}C(CH_3)CH_2Cl \longrightarrow RC(=O)NHC(CH_3)_2CH_2Cl \longrightarrow \text{RC=N–C(CH}_3)_2\text{–CH}_2\text{–O (ring)}$$

Dihydro-1,3-thiazines and Δ^2-Thiazolines. Substitution of mercapto alcohols[59, 87, 95, 96] or methallyl mercaptan[90] for the diols used in the oxazine synthesis provided a new route to dihydro-1,3-thiazines and Δ^2-thiazolines, respectively. The addition of 2-methyl-2-hydroxypropanethiol or methallyl mercaptan to a cold solution of acetonitrile in concentrated sulfuric acid leads to the formation of 2,4,4-trimethyl-Δ^2-thiazoline.

$$(CH_3)_2C(OH)CH_2SH \xrightarrow[(50\%)]{CH_3CN,\ H_2SO_4} \text{2,4,4-trimethyl-}\Delta^2\text{-thiazoline} \xleftarrow[(23\%)]{CH_3CN,\ H_2SO_4} CH_2{=}C(CH_3)CH_2SH$$

The higher yield obtained by using the mercapto alcohol was attributed to the fact that methallyl mercaptan polymerizes more readily than the alcohol in concentrated sulfuric acid.[59]

A stereospecific synthesis of Δ^2-thiazolines from episulfides has been reported.[93,94] The proposed mechanism for this reaction involves protonation of the episulfide, ring opening by nucleophilic attack by the nitrile, and ring closure to form the thiazoline. The preparation of (4R:5S)-(−)*cis*-2,4,5-trimethyl-Δ^2-thiazoline **(16)** from (SS)-(−)-*trans*-2-butene episulfide **(15)** demonstrated that the reaction proceeds without racemization.[93]

$$\underset{\mathbf{15}}{\text{trans-2-butene episulfide}} \xrightarrow[H^{\oplus}]{CH_3CN} CH_3CH(\overset{\oplus}{N}{\equiv}CCH_3)CH(SH)CH_3 \longrightarrow \underset{\mathbf{16}}{\text{cis-2,4,5-trimethyl-}\Delta^2\text{-thiazoline}}$$

When butanethiol is employed, a dihydro-1,3-thiazine is obtained.[59, 87] *p*-Aminobenzonitrile reacted with 3-methyl-3-hydroxy-*n*-butanethiol to give 2-(*p*-aminophenyl)-4,4-dimethyl-5,6-dihydro-1,3-thiazine. With

$$(CH_3)_2C(OH)CH_2CH_2SH + NCC_6H_4NH_2\text{-}p \xrightarrow{H_2SO_4} \text{2-}(C_6H_4NH_2\text{-}p)\text{-4,4-dimethyl-5,6-dihydro-1,3-thiazine}\ (53\%)$$

boron trifluoride etherate as catalyst for the formation of the thiazine, yields of 75% were reported.[95]

Benzylmercaptobutanols have also been used to give the dihydro-1,3-thiazine with concomitant debenzylation of the sulfide. The reaction probably involves attack by the nitrilium ion on the sulfide with expulsion of the benzyl cation. The existence of this cation has been verified by the isolation of N-benzylbenzamide.[95]

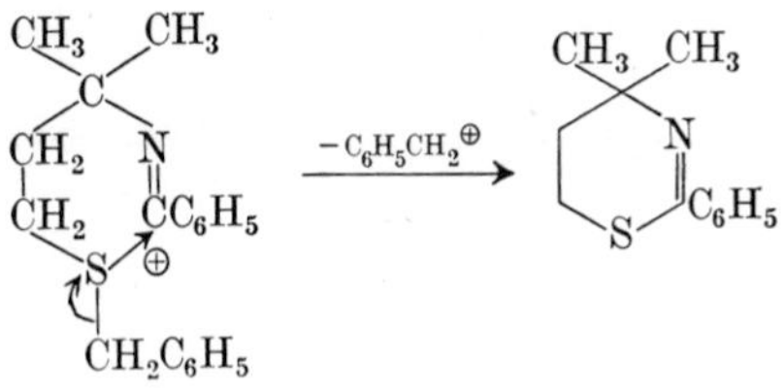

Bis(heterocyclyl)alkanes. The reactions leading to the heterocycles described in the preceding paragraphs can be made to furnish *bis*(heterocyclyl)alkanes by employing dinitriles. The ring closure may be directed to the formation of a mono- or di-cyclic product by utilizing the proper quantities of dinitrile and alcohol.

The addition of 2,5-dimethyl-2,5-hexanediol, 2-methyl-2,4-pentanediol, or 4-mercapto-2-methyl-2-butanol to a dinitrile in sulfuric acid yielded N-heterocyclic bases of the following type.[104]

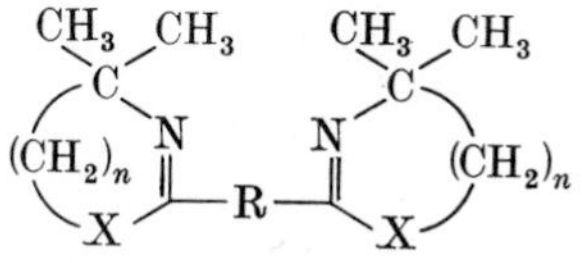

X = O, S;
R = alkylene

1-Pyrrolines and 5,6-Dihydropyridines. A series of 1-pyrrolines was prepared by heating 1,4-diols with various nitriles.[90, 91] The yields

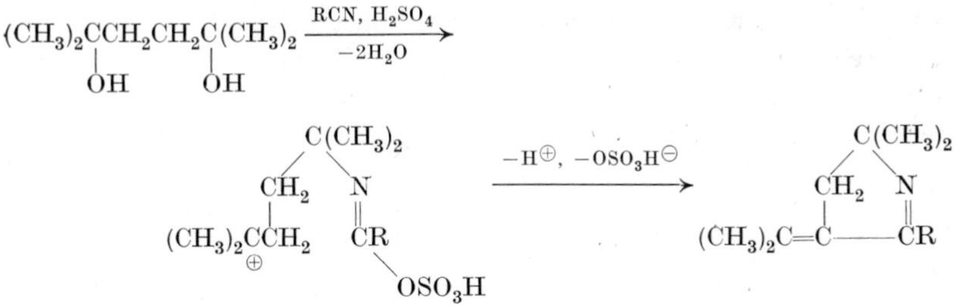

are 60–80% with the diols. Under the same conditions, however, 2,5-dimethyl-2,5-hexadiene gave the corresponding pyrroline in a much smaller yield (28%) as a consequence of a competing polymerization.[90]

Replacement of the 1,4-diols by 1,3-diol 2,4-dimethylpentane-2,4-diol permits the preparation of a variety of 5,6-dihydropyridines. The yields are considerably lower (∼20%) because of cleavage of the diol to acetone and isobutene. From the 1,3-diol and acetonitrile the N-*t*-butylacetamide was isolated in 50–55% yield.[90]

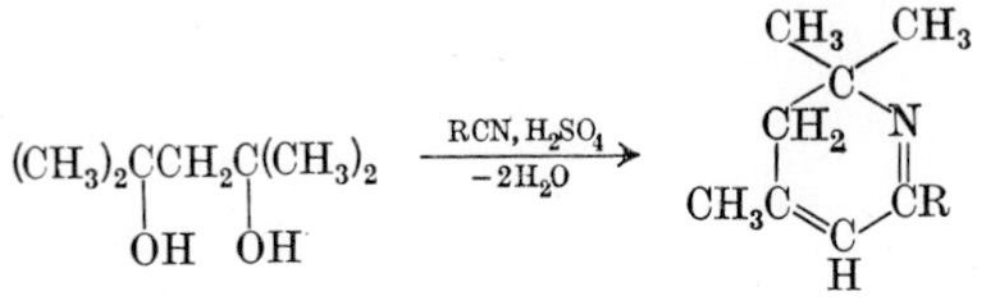

$(CH_3)_2C(OH)CH_2C(OH)(CH_3)_2$ $\xrightarrow[-2H_2O]{RCN,\ H_2SO_4}$

CH₃ CH₃ C CH₂ N CH₃C CR C H

In the course of their studies on oxazine preparation Meyers and co-workers discovered that by controlling the water activity the reaction could be modified to give dihydropyridines exclusively.[86] Treatment of α-(3-cyclopentenyl)-*t*-butyl alcohol **(13b)** with acetonitrile in sulfuric acid concentrations above 93% led to an azocarbonium ion that lost a proton to give 3,4-cyclopenteno-5,6-dihydropyridine.

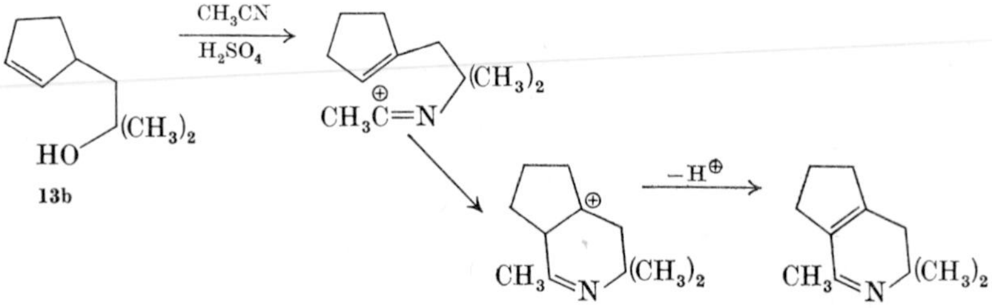

HO $(CH_3)_2$ **13b** $\xrightarrow[H_2SO_4]{CH_3CN}$ $CH_3\overset{\oplus}{C}{=}N$ $(CH_3)_2$ → $\oplus$ CH_3 N $(CH_3)_2$ $\xrightarrow{-H^{\oplus}}$ CH_3 N $(CH_3)_2$

A similar reaction with α-(1-cyclohexenyl)-*t*-butyl alcohol **(15a)** gave the spirodihydropyridine **18** (see Chart I) along with an isoquinoline **17**.[92]

Isoquinolines. 3,4-Dihydroisoquinolines have been obtained by the reaction of methyleugenol with alkoxyarylnitriles[34] and of isosafrol or methylisoeugenol with a variety of nitriles.[54] For example, veratronitrile and methyleugenol give 1-(3′,4′-dimethoxyphenyl)-3-methyl-6,7-dimethoxy-3,4-dihydroisoquinoline in 53% yield.[34]

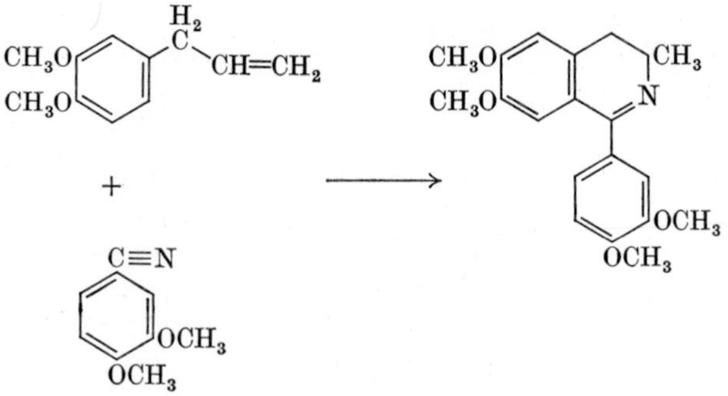

CH_3O CH_3O $\overset{H_2}{C}$ $CH{=}CH_2$ + $C{\equiv}N$ OCH_3 OCH_3 ⟶ CH_3O CH_3O CH_3 N OCH_3 OCH_3

CHART I

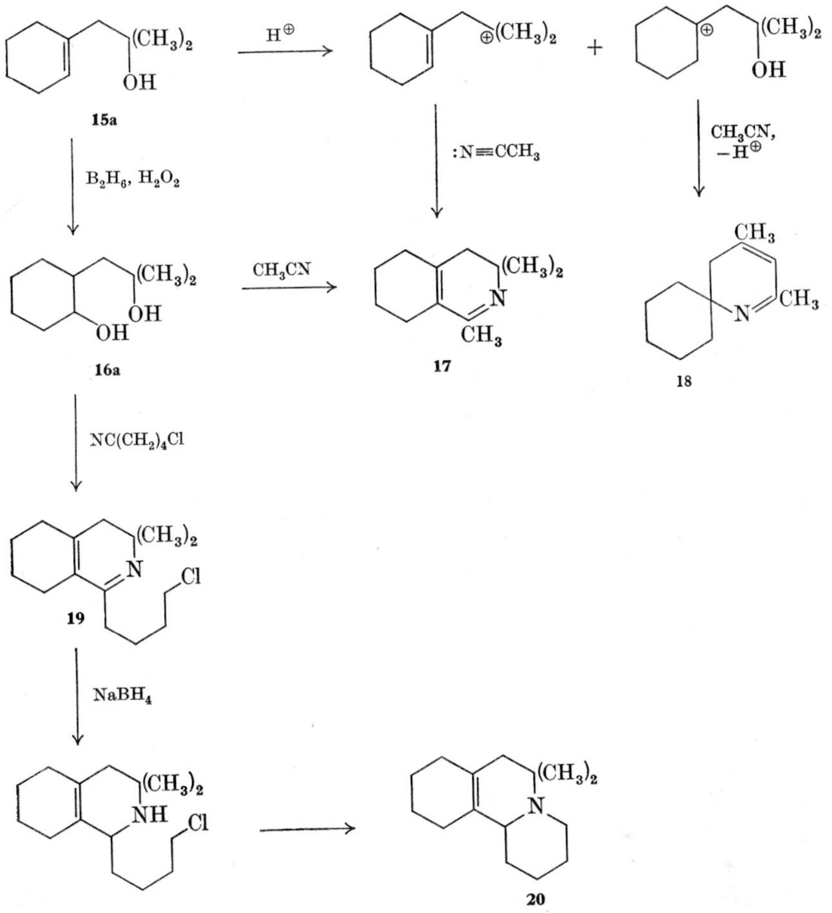

In an attempt to circumvent the formation of the spiro base **18** (Chart I) Meyers utilized the glycol **16a**; however, the desired isoquinoline **17** was obtained in only 30% yield.

In another modification the same glycol, **16a**, was treated with δ-chlorovaleronitrile to give the hexahydroisoquinoline **19**. Reduction followed by cyclization afforded the quinolizine **20**.

2-Pyridones and 2-Quinolones. The use of keto nitriles in the Ritter reaction provides a direct route to 3,4-dihydro-2-pyridones[98, 99] and 2-quinolones.[97] (The synthesis of 5,6-dihydro-2-pyridones is discussed on p. 246.) An example of the dihydropyridone synthesis is given in the accompanying equation.[99] A similar approach was applied to 2-(2-cyano-

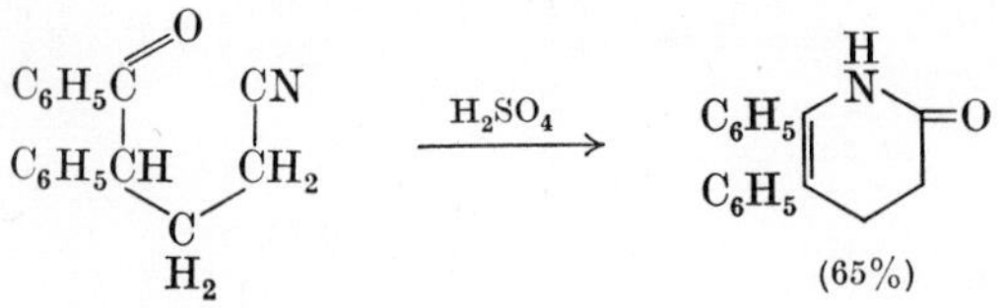

$$C_6H_5COCH(C_6H_5)CH_2CH_2CN \xrightarrow{H_2SO_4} \text{5,6-diphenyl-3,4-dihydro-2-pyridone} \; (65\%)$$

alkyl)cyclohexanones, which formed the corresponding cycloalkano[e]-2-pyridones by means of a consecutive cyclization-aromatization. When 2-(2-cyanoethyl)cyclohexanone was allowed to stand in sulfuric acid for 3 hours at room temperature, 5,6,7,8-tetrahydro-2-quinolone was formed.[97]

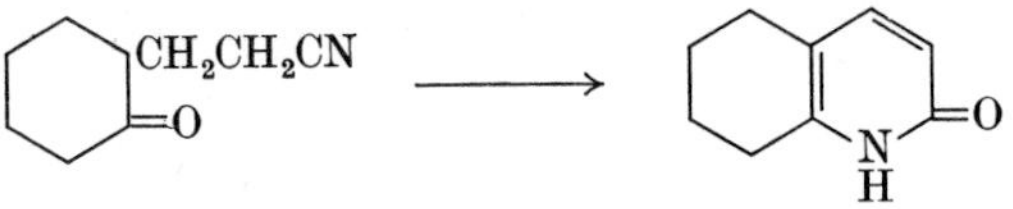

$$\text{2-}(CH_2CH_2CN)\text{cyclohexanone} \longrightarrow \text{5,6,7,8-tetrahydro-2-quinolone}$$

Aromatization also occurred when the corresponding cyclopentanone and cycloheptanone nitriles were cyclized in sulfuric acid.[97]

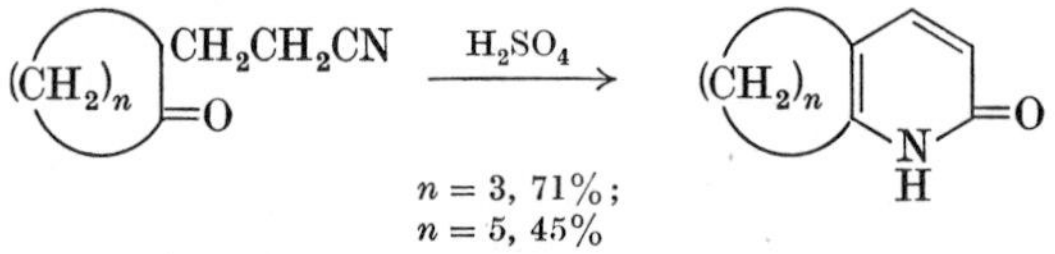

$$(CH_2)_n\text{-cycloalkanone-}CH_2CH_2CN \xrightarrow{H_2SO_4} (CH_2)_n\text{-cycloalkano-2-pyridone}$$

$n = 3$, 71%; $n = 5$, 45%

1-Azabicycloalkanes. A new approach to polycyclic bases utilizing the Ritter reaction was investigated by Meyers and co-workers.[101–103, 106] The method involves the treatment of a ditertiary glycol with an ω-chloronitrile to give an ω-chloroalkyl-1-pyrroline. Subsequent reduction and cyclization via intramolecular alkylation yields the 1-azabicycloalkane.

The formation of 1-azabicyclo[3.2.0]heptane in 30% yield is illustrative of this unique reaction which takes place in a single step without isolation of the intermediate 2-chloroethyl-1-pyrroline. By substituting 4-chlorobutyronitrile or 5-chlorovaleronitrile the analogous 1-azabicyclo[3.3.0]- or [4.3.0]-cycloalkanes can be obtained in equally good yields[101] (p. 243).

Cyclopenteno[d]-1-azabicycloalkanes have also been synthesized by this procedure from α-(2-hydroxycyclopentyl)-*t*-butyl alcohol and the appropriate chloronitrile.[102] The route outlined below affords the final tricyclic

[106] A. I. Meyers and H. Singh, to be published.

CH_3 CH_3 C CH_2 OH CH_2 OH C CH_3 CH_3 —(Conc. H_2SO_4; $NC(CH_2)_2Cl$)→ CH_3 CH_3 C CH_2 OH CH_2 $N{=}\overset{\oplus}{C}CH_2CH_2Cl$ C CH_3 CH_3

$-H_2O$

$(CH_3)_2$ N H ⊕ $=C(CH_3)_2$ CH_2CH_2Cl —($NaBH_4$; pH 3–4)→ $(CH_3)_2$ N H $=C(CH_3)_2$ CH_2 CH_2 Cl

pH 8–9

$(CH_3)_2$ N $=C(CH_3)_2$

compound in 46% overall yield and represents the formation of rings B, C, and D of an azasteroid nucleus.

$(CH_3)_2$ OH OH —($Cl(CH_2)_4CN$; H_2SO_4)→ $(CH_3)_2$ N $(CH_2)_4Cl$

$NaBH_4$, pH 3–4

$(CH_3)_2$ HN $Cl(CH_2)_4$ —(pH 8–9)→ $(CH_3)_2$ N B C D

1,4,5,6-Tetrahydro-*as*-triazines.[100] Hydrazino alcohols containing a tertiary aliphatic or secondary benzylic hydroxyl group condense with a variety of nitriles to give triazines. N-Amino-(—)-ephedrine reacted with benzonitrile to give *trans*-(+)-1,6-dimethyl-3,5-diphenyl-1,4,5,6-tetrahydro-*as*-triazine **(21)**. The reaction appears to proceed by attack of the nitrilium sulfate on the carbonium ion resulting from the dehydration of the hydrazino alcohol.

2,4,4-Triphenyl-5-oxazolone. The treatment of benzilic acid and benzonitrile with concentrated sulfuric acid by Japp and Findlay represents the earliest report of a reaction of the Ritter type.[107] Reinvestigation of this reaction confirmed the participation of the $(C_6H_5)_2\overset{\oplus}{C}CO_2H$ cation to give 2,4,4-triphenyl-5-oxazolone.[105]

The Nitrile Source

The general applicability of the reaction to nitriles of various structures is evident from Ritter's original investigation.[1] The reactions described in the preceding sections and those referred to in the tables further exemplify the great variety of compounds which have been utilized as a source of the C≡N group.

[107] F. R. Japp and A. Findlay, *J. Chem. Soc.*, **75**, 1027 (1899).

Although the yields of *t*-alkylamides are not markedly affected by the nitrile structure, there is a decrease in the sequence, $CH_2{=}CHCN >$ $C_6H_5CN > CH_3CN$ in the reaction with isobutene,[28] methylbutene, and isooctene.[1] Thus a phenyl group or a double bond in conjugation with the nitrile exerts a favorable but not a major influence.[28] The effect of strongly electronegative substituents in the α position was described for camphene (p. 221); however, there is insufficient experimental evidence to extend this to other alkenes that undergo Wagner rearrangements.

An unexpected result was obtained with α-morpholinyl nitriles.[108] When the β-carbon atom is tertiary, an α-hydroxyamide is produced.

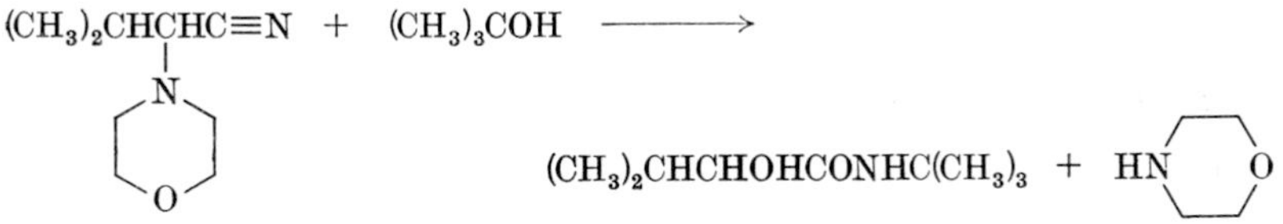

$$(CH_3)_2CHCH(NC_4H_8O)C{\equiv}N + (CH_3)_3COH \longrightarrow (CH_3)_2CHCHOHCONHC(CH_3)_3 + HNC_4H_8O$$

Similar products are obtained if the α-amino substituent contains groups larger than methyl.

Some unusual sources of the nitrile group have been reported. In an attempt to prepare alkyl isonitriles, Heldt obtained the corresponding N-alkylformamides from hexacyanoferric(II) acid and *t*-butyl alcohol, *t*-amyl alcohol, or 1-butene. It is not certain that this acid furnished the —C≡N group, since gaseous hydrogen cyanide was present during the reaction.[109]

The acid stability of 6-deoxytetracyclines has made possible the conversion of the 2-cyano compound to the corresponding carboxamide by means of the Ritter reaction. For example, the reaction of isobutene with 6-demethyl-6-deoxytetracyclinonitrile gave N[2]-*t*-butyl-6-demethyl-6-deoxytetracycline.

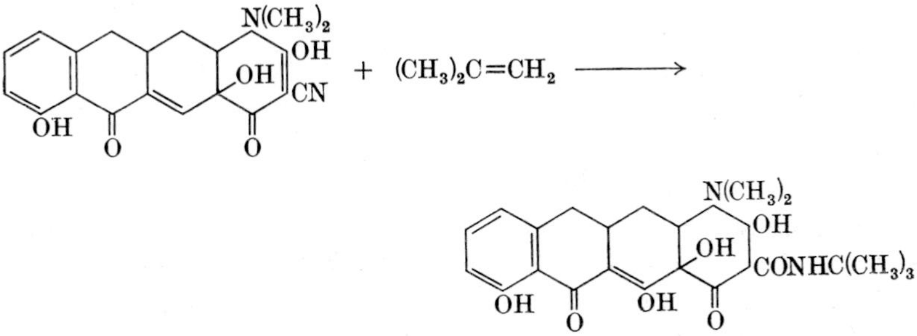

$$\text{[tetracyclinonitrile: } N(CH_3)_2, OH, OH, CN, OH, O, O] + (CH_3)_2C{=}CH_2 \longrightarrow \text{[tetracycline: } N(CH_3)_2, OH, OH, CONHC(CH_3)_3, OH, OH, O, O]$$

[108] D. Giraud-Clenet and J. Anatol, *Compt. Rend.*, **262**, 224 (1966).
[109] W. Z. Heldt, *J. Org. Chem.*, **26**, 3226 (1961).

Eugster and co-workers employed 5-methyl- and 5-phenyl-isoxazole in an unusual extension of the Ritter reaction.[110] Under acid conditions these compounds behave as cyanoacetones and give the corresponding Ritter products. The reaction of 5-methylisoxazole with *t*-butyl alcohol yields N-*t*-butylacetoacetamide.

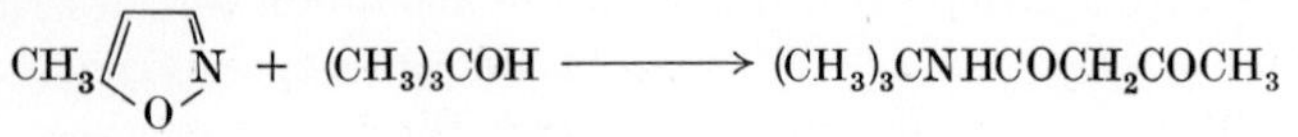

$$\text{5-methylisoxazole} + (CH_3)_3COH \longrightarrow (CH_3)_3CNHCOCH_2COCH_3$$

With α,β-unsaturated ketones, the isoxazole forms 5,6-dihydro-2-pyridones. The proposed mechanism follows.

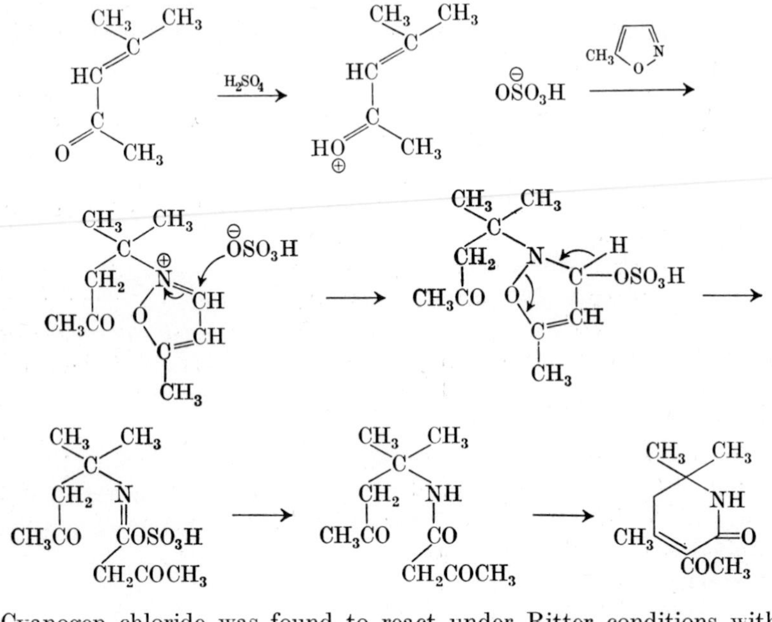

$$(CH_3)_2C{=}CHCOCH_3 \xrightarrow{H_2SO_4} (CH_3)_2C{=}CHC(CH_3){=}\overset{\oplus}{O}H \;\; \overset{\ominus}{O}SO_3H \xrightarrow{\text{5-methylisoxazole}}$$

$$\longrightarrow \longrightarrow (CH_3)_2C(CH_2COCH_3)N{=}C(OSO_3H)CH_2COCH_3 \longrightarrow (CH_3)_2C(CH_2COCH_3)NHCOCH_2COCH_3 \longrightarrow \text{6,6-dimethyl-4-methyl-3-acetyl-5,6-dihydro-2-pyridone}$$

Cyanogen chloride was found to react under Ritter conditions with olefins containing internal or substituted terminal double bonds. Although the yields are considerably lower, the amine is obtained directly without the hydrolysis required when hydrogen cyanide is used.[111]

RELATED SYNTHETIC PROCESSES

Halogens and Halogen Acids as Catalysts. A similar reaction has been conducted in media other than strong acids. These modifications of the Ritter reaction conditions have added to the general applicability of this procedure for the preparation of N-substituted amides.

[110] C. H. Eugster, L. Leichner, and E. Jenny, *Helv. Chim. Acta.*, **46,** 543 (1963).
[111] E. M. Smolin, *J. Org. Chem.*, **20,** 295 (1955).

Cairns and co-workers showed that the interaction of chlorine and an olefin in the presence of a nitrile yielded an imidoyl chloride which could be hydrolyzed readily to an N-(2-chloroalkyl)amide.[112] 2-Haloalkylamines were obtained when hydrogen cyanide was the nitrile source. A comparable reaction was described by Theilacker who added hypochlorous

$$RCH{=}CH_2 + Cl_2 \rightarrow R\overset{\oplus}{C}HCH_2Cl + Cl^{\ominus}$$

$$R\overset{\oplus}{C}HCH_2Cl \xrightarrow{R'CN} \underset{\underset{\oplus}{N{=}CR'}}{RCHCH_2Cl} \xrightarrow{Cl^{\ominus}} \underset{N{=}CClR'}{RCHCH_2Cl} \xrightarrow{H_2O} \underset{NHCOR'}{RCHCH_2Cl} + HCl$$

acid to cyclohexene and used aqueous acetonitrile as a solvent.[113] The unexpected participation of acetonitrile yielded N-(2-chlorocyclohexyl)-acetamide. When α,β-unsaturated acids are used in place of olefins,

$$\text{cyclohexene} + ClOH + N{\equiv}CCH_3 \longrightarrow \left[\text{2-chlorocyclohexyl-}N{=}\overset{OH}{C}CH_3 \right] \longrightarrow \text{2-chlorocyclohexyl-}NHCOCH_3$$

hydrolysis yields the β-amino α-hydroxy-carboxylic acids in yields up to 70%.

The use of bromine has also been reported.[114] In this case the nitrile opens the intermediate cyclic bromonium ion and incorporates itself into the molecule in a stereospecific manner. The reaction proceeds only when the halide anion liberated from the original halogen is removed from solution.

$$>C{=}C< \xrightarrow{Br_2} -C\underset{\overset{\oplus}{Br}}{\diagdown\diagup}C- + Br^{\ominus} \xrightarrow{Ag^{\oplus}} AgBr$$

$$\xrightarrow{N{\equiv}CR} -\underset{Br}{C}-\overset{N{=}\overset{\oplus}{C}R}{C}- \xrightarrow[-H^{\oplus}]{H_2O} -\underset{Br}{C}-\overset{HNCOR}{C}-$$

[112] T. L. Cairns, P. J. Graham, P. L. Barrick, and R. S. Schreiber, *J. Org. Chem.*, **17,** 751 (1952).

[113] W. Theilacker, *Angew. Chem., Intern. Ed. Engl.*, **6,** 94 (1967).

[114] A. Hassner, L. Levy, and R. Gault, *Chem. Eng. News*, **44,** 44 (April 11, 1966).

Friedel-Crafts Catalysts. *N*-Cyclopentyl- and cyclohexyl-amide were prepared in 5–55% yields by the reaction of nitriles with cyclopentene and cyclohexene under conditions of the Gattermann aldehyde reaction.[115] Bromine and iodine were also used to facilitate the aluminum chloride-catalyzed reaction.

$$\text{cyclohexene} \xrightarrow{H^{\oplus}} \text{cyclohexyl cation}^{\oplus} \xrightarrow{RCN} R\overset{\oplus}{C}{=}NC_6H_{11}$$

$$\text{cyclohexene} \xrightarrow{HCl} \text{cyclohexyl chloride} \xrightarrow{AlCl_3} \text{cyclohexyl cation}^{\oplus}$$

$$R\overset{\oplus}{C}{=}NC_6H_{11} + AlCl_4^{\ominus} \longrightarrow R\overset{Cl}{\overset{|}{C}}{=}NC_6H_{11} + AlCl_3$$

$$R\overset{Cl}{\overset{|}{C}}{=}NC_6H_{11} + H_2O \longrightarrow R\overset{OH}{\overset{|}{C}}{=}NC_6H_{11} \rightleftharpoons RCONHC_6H_{11}$$

The results indicated that the Ritter method is superior with respect to both yields and simplicity. It is noteworthy that primary and tertiary alkyl halides failed to give isolable amides.[115] The reactions of trialkyl oxonium fluoborate salts with nitriles to give nitrilium salts have been described.[116] Subsequent hydrolysis yields the corresponding N-alkyl acid amides.

$$R_3O^{\oplus}BF_4^{\ominus} + R'C{\equiv}N \rightarrow \underset{(60\text{–}90\%)}{[R'C{\equiv}\overset{\oplus}{N}R]BF_4^{\ominus}} + R_2O$$

The synthesis of heterocyclic compounds from nitrilium salts has been investigated by Lora-Tamayo and co-workers.[117–122] They were able to prepare 3,4-dihydro-isoquinolines,[118] -papaverines,[119] -quinazolines,[120] 4,5-diphenyloxazoles,[121] and 4,4,6-trimethyl-4H-1,3-oxazines[122] by cyclization of suitable nitrilium intermediates. The nitrilium salts were formed

[115] G. W. Cannon, K. K. Grebker, and Y. Hsu, *J. Org. Chem.*, **18**, 516 (1953).

[116] H. Meerwein, P. Laasch, R. Mersch, and J. Spille, *Chem. Ber.*, **89**, 209 (1956).

[117] M. Lora-Tamayo, R. Madroñero, and G. G. Muñoz, *Chem. Ind.*, (*London*) 657 (1959).

[118] M. Lora-Tamayo, R. Madroñero, and G. G. Muñoz, *Chem. Ber.*, **93**, 289 (1960).

[119] M. Lora-Tamayo, R. Madroñoro, G. G. Muñoz, J. M. Marzal, and M. Stud, *Chem. Ber.*, **94**, 199 (1961).

[120] M. Lora-Tamayo, R. Madroñero, and G. G. Muñoz, *Chem. Ber.*, **94**, 208 (1961).

[121] M. Lora-Tamayo, R. Madroñero, and H. Leipprand, *Chem. Ber.*, **97**, 2230 (1964).

[122] M. Lora-Tamayo, R. Madroñero, G. G. Muñoz, and H. Leipprand, *Chem. Ber.*, **97**, 2234 (1964).

by the action of a metal halide-nitrile complex. Thus, by heating equimolecular amounts of β-chloroethylbenzene and the tin tetrachloride-acetonitrile complex, a 91% yield of 1-methyl-3,4-dihydroisoquinoline

$$C_6H_5CH_2CH_2Cl + CH_3C{\equiv}N{\rightarrow}SnCl_4 \rightarrow \left[C_6H_5CH_2CH_2\overset{\oplus}{N}{\equiv}C{-}CH_3 \leftrightarrow C_6H_5CH_2CH_2N{=}\overset{\oplus}{C}{-}CH_3 \right] SnCl_5^{\ominus} \rightarrow \text{1-methyl-3,4-dihydroisoquinoline}$$

was obtained.[118] Although the yields vary considerably, the method provides a facile route to the heterocycles cited.

Koch-Haaf Reaction. The reaction of carbonium ions with carbon monoxide in sulfuric acid followed by hydrolysis to give carboxylic acids is known as the Koch-Haaf reaction. Carbon monoxide, either generated *in situ* from formic acid[123] or supplied as the gas under pressure,[124] has been used to produce branched carboxylic acids in good yields from alkenes.

$$>C{=}C< \xrightarrow{+H^{\oplus}} -\underset{H}{C}-\overset{\oplus}{C}- \xrightarrow{+CO} -\underset{H}{C}-\underset{\overset{\oplus}{CO}}{C}- \xrightarrow[-H^{\oplus}]{H_2O} -\underset{H}{C}-\underset{CO_2H}{C}-$$

The reaction has been extended to include alcohols, alkyl chlorides, esters, and paraffins as the carbonium ion source.[123, 124]

In a comparative study of the Koch-Haaf and Ritter reactions Christol and Solladie furnished evidence that the nucleophilic reagent, carbon monoxide, has greater reactivity than its counterpart, the nitrile, particularly with the compounds containing a phenyl group.[12] Indeed, whereas the Ritter reaction failed with diphenylmethylcarbinol,[40] the Koch-Haaf reaction furnishes the corresponding carboxylic acid in 18% yield.[12]

Thiocarbamate Synthesis. The preparation of *N*-substituted thiocarbamates from thiocyanates, using methods analogous to that of the Ritter reaction, was described by Riemschneider.[125] He found that thiocyanates react with alcohols and olefins in the presence of sulfuric acid to give products which, on hydrolysis, yield N-substituted thiocarbamates.

$$RSCN \xrightarrow[(H_2SO_4)]{R'OH} RSCONHR'$$

[123] H. Koch and W. Haaf, *Ann.*, **618**, 251 (1958).
[124] W. Haaf and H. Koch, *Ann.* **638**, 122 (1960).
[125] R. Riemschneider, *J. Am. Chem. Soc.*, **78**, 844 (1956).

The formation of these thiocarbamates from a carbonium ion may be formulated in the following way. Primary alkyl thiocyanates react with

$$(CH_3)_2\overset{\oplus}{C}H + \overset{\ominus}{N}{=}\overset{\oplus}{C}SCH_3 \rightleftharpoons (CH_3)_2CHN{=}\overset{\oplus}{C}SCH_3$$

$$(CH_3)_2CHN{=}\overset{\oplus}{C}SHC_3 + H_2O \rightleftharpoons (CH_3)_2CHNHC({=}O)SCH_3 + H^{\oplus}$$

alcohols and olefins to give the desired thiocarbamates in good yields; 1,4-dithiocyanatobutane and 1,5-dithiocyanatopentane yielded the expected N,N′-substituted *bis*-thiocarbamates.

EXPERIMENTAL CONDITIONS

Nature of the Acid. Perchloric,[19] phosphoric,[57] polyphosphoric[45, 46, 49–51] formic,[68, 69] substituted sulfonic acids,[5, 26, 111] and boron trifluoride[96] have been used as reagents in the Ritter reaction. The best yields have been reported for the reactions which were conducted in 85–90% sulfuric acid.

A detailed study of the effect of various mineral acids on the formation of N-*t*-butylacrylamide in acetic anhydride showed that the highest yields resulted from the use of sulfuric acid.[28] This report and others[19, 24, 26, 68, 70] describe lower yields when other acids are substituted for sulfuric acid, thus supporting the particular role of sulfuric acid or the bisulfate anion in the Ritter reaction.

Glacial acetic acid has frequently been employed as a diluent; however, Benson and Ritter reported that secondary alcohols would not react under these conditions.[55] It is noteworthy that esterification competes with amide formation yielding acetates and giving lower yields.[28, 57]

Nature of the Solvent. In addition to glacial acetic acid, other polar and non-polar solvents have been used. They include acetic anhydride, di-*n*-butyl ether, chloroform, carbon tetrachloride, hexane, and nitrobenzene.

Christol and Solladie made a comparison of products using 96% sulfuric acid with and without solvent.[11] They concluded that in highly polar solvents the attacking nitrile is strongly polarized and becomes solvated, resulting in enhancement of its nucleophilic reactivity. On the other hand, in non-polar solvents such as carbon tetrachloride the solvation phenomena are less important since the medium is less polar. In this case the yields are generally lower.

A related study by Glikmans and co-workers employing di-*n*-butyl ether and dioxane gave similar results, and the Ritter reaction failed completely in the presence of methanol or ethanol.[28]

Reaction Time and Temperature. An examination of the experimental data clearly demonstrates that there is no rigid pattern, but a great number of variations both in reaction time and temperatures. Generally the reaction temperatures are mild; that is, in the 25–50° range. The use of liquid hydrogen cyanide, of course, necessitates temperature in the 0–10° range, and there are a few examples of the reaction proceeding at lower temperatures.[36]

The reaction time is also quite variable. Swern reported times as brief as 15 minutes, and related time to reaction size.[19, 42] With lower temperatures, longer reaction times have been used. A study of the influence of time versus yield was reported for the reaction of isobutene with acrylonitrile at 25°. The results show that the reaction is 96.5% complete after 3 hours and that extending the reaction time to 16 hours provides only an increase of 1.9% in the yield of N-*t*-butylacrylamide.[28]

EXPERIMENTAL PROCEDURES

N-Benzylacrylamide. Preparation from acrylonitrile and benzyl alcohol in 59–62% yield is described in *Organic Syntheses*.[56]

N-*t*-Butylacetamide.[1] Gaseous isobutene (12 g., 0.21 mole) is led into a solution of 4.5 g. (0.11 mole) of acetonitrile in 50 ml. of glacial acetic acid containing 10 g. (0.1 mole) of concentrated sulfuric acid while the temperature is maintained at about 20°. The reaction vessel is then stoppered loosely and allowed to stand overnight. The reaction mixture is poured into 200 ml. of water, neutralized with sodium carbonate, and extracted with five 50-ml. portions of diethyl ether. The combined extracts are dried over anhydrous potassium carbonate, most of the ether is removed at 50°, and the remainder is evaporated at room temperature. The residue (11 g., 85%) is recrystallized from hexane as needles, m.p. 97–98°, b.p. 194° (cor.).

1-Methylcyclohexylamine.[60] 1-Methylcyclohexanol (114 g., 1.0 mole) is slowly added to a previously cooled mixture of 122 g. (4.5 moles) of liquid hydrocyanic acid in 131 g. (1.34 moles) of concentrated sulfuric acid, the temperature being maintained below 0°. This temperature is maintained for 1 hour after completion of the addition. The reaction mixture is then allowed to stand at 25–30° for 24 hours, after which it is cooled and 250 g. of ice and 250 g. of 50% aqueous sodium hydroxide are added. The mixture should be at pH 9; if not, additional sodium hydroxide solution is added.

The reaction mixture is then heated at 50–60° for 4 hours. Sodium hydroxide pellets (50 g.) are added, and the mixture is steam-distilled. The organic distillate is separated and dried over potassium hydroxide. Distillation at 31–38°/8–10 mm. gives 70 g. (60%) of 1-methylcyclohexylamine.

1-Phenylethylamine.[8] Concentrated sulfuric acid (30 ml.) is added dropwise with stirring to a solution of 12.2 g. (0.1 mole) of 1-phenylethanol in 75 ml. of di-*n*-butyl ether in which is suspended 16.25 g. (0.25 mole) of potassium cyanide. The temperature rises rapidly but is maintained between 60° and 80° by external cooling. After the addition, agitation is continued for 1 hour at 50°. The reaction mixture is then poured onto 300 g. of ice, made alkaline with 20% aqueous sodium hydroxide, and extracted with diethyl ether. The ethers are removed, and the resulting oil is heated under reflux for 3 hours with twice its volume of concentrated hydrochloric acid. After the solution has cooled, it is extracted with ether to remove any impurities. The water layer is made alkaline, and the amine extracted with ether. The ether extracts are dried over anhydrous potassium carbonate and fractionally distilled; the yield is 70.2 g. (60%) of a colorless oil, b.p. 91°/20 mm.

Ethyl 3-propionamido-3-phenylbutyrate.[62] Concentrated sulfuric acid (10 ml.) is added dropwise with stirring to a mixture of 0.05 mole of ethyl 3-hydroxy-3-phenylbutyrate and 0.06 mole of propionitrile at 0° over a period of 30 minutes. The mixture, which becomes viscous after about half of the sulfuric acid is added, is left standing overnight at room temperature and is then poured into 100 ml. of ice water. The amido ester is collected by filtration and crystallized from methylcyclohexane. The yield of crystals, m.p. 83°, is 73%.

Di-N-*t*-butyladipamide.[71] A solution of 0.05 mole of adiponitrile and 0.1 mole of methyl *t*-butyl ether in 10 ml. of glacial acetic acid is added with stirring and cooling (30°) to a solution of 10 g. of 100% sulfuric acid in 20 ml. of glacial acetic acid. After 3 hours the reaction mixture is poured into water. The di-N-*t*-butyladipamide that precipitates is isolated by filtration. The yield of the diamide, m.p. 212°, is 75%.

Methylene-*bis*-*p*-toluamide.[68] A solution of 1.5 g (0.05 mole) of trioxane in 11.7 g. (0.1 mole) of *p*-tolunitrile is added slowly with stirring to 38 ml. of 85% sulfuric acid in a 125-ml. three-necked flask. The temperature is maintained at 30° by cooling with an ice bath. After 3 hours the solution is poured into 300 ml. of ice and water. Methylene-*bis*-*p*-toluamide separates as white crystals which are filtered and recrystallized from 95% ethanol. The yield is 12.4 g. (83%) of crystals melting at 209–210°.

N-(Benzamidomethyl)acrylamide.[75] A mixture of 7.6 g. (0.05

mole) of N-methylolbenzamide and 3.7 g. (0.07 mole) of acrylonitrile is added slowly to 50 ml. of 96% sulfuric acid with stirring. The temperature is kept at 30° by cooling. At the end of 1 hour the reaction mixture is poured into 200 g. of ice and water. The crystalline product which precipitates is filtered and recrystallized from water to give 6.0 g. (60%) of N-(benzamidomethyl)acrylamide. This compound polymerizes in the melting-point tube when it is heated above 150°.

N-*t*-Butylphenylacetamide.[66] A mixture of 11.7 g. (0.1 mole) of benzyl cyanide, 9.25 g. (0.1 mole) of *t*-butyl chloride, and 24 g. of 90% formic acid is heated under reflux for 5 hours and poured into a mixture of ice and water. The solid that precipitates is filtered and dried. The yield of amide melting at 111–113° is 4.3 g. (23%).

Acetamidostearic Acid.[19] A mixture of 282 g. (1.0 mole) of oleic acid and 123 g. (3.0 moles) of acetonitrile in a cylindrical dropping funnel is stirred vigorously and added during 35 minutes to 338 ml. (6 moles) of 95% sulfuric acid in a 2-l. three-necked flask fitted with a thermometer and an efficient stirrer. The reaction temperature is maintained between 27° and 30° by external cooling. Fifteen minutes after addition is complete, the mixture is poured with stirring onto approximately 2 l. of chopped ice and water. The soft, sirupy, insoluble mass is stirred occasionally and then allowed to stand overnight in the dilute acid. The following morning, stirring is continued until the product hardens to a crumbly, wax-like solid which is filtered, washed several times with cold water, and dried; the yield of crude acetamidostearic acid, m.p. 56–58°, is 339 g. (99%).

4-Methyl-4-benzamido-2-pentanone.[72] To a solution of 19.6 g. (0.2 mole) of mesityl oxide and 22.0 g. (0.21 mole) of benzonitrile is added 20 ml. of concentrated sulfuric acid. The temperature is kept below 30° by means of an ice bath. After addition is complete, the reaction mixture is warmed to 50° and is kept at this temperature for 1 hour. The dark, viscous liquid is poured into 300 ml. of ice cold water. The resulting solid is filtered and washed with 10% aqueous potassium carbonate, then with water. The yield, m.p. 98–100°, is 20–24 g. (45–55%). Recrystallization from cyclohexene and dilute ethanol furnishes white needles, m.p. 100–101°.

3-Dichloroacetamidoisocamphane.[36] Sulfuric acid (20 ml., 98%) is added with stirring, at a temperature below −15°, over a period of 1 to 1½ hours to a solution of 44.3 g. of camphene and 43.0 g. of dichloroacetonitrile in 20 ml. of propionic acid. The reaction mixture is kept for 2 days at −20° to −15°; then it is poured onto 300 ml. of ice water, the aqueous solution neutralized with sodium carbonate, the crystalline precipitate collected by filtration and dried in a vacuum desiccator over phosphorus pentoxide. After recrystallization from 150 ml. of petroleum ether (b.p. 70–100°), 50.2 g. of material (m.p. 106–108°) is obtained; a further 10 g. of the material is isolated from the mother liquor by evaporation of part of the solvent and recrystallization. The total yield is 60.2 g. (69.7%).

4,4,6-Trimethyl-2-phenyl-5,6-dihydro-1,3-oxazine.[83] Benzonitrile (20.6 g., 0.2 mole) is added dropwise with stirring to 100 g. of 92% sulfuric acid at 2–4° over 20 minutes. Then 23.6 g. (0.2 mole) of 2-methyl-2,4-pentanediol is added dropwise with stirring at 3–6° during 2 hours. The resulting solution is poured with stirring on 1 kg. of cracked ice, and the mixture is half-neutralized with 40% aqueous sodium hydroxide, then extracted with several portions of diethyl ether. The combined ether extracts are dried over anhydrous potassium carbonate and, after removal of the ether, the residual oil is distilled through a 30-cm. vacuum-jacketed Vigreux column. Two recrystallizations from ethanol-water (the compound is dissolved at room temperature, and the solution is then strongly cooled) give colorless crystals, m.p. 34–35°.

2-(*p*-Aminophenyl)-4,4-dimethyl-Δ^2-thiazoline.[59] To a solution of 3.5 g. (0.03 mole) of *p*-aminobenzonitrile in 25 ml. of concentrated sulfuric acid previously cooled to 3° is added with stirring 2.1 g. (0.02 mole) of 2-hydroxy-2-methylpropanethiol during 30 minutes. The reaction mixture, which is golden yellow, is stirred at 3–5° for an additional hour; then it is poured on 300 g. of chipped ice. The cold, aqueous acid solution is then extracted with chloroform until the chloroform layer is colorless. After the aqueous solution has been poured through fluted filter paper to remove the excess chloroform, it is carefully neutralized with 30% aqueous sodium hydroxide. The heterocyclic base appears as a crude brown solid which is collected in a Buchner funnel and then washed several times with hot water to remove any unchanged nitrile. Recrystallization from aqueous ethanol yields 2.3 g. (55%) of a very light yellow crystalline material, m.p. 162–164°. The picrate melts at 91–93°.

2-Vinyl-1-pyrroline.[104] To a cold solution of 14.0 g. (0.10 mole) of 3,3′-thiodipropionitrile[126] in 100.0 ml. of concentrated sulfuric acid is added with stirring 29.6 g. (0.20 mole) of 2,5-dimethyl-2,5-hexanediol. The temperature of the reaction is kept below 10° by employing an ice bath. After the glycol addition has been completed, the mixture is stirred for an additional hour at 4–6° and then slowly poured over 300 g. of chipped ice. The aqueous acid solution, after it has been shaken several times with 75-ml. portions of chloroform to extract polymeric material, is cautiously neutralized with 30% aqueous sodium hydroxide. The heterocyclic base, which separates, is taken up in diethyl ether and dried overnight with potassium carbonate. After the ether is removed at atmospheric pressure,

[126] F. M. Cowen, *J. Org. Chem.*, **20**, 287 (1955).

the residual oil is distilled under reduced pressure to give a light yellow oil, b.p. 86–87°/1.25 mm.; n_D^{30} 1.5019.

1-(3′,4′-Dimethoxyphenyl)-3-methyl-6,7-dimethoxy-3,4-dihydroisoquinoline.[34] Veratronitrile (9.8 g., 0.06 mole) is added portionwise to concentrated sulfuric acid (15 ml.) in an ice bath with rapid stirring. Methyleugenol (10.2 ml., 0.06 mole) is then added during 2 minutes. The cooling bath is removed before addition of the methyleugenol, and the temperature, which rises rapidly during the addition, is maintained around 80° by intermittent cooling. After it has stood at room temperature for 3 days, the reaction mixture is poured into water and neutralized with sodium carbonate. A yellow mass separates which gradually crystallizes on stirring in warm water. It is filtered, washed with water, and dried. The crude dihydroisoquinoline (11.5 g., 53%) is crystallized from aqueous ethanol as yellow crystals (6.7 g.) melting at 134–137°. Recrystallization yields a product melting at 138–139°.

5,6-Diphenyl-1,2,3,4-tetrahydro-2-pyridone.[99] In a flask fitted with a mechanical stirrer is placed 20 ml. of 90% sulfuric acid, and 10 g. (0.04 mole) of γ-benzoyl-γ-phenylbutyronitrile is added in 5 minutes at 25–27°. Solution is quite rapid with the evolution of a little heat. The stirring is continued for 3 hours, after which the mixture is allowed to stand overnight. The next day, the viscous, brownish-red solution is poured with stirring over 200 g. of cracked ice. The precipitate forms as a whitish lump. The dilute acid is decanted, the lump is treated with 150 ml. of water, and the mixture is heated on the steam bath. The mass disintegrates and becomes crystalline. It is crushed, filtered, washed with water to remove the acid, treated with 3% aqueous ammonia; after it has stood in the ammonia solution for an hour, it is filtered, washed with water, and dried. Recrystallization from ethyl Cellosolve gives 6.0 g., and an additional 0.5 g. is isolated from the mother liquors. The yield, m.p. 218–219°, is 65%.

2,2-Dimethylcyclopenteno(*d*)-1-azabicyclo[4.4.0]decane.[102] To a cold (0–3°) solution of 9.05 g. (0.077 mole) of δ-chlorovaleronitrile in 150 ml. of 98% sulfuric acid is added dropwise 11.06 g. (0.07 mole) of α-(2-hydroxycyclopentyl)-*t*-butyl alcohol. During the addition efficient stirring is maintained and the temperature of the mixture is kept below 10°. The time required for the complete addition of the glycol under these conditions is approximately 1½ hours.

The deep reddish reaction mixture is slowly poured over 500 g. of chipped ice in a 2-l. beaker. (The following two reactions, *i.e.*, the sodium borohydride reduction and the intramolecular alkylation, are carried out in this vessel.) The aqueous acid solution is extracted several times with chloroform to remove insoluble polymeric material and then partially neutralized (to pH 2–4) with 35% aqueous sodium hydroxide. The temperature is maintained below 40° during the neutralization with the aid of external cooling. The electrodes of a pH meter (Beckman Zeromatic) are inserted into the solution and the acidity is adjusted to pH 3–4 by the addition of 4*M* sulfuric acid and 6*M* sodium hydroxide which are contained in burets situated above the beaker. The clear solution is cooled to room temperature, and a freshly prepared solution of sodium borohydride (2.66 g., 30 ml. of water, and one drop of 35% sodium hydroxide) is added dropwise with stirring supplied by a magnetic stirrer.

The pH of the reaction during the borohydride addition is constantly kept between 3 and 4 by the periodic addition of sulfuric acid or sodium hydroxide. The sodium borohydride addition is complete after 1 hour at 25°. The solution is allowed to stir at room temperature overnight and is then acidified to pH 1 and stirred for 1 hour to destroy the excess sodium borohydride. After addition of 300 ml. of water the pH is again adjusted to 8–8.5 and the solution is stirred for 5 hours. The oil that separates is extracted with diethyl ether, and the extract is dried over potassium carbonate. Removal of the ether on a steam bath and distillation of the residual oil given 6.6 g. (46%) of the tricyclic base, b.p. 100–102°/1.0 mm., n_D^{30} 1.5062. The infrared spectrum (carbon tetrachloride) shows only strong absorption at 3.4 μ (CH), 6.9 μ (CH_2 bending), and ring skeletal vibrations above 8 μ.

***trans*-1,6-Dimethyl-3,5-diphenyl-1,4,5,6-tetrahydro-*as*-triazine.**[100] A solution of 36 g. (0.2 mole) of *N*-amino-(+)-pseudoephedrine in 50 ml. of chloroform is added dropwise to cooled (5°), stirred, concentrated sulfuric acid (250 ml.). After the addition is complete, 20.6 g. (0.2 mole) of benzonitrile is added, and the mixture is stirred at ambient temperature overnight. The mixture is poured onto crushed ice, washed with chloroform, made basic with sodium carbonate solution, and the precipitated solid is removed by suction filtration. The solid is recrystallized from isopropyl alcohol to yield 18.5 g. (35%) of white crystalline product, m.p. 146–147°.

TABULAR SURVEY

The tables include reactions compiled from the literature through December 1966. The authors feel that the data are reasonably complete, but some publications were undoubtedly missed.

The tables are arranged in the order in which different sources of the carbonium ion are discussed in the text. The reactants within a table are, in general, listed in order of increasing size and complexity.

TABLE I. ALKANES

Alkane	Nitrile/Hydride Acceptor	Reaction Medium	Product(s)	Yield (%)	Refs.
$(CH_3)_2CHCH_2CH_3$	$HCN/(CH_3)_3COH$	96% H_2SO_4	$(CH_3)_3CNH_2$	27 (31)*	30
			$(CH_3)_2C(NH_2)CH_2CH_3$	33 (45)*	
			$(CH_3)_2CHCH(NH_2)CH_3$	1.5 (2)*	
$(CH_3)_2CHCH(CH_3)_2$	$HCN/(CH_3)_3COH$	96% H_2SO_4	$(CH_3)_3CNH_2$	40 (43)*	30
			$(CH_3)_2C(NH_2)CH(CH_3)_2$	14 (22)*	
			$(CH_3)_3CCH(NH_2)CH_3$	5 (8)*	
			$(CH_3)_2CHCH(NH_2)C_2H_5$	3 (4)*	
			$CH_3CH(NH_2)CH(CH_3)C_2H_5$	0.6 (1)*	
 [cyclopentane ring]–CH_3	$HCN/(CH_3)_3COH$	96% H_2SO_4	$(CH_3)_3CNH_2$	27 (43)*	30
			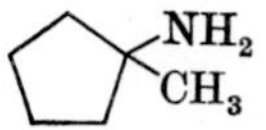 [cyclopentane ring] with NH_2 and CH_3 on the same carbon	6.5 (14)*	
			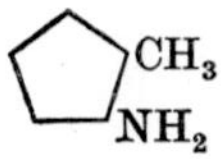 [cyclopentane ring] with CH_3 and NH_2 on adjacent carbons (*cis*)	0.5 (1)*	

Note: References 127–149 are on p. 325.

* This figure is the yield of the product as a percentage of the mixture of products.

TABLE I. ALKANES (*Continued*)

Alkane	Nitrile/Hydride Acceptor	Reaction Medium	Product(s)	Yield (%)	Refs.
 CH_3	$HCN/(CH_3)_3COH$	96% H_2SO_4	$(CH_3)_3CNH_2$	49 (52)*	30
			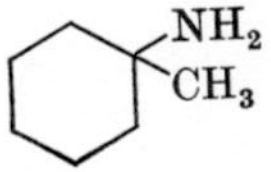 NH_2, CH_3	14 (23)*	
			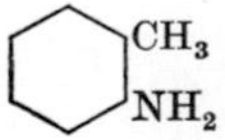 CH_3, NH_2 (*cis*)	2.4 (3)*	
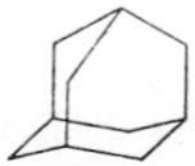 (Adamantane)	$HCN/(CH_3)_3COH$	96% H_2SO_4/*n*-hexane	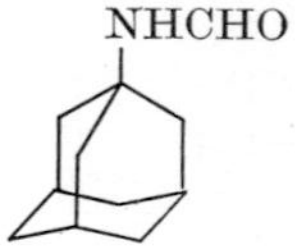 NHCHO	78	30, 31
	$HCN/CH_3CH_2CHOHCH_3$	100% H_2SO_4/*n*-hexane	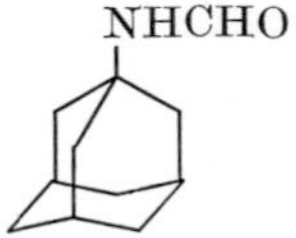 NHCHO	75	30
	$HCN/(CH_3)_2C{=}CH_2$	96% H_2SO_4/cyclohexane	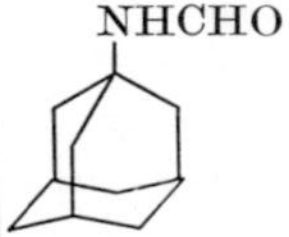 NHCHO	60	30

$HCN/(CH_3)_3CCl$	96% H_2SO_4/CCl_4	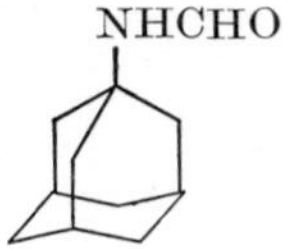 NHCHO (adamantane structure)	42	30
$CH_3CN/(CH_3)_3COH$	96% H_2SO_4/*n*-hexane	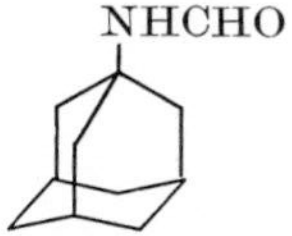 NHCHO (adamantane structure)	36	30

Note; References 127–149 are on p. 325.

* This figure is the yield of the product as a percentage of the mixture of products.

TABLE II-A. 1-ALKENES

$$CH_3(CH_2)_nCH{=}CH_2 + RCN \xrightarrow{H_2SO_4} CH_3(CH_2)_nCH(CH_3)NHCOR$$

n	R	Yield (%)	Refs.
2	$CH_2ClCHCl$	25	26
5	CCl_3	95	26
7	CH_2CO_2H	—	26
9	H	—	26
9	CH_3	70	26
9	$CH_2{=}CH$	70	26
9	CH_2CO_2H	90	26
11	H	37	26
11	CH_3	80	26
11	$CH_2{=}CH$	68	26
11	CH_2CO_2H	84	26
13	H	44	26
13	CH_3	67	24, 26
13	$CH_2{=}CH$	58	26
13	CH_2CO_2H	53	26
15	H	45	26
15	CH_3	66	26
15	CCl_3	56	26
15	$CH_2{=}CH$	73	26
15	$CH_2ClCHCl$	75	26
15	CH_2CO_2H	59	26

Note: References 127–149 are on p. 325.

TABLE II-B. Isobutene

$(CH_3)_2C{=}CH_2 + RCN \rightarrow (CH_3)_3CNHCOR$

[The product is $(CH_3)_3CNHCOR$ unless a structural formula is shown.]

R	Reaction Medium	Product	Yield (%)	Refs.
CH_3	H_2SO_4/CH_3CO_2H		85	1, 5, 28
$CH_2{=}CH$	H_2SO_4/CH_3CO_2H		98	28
	$HClO_4/CH_3CO_2H$		76	28
	H_3PO_4/CH_3CO_2H		4	28
	HCl/CH_3CO_2H		0	28
	$H_2SO_4/(n\text{-}C_4H_9)_2O$		78	28
	H_2SO_4/dioxane		86	28
	H_2SO_4/CH_3OH		0	28
C_2H_5	H_2SO_4/CH_3CO_2H		79	28
$(C_2H_5)_2NCH_2$	H_2SO_4/CH_3CO_2H		—	127
C_6H_5	H_2SO_4/CH_3CO_2H		93	1, 28
$C_6H_5CH_2$	H_2SO_4/CH_3CO_2H		80	1
$2\text{-}CH_3C_6H_4$	H_2SO_4/CH_3CO_2H		32	28
$3\text{-}CH_3C_6H_4$	H_2SO_4/CH_3CO_2H		50	28
$4\text{-}CH_3C_6H_4$	H_2SO_4/CH_3CO_2H		57	28
$n\text{-}C_9H_{19}$	H_2SO_4/CH_3CO_2H		75	1
$n\text{-}C_{10}H_{21}$	H_2SO_4/CH_3CO_2H		75	5
$n\text{-}C_{13}H_{27}$	H_2SO_4/CH_3CO_2H		70	1
$n\text{-}C_{14}H_{29}$	H_2SO_4/CH_3CO_2H		70	5
$n\text{-}C_{17}H_{35}$	H_2SO_4/CH_3CO_2H		85	1
$n\text{-}C_{18}H_{37}$	H_2SO_4/CH_3CO_2H		85	5
CN	H_2SO_4	$CONHC(CH_3)_3$* – $CONHC(CH_3)_3$	45	5

Note: References 127–149 are on p. 325.

* This is a complete structural formula.

TABLE II-B. Isobutene (*Continued*)

R	Reaction Medium	Product	Yield (%)	Refs.
CH_2CN	H_2SO_4/CH_3CO_2H	$[CONHC(CH_3)_3]_2CH_2$*	50	55
$(CH_2)_3CN$	H_2SO_4	$CONHC(CH_3)_3$* — $(CH_2)_3$ — $CONHC(CH_3)_3$	25	55
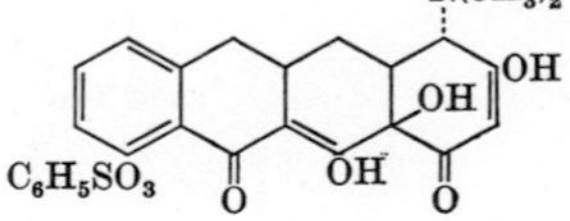 [structure: $N(CH_3)_2$, OH, OH, $C_6H_5SO_3$, O, OH, O]	H_2SO_4/CH_3CO_2H	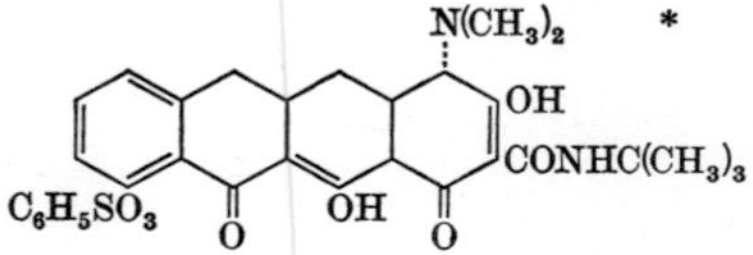 [structure: $N(CH_3)_2$, OH, $CONHC(CH_3)_3$, $C_6H_5SO_3$, O, OH, O]*	53	128
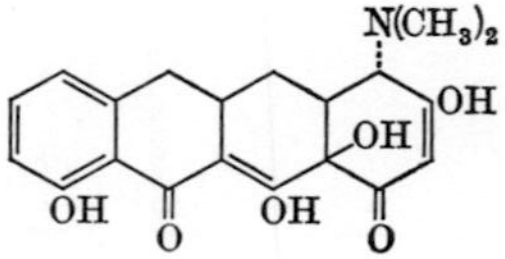 [structure: $N(CH_3)_2$, OH, OH, OH, O, OH, O]	H_2SO_4/CH_3CO_2H	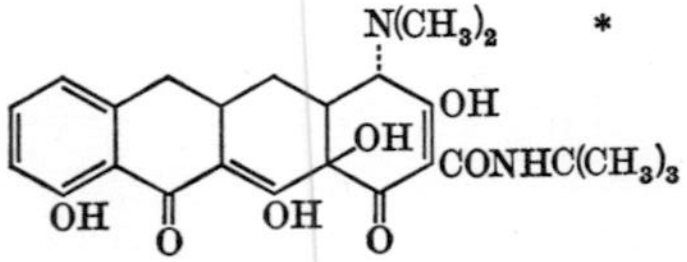 [structure: $N(CH_3)_2$, OH, OH, $CONHC(CH_3)_3$, OH, O, OH, O]*	71	128

Note: References 127–149 are on p. 325.

* This is a complete structural formula.

TABLE II-C. BRANCHED OR SUBSTITUTED 1-ALKENES
(See also Tables II-B and II-F.)

Alkene	Nitrile	Reaction Medium	Product	Yield (%)	Refs.
$C_2H_5C(CH_3)=CH_2$	$ClCN$	H_2SO_4	$C_2H_5C(CH_3)_2NH_2$	33	111
	$CH_2=CHCN$	H_2SO_4/CH_3CO_2H	$C_2H_5C(CH_3)_2NHCOCH=CH_2$	90	28
	$(C_2H_5)_2NCH_2CN$	H_2SO_4	$C_2H_5C(CH_3)_2NHCOCH_2N(C_2H_5)_2$		127
	$O(CH_2CH_2)_2NCH_2CN$	H_2SO_4	$C_2H_5C(CH_3)_2NHCOCH_2N(CH_2CH_2)_2O$	—	127
	$(CH_2)_5NCH_2CN$	H_2SO_4	$C_2H_5C(CH_3)_2NHCOCH_2N(CH_2)_5$	—	127
	$(CH_2)_5NCH_2CH_2CN$	H_2SO_4	$C_2H_5C(CH_3)_2NHCOCH_2CH_2N(CH_2)_5$	—	127
	$(C_2H_5)_2NCH(C_2H_5)CN$	H_2SO_4	$C_2H_5C(CH_3)_2NHCOCH(C_2H_5)N(C_2H_5)_2$	—	127
	$(C_2H_5)_2NCH(C_5H_{11}\text{-}n)CN$	H_2SO_4	$C_2H_5C(CH_3)_2NHCOCH(C_5H_{11}\text{-}n)N(C_2H_5)_2$	—	127
$(CH_3)_3CC(CH_3)=CH_2$	$NaCN$	H_2SO_4/CH_3CO_2H	$(CH_3)_3CC(CH_3)_2NHCHO$	83	37
$(CH_3)_3CCH(CH_3)CH=CH_2$	$CH_2=CHCN$	H_2SO_4/CH_3CO_2H	$(CH_3)_3CCH(CH_3)CH(CH_3)NHCOCH=CH_2$	80	28
$(CH_3)_3CCH_2C(CH_3)=CH_2$	$ClCN$	H_2SO_4	$(CH)_3CCH_2C(CH_3)_2NH_2$	27	111
$C_6H_5CH=CH_2$	HCN	H_2SO_4/CH_3CO_2H	$C_6H_5CH(CH_3)NHCHO$	—	5
	H_2NCN	H_2SO_4/CH_3CO_2H	$C_6H_5CH(CH_3)NHCONH_2$	—	5
	CH_3CN	H_2SO_4/CH_3CO_2H	$C_6H_5CH(CH_3)NHCOCH_3$	43	1
		$C_6H_5SO_3H/CH_3CO_2H$		40	5
	$C_6H_5CH_2CN$	H_2SO_4/CH_3CO_2H	$C_6H_5CH(CH_3)NHCOCH_2C_6H_5$	—	5
$C_6H_5CH_2CH=CH_2$	KCN	$H_2SO_4/(n\text{-}C_4H_9)_2O$	$C_6H_5CH_2CH(CH_3)NH_2$	60	8
	CH_3CN	H_2SO_4	$C_6H_5CH_2CH(CH_3)NHCOCH_3$	40	5
	C_6H_5CN	H_2SO_4	$C_6H_5CH_2CH(CH_3)NHCOC_6H_5$	80	8, 34
	$C_6H_5CH_2CN$	H_2SO_4	$C_6H_5CH_2CH(CH_3)NHCOCH_2C_6H_5$	53	34
	$3,4\text{-}(CH_3O)_2C_6H_3CN$	$H_2SO_4/(n\text{-}C_4H_9)_2O$	$C_6H_5CH_2CH(CH_3)NHCOC_6H_3(OCH_3)_2\text{-}3,4$	53	34
	$3,4\text{-}(C_2H_5O)_2C_6H_3CN$	$H_2SO_4/(n\text{-}C_4H_9)_2O$	$C_6H_5CH_2CH(CH_3)NHCOC_6H_3(OC_2H_5)_2\text{-}3,4$	97	34
$C_6H_5C(CH_3)=CH_2$	CH_3CN	H_2SO_4	$C_6H_5C(CH_3)_2NHCOCH_3$	28	1
		$C_6H_5SO_3H/CH_3CO_2H$		—	5
	$CH_2=CHCN$	H_2SO_4/THF	$C_6H_5C(CH_3)_2NHCOCH=CH_2$	87	23
$4\text{-}CH_3C_6H_4C(CH_3)=CH_2$	CH_3CN	H_2SO_4	$4\text{-}CH_3C_6H_4C(CH_3)_2NHCOCH_3$	39	1
$CH_3O_2C(CH_2)_8CH=CH_2$	C_6H_5CN	H_2SO_4	$HO_2C(CH_2)_{10}NHCOC_6H_5$*	16	63

Note: References 127–149 are on p. 325.

* The position of the C_6H_5CONH group was not determined.

TABLE II-D. 2-Alkenes
(See also Tables II-E and II-F.)

Alkene	Nitrile	Reaction Medium	Product	Yield (%)	Ref.
$CH_3CH{=}CHCH_3$	ClCN	H_2SO_4	$C_2H_5CH(NH_2)CH_3$	23	111
$CH_3CH{=}C(CH_3)_2$	$NC(CH_2)_2CN$	H_2SO_4/CH_3CO_2H	$CH_2CONHC(CH_3)_2C_2H_5$ — $CH_2CONHC(CH_3)_2C_2H_5$	40	55
	$NC(CH_2)_3CN$	H_2SO_4/CH_3CO_2H	$CH_2(CH_2CONHC(CH_3)_2C_2H_5)_2$	35	55
$CH_3CH{=}C(CH_3)C_2H_5$	$(C_2H_5)_2NCH_2CN$	H_2SO_4	$(C_2H_5)_2C(CH_3)NHCOCH_2N(C_2H_5)_2$	—	127
	[piperidine ring drawn]NCH_2CN	H_2SO_4	$(C_2H_5)_2C(CH_3)NHCOCH_2N$[piperidine ring drawn]	—	127
	[morpholine ring drawn, O]NCH_2CN	H_2SO_4	$(C_2H_5)_2C(CH_3)NHCOCH_2N$[morpholine ring drawn, O]	—	127
$(CH_3)_2C{=}CHC_2H_5$	$C_6H_5CH_2CN$	H_2SO_4	$(CH_3)_2C(C_3H_7\text{-}n)NHCOCH_2C_6H_5$	69	127
$(CH_3)_2C{=}CHC_3H_7\text{-}n$	H_2NCN	H_2SO_4	$(CH_3)_2C(C_4H_9\text{-}n)NHCONH_2$	12	1
$C_6H_5CH{=}CHCH_3$	CH_3CN	H_2SO_4/CH_3CO_2H	$C_6H_5CH_2CH(CH_3)NHCOCH_3$	40	4
$C_6H_5CH{=}C(CH_3)_2$	KCN	$H_2SO_4/(n\text{-}C_4H_9)_2O$	$C_6H_5CH_2C(CH_3)_2NH_2$	60	8
$3,4\text{-}Cl_2C_6H_3CH{=}C(CH_3)_2$	KCN	$H_2SO_4/(n\text{-}C_4H_9)_2O$	$3,4\text{-}Cl_2C_6H_3CH_2C(CH_3)_2NHCHO$	—	41
	HCN	$H_2SO_4/(n\text{-}C_4H_9)_2O$	$3,4\text{-}Cl_2C_6H_3CH_2C(CH_3)_2NH_2$	—	119
$C_6H_5C(CH_3){=}C(CH_3)_2$	HCN	$H_2SO_4/(n\text{-}C_4H_9)_2O$	$C_6H_5C(CH_3)(NH_2)CH(CH_3)_2$	60	9
$4\text{-}CH_3OC_6H_4CH{=}C(CH_3)_2$	HCN	—	$4\text{-}CH_3OC_6H_4CH_2C(CH_3)_2NHCHO$	19	129
$C_6H_5CH{=}CHCH(CH_3)_2$	KCN	$H_2SO_4/(n\text{-}C_4H_9)_2O$	$C_6H_5CH(NH_2)CH_2CH(CH_3)_2$	35	8
[quinolinone structure: OH; $CH_2CH{=}C(CH_3)_2$; =O; N; CH_3; H_3CO]	CH_3CN	H_2SO_4	[quinolinone structure: OH; =O; N; CH_3; H_3CO; $C(CH_3)_2NHCOCH_3$]	28	130
[quinolinone structure: OCH_3; $CH_2CH{=}C(CH_3)_2$; =O; N; CH_3; H_3CO]	CH_3CN	H_2SO_4	[quinoline structure: OCH_3; N; CH_3; H_3CO; $C(CH_3)_2NHCOCH_3$]	44	130

Note: References 127–149 are on p. 325.

TABLE II-E. 2-METHYL-2-BUTENE

$$(CH_3)_2C{=}CHCH_3 + RCN \xrightarrow{H_2O} (CH_3)(C_2H_5)CNHCOR$$

R	Reaction Medium	Yield (%)	Refs.
H_2N	H_2SO_4/CH_3CO_2H	60	1, 5
CH_2Cl	H_2SO_4	21	1
CH_3	H_2SO_4	70	1, 5
H_2NCONH	H_2SO_4	75	1
$CH_2{=}CH$	H_2SO_4/CH_3CO_2H	92	1, 5, 28
H_2NCOCH_2	H_2SO_4/CH_3CO_2H	65	1
C_2H_5	H_2SO_4	50	1
$CH_2{=}C(CH_3)$	H_2SO_4	65	1
$CH_3O_2CCH_2$	H_2SO_4	57	1
$(CH_3)_2CCl$	H_2SO_4	35	1
n-C_3H_7	H_2SO_4	50	1
$C_2H_5O_2CCH_2$	H_2SO_4	45	1
$CH_3CO_2CH(CH_3)$	H_2SO_4	56	1
$CH_3CO_2CH_2CH_2$	H_2SO_4	76	1
n-C_4H_9	H_2SO_4	58	1
C_6H_5	H_2SO_4	70	1
$C_6H_5CH_2$	H_2SO_4	70	1, 5
4-$O_2NC_6H_4CH_2$	H_2SO_4	89	1
2-$CH_3C_6H_4$	H_2SO_4	35	1
4-$CH_3C_6H_4$	H_2SO_4	45	1
n-C_7H_{15}	H_2SO_4	52	1
$C_6H_5CHC{=}H_2$	H_2SO_4	70	1
n-C_9H_{19}	H_2SO_4	45	1
n-$C_{11}H_{23}$	H_2SO_4	39	1
n-$C_{13}H_{27}$	H_2SO_4	25	1
n-$C_{17}H_{35}$	H_2SO_4	22	1

Note: References 127–149 are on p. 325.

TABLE II-F. Diisobutylene

$$\left.\begin{array}{c}(CH_3)_3CCH{=}C(CH_3)_2\\+\\(CH_3)_2CCH_2C(CH_3){=}CH_2\end{array}\right\} + RCN \rightarrow (CH_3)_3CCH_2C(CH_3)_2NHCOR$$

$$\left.\begin{array}{c}(CH_3)_3CCH{=}C(CH_3)_2\\+\\(CH_3)_3CCH_2C(CH_3){=}CH_2\end{array}\right\} + NC{-}R{-}CN \rightarrow$$

$$(CH_3)_3CCH_2C(CH_3)_2NHCORCONH(CH_3)_2CCH_2C(CH_3)_3$$

R	Reaction Medium	Yield (%)	Refs.
H	H_2SO_4/CH_3CO_2H	50–70	1, 4, 5
H_2NCO	H_2SO_4/CH_3CO_2H	55	131
CH_3	H_2SO_4/CH_3CO_2H	85	1, 5
	92% H_2SO_4	70*	132
	98% H_2SO_4	60*	1
	98% H_2SO_4	50	5
	$C_6H_5SO_3H/CH_3CO_2H$	90	5
	$SnCl_2/CHCl_3$	—	5
H_2NCONH	H_2SO_4/CH_3CO_2H	50	1
$H_2NC(=NH)NH$	H_2SO_4/CH_3CO_2H	50	5
$CH_2{=}CH$	H_2SO_4/CH_3CO_2H	71	133
H_2NCOCH_2	H_2SO_4/CH_3CO_2H	65	1, 5
$CH_3CH(OH)$	H_2SO_4/CH_3CO_2H	65	133
$HOCH_2CH_2$	H_3SO_4/CH_3CO_2H	85	1
CH_2†	4-$CH_3C_6H_4SO_3H/CH_3CO_2H$	60	5
	H_2SO_4/CH_3CO_2H	50	55
$CH_2{=}C(CH_3)$	H_2SO_4/CH_3CO_2H	60	133
$CH_3C(OH)CH_2$	H_2SO_4/CH_3CO_2H	85	5
$(CH_3)_2NCH_2$	H_2SO_4/CH_3CO_2H	60	32, 127
$(CH_2)_2$†	H_2SO_4/CH_3CO_2H	50	55
$C_2H_5O_2CCH_2$	H_2SO_4/CH_3CO_2H	70	1, 5
$(CH_3)_2NCH_2CH_2$	H_2SO_4/CH_3CO_2H	80	32, 127
$(CH_2)_4NCH_2$ (pyrrolidine ring)	H_2SO_4/CH_3CO_2H	65	32, 127
$O(CH_2CH_2)_2NCH_2$ (morpholine ring)	H_2SO_4/CH_3CO_2H	65	32, 127
$(C_2H_5)_2NCH_2$	H_2SO_4/CH_3CO_2H	60	32, 127
C_6H_5	H_2SO_4/CH_3CO_2H	90	1
$(CH_2)_5NCH_2$ (piperidine ring)	H_2SO_4/CH_3CO_2H	35	32, 127
$O(CH_2CH_2)_2NCH_2CH_2$ (morpholine ring)	H_2SO_4/CH_3CO_2H	80	32, 127
$O(CH_2CH_2)_2NCH(CH_3)$ (morpholine ring)	H_2SO_4/CH_3CO_2H	30	32, 127

Note: References 127–149 are on p. 325.

* The product was $(CH_3)_3CCH_2C(OH)(CH_3)_2$.

† This is a dinitrile.

TABLE II-F. DIISOBUTYLENE (*Continued*)

R	Reaction Medium	Yield (%)	Refs.
$(C_2H_5)_2NCH_2CH_2$	H_2SO_4/CH_3CO_2H	80	32, 127
$(C_2H_5)_2NCH(CH_3)$	H_2SO_4/CH_3CO_2H	42	32, 127
$C_6H_5CH_2$	H_2SO_4/CH_3CO_2H	75	1, 5
(piperidino)NCH_2CH_2	H_2SO_4/CH_3CO_2H	70	32, 127
(2-methylpiperidino)NCH_2, CH_3	H_2SO_4/CH_3CO_2H	70	32, 127
n-C_7H_{15}	H_2SO_4/CH_3CO_2H	70	1
$(C_2H_5)_2NCH(C_2H_5)$	H_2SO_4/CH_3CO_2H	5	32, 127
$C_6H_5CH{=}CH$	H_2SO_4/CH_3CO_2H	77	133
$CH_3(CH_2)_6CH_2$	H_2SO_4/CH_3CO_2H	—	5
$(C_2H_5)_2NCH(C_3H_7\text{-}n)$	H_2SO_4/CH_3CO_2H	7	32, 127
n-C_9H_{19}	H_2SO_4/CH_3CO_2H	70	1
n-$C_{10}H_{21}$	H_2SO_4/CH_3CO_2H	70	5
$(C_2H_5)_2NCH(C_5H_{11}\text{-}n)$	H_2SO_4/CH_3CO_2H	8	32, 127
$(CH_3)_3CCH_2CH(CH_3)CH_2CH(OH)$	H_2SO_4/CH_3CO_2H	—	134
n-$C_{11}H_{23}$	H_2SO_4/CH_3CO_2H	60	1
n-$C_{13}H_{27}$	H_2SO_4/CH_3CO_2H	50	1
n-$C_{18}H_{37}$	$CH_3SO_3H/(n\text{-}C_4H_9)_2O$	—	5
$C_{10}H_{21}CH(OH)$	80% H_2SO_4	—	134

Note: References 127–149 are on p. 325.

TABLE II-G. CAMPHENE

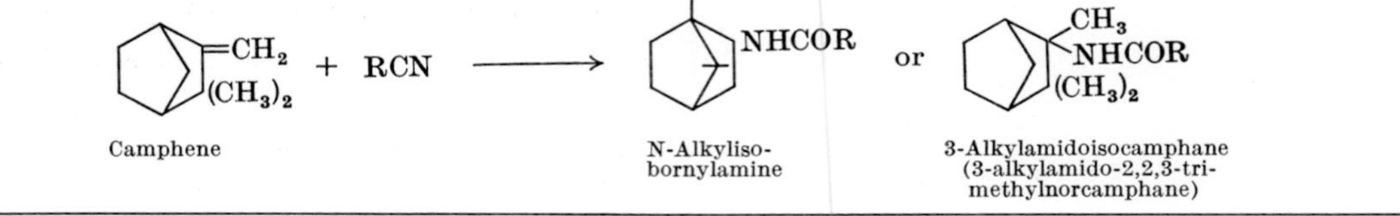

R	Temp. (°C)	Reaction Medium	Product	Yield (%)	Refs.
H	20	H_2SO_4/CH_3CO_2H	N-Formylisobornylamine	40	1, 5
	0	CH_3CO_2H	3-Formamidoisocamphane	73	37
	−15	$H_2SO_4/C_2H_5CO_2H$*	3-Formamidoisocamphane	76	36
CH_3	20	H_2SO_4/CH_3CO_2H	N-Acetylisobornylamine	75	1, 5
$CHCl_2$	−15	$H_2SO_4/C_2H_5CO_2H$	3-Dichloroacetamidoisocamphane	70	36
CCl_3	−15	H_2SO_4/CH_3CO_2H	3-Trichloroacetamidoisocamphane	60	36
$NC(CH_2)_2$	25	H_2SO_4/CH_3CO_2H	N-Isobornylsuccinamide	24	55
	20	$4\text{-}CH_3C_6H_4SO_3H/CH_3CO_2H$	N,N′-Diisobornylsuccinamide	—	5
C_6H_5	20	H_2SO_4/CH_3CO_2H	N-Benzoylisobornylamine	90	1, 5
$C_6H_5CH_2$	20	H_2SO_4/CH_3CO_2H	N-Phenylacetylisobornylamine	67	1
$C_6H_5CH{=}CH$	20	H_2SO_4/CH_3CO_2H	N-Cinnamoylisobornylamine	—	5
$(CH_3)_3CCH_2CH(CH_3)$-$CH_2CH(OH)CN$	35	$H_2SO_4/(n\text{-}C_4H_9)_2O$	N-Dehydrocamphenyl-2-hydroxydecanamide†	—	134

Note: References 127–149 are on p. 325.

* Potassium cyanide was used as the nitrile source.

† The product corresponded in composition to this formula; its structure was not established.

TABLE II-H. POLYPROPYLENES

Polymer Reactant	Nitrile	Reaction Medium	Product*	Yield (%)	Ref.
Propylene tetramer	$(CH_3)_3CCH_2CH(CH_3)CH_2CHOHCN$	H_2SO_4	n-$C_3H_7[CH(CH_3)CH_2]_3NHCO$ / $(CH_3)_3CCH_2CH(CH_3)CH_2CHOH$	—	111
	n-$C_{18}H_{37}CHOHCN$	H_2SO_4	$ANHCOCHOH(CH_2)_{17}CH_3$	59	134
	n-$C_3H_7CH{=}C(C_2H_5)CHOHCN$	$H_2SO_4/(n\text{-}C_4H_9)_2O$	$ANHCOCHOHC(C_2H_5){=}CHC_3H_7$-$n$	75	134
	n-$C_4H_9CH(C_2H_5)CHOHCN$	$H_2SO_4/(n\text{-}C_4H_9)_2O$	$ANHCOCHOHCH(C_2H_5)C_4H_9$-$n$	89	134
	n-$C_6H_{13}CHOHCN$	$H_2SO_4/(n\text{-}C_4H_9)_2O$	$ANHCOCHOHC_6H_{13}$-n†	—	134
	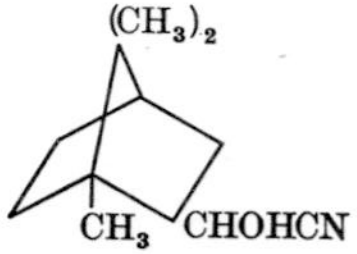 $(CH_3)_2$ CH_3 CHOHCN	$H_2SO_4/(n\text{-}C_4H_9)_2O$	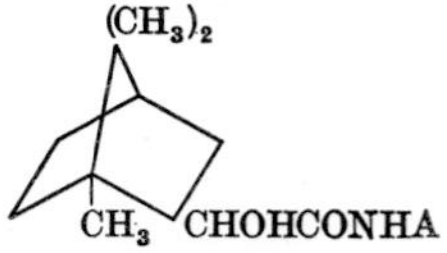 $(CH_3)_2$ CH_3 CHOHCONHA	—	134
	n-$C_8H_{17}CHOHCN$	80% H_2SO_4	ANHCOCHOHCONHA	91	134
	$(CH_3)_3CCH_2CH(CH_3)CH_2CHOHCN$	$BF_3 \cdot H_2O$	$ANHCOCHOHCH_2CH(CH_3)CH_2$-$C(CH_3)_3$†	—	134
Propylene polymer	ClCN	H_2SO_4	Amine (primary)	27	134
	$(CH_3)_3CCH_2CH(CH_3)CH_2CHOHCN$	H_2SO_4	$ANHCOCHOHCH_2CH(CH_3)CH_2$-$C(CH_3)_3$†	91	134
	n-$C_6H_{13}CHOHCN$	$BF_3 \cdot H_2O$	$C_{14}H_{29}NHCOCHOHC_6H_{13}$-$n$†	—	134
	$(CH_3)_3CCH_2CH(CH_3)CH_2CHOHCN$	$BF_3 \cdot H_2O$	$C_{12}H_{25}NHCOCHOHCH_2$-$CH(CH_3)CH_2C(CH_3)_3$†	—	134

Note: References 127–149 are on p. 325.

* The A in the product is the portion coming from the polypropylene.

† The product corresponded in composition to this formula; its structure was not established.

TABLE II-I. CHLOROALKENES

$$CH_2{=}C(CH_3)CH_2Cl + RCN \xrightarrow{H_2SO_4} (CH_3)_2C(CH_2Cl)(NHCOR)$$

R	Yield (%)	Refs.
CH_3	—	89
$CH_2{=}CH$	92	89, 133
H_2NCOCH_2	—	89
$C_2H_5O_2CCH_2$	—	89
C_6H_5	—	89
$C_6H_5CH_2$	—	89

Note: References 127–149 are on p. 325.

TABLE III. CYCLOALKENES

Cycloalkene	Nitrile	Reaction Medium	Product(s)	Yield (%)	Refs.
	ClCN	H_2SO_4	 NH_2	14	111
	$CH_2{=}CHCN$	H_2SO_4	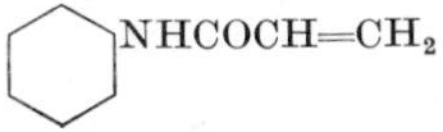 $NHCOCH{=}CH_2$	81	28
	C_2H_5CN	Polyphosphoric acid	 $NHCOC_2H_5$	37	46
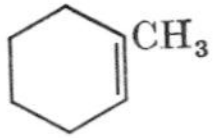 CH_3	HCN	$HBF_4/(C_2H_5)_2O$	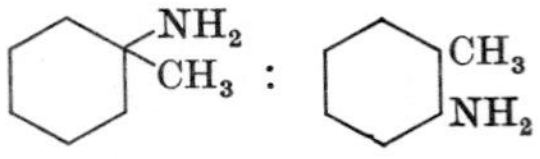 NH_2, CH_3 : CH_3, NH_2 (19.6:1)	37	25
	NaCN	H_2SO_4/CH_3CO_2H	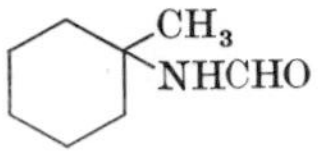 CH_3, NHCHO	65	37
 CH_3, CH_3	NaCN	H_2SO_4/CH_3CO_2H	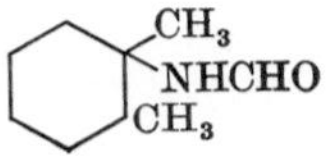 CH_3, NHCHO, CH_3	65	37
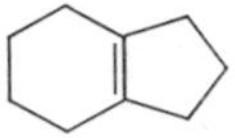	KCN	$H_2SO_4/(n\text{-}C_4H_9)_2O$	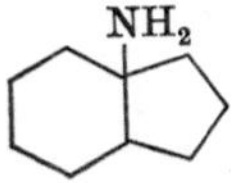 NH_2	35	6
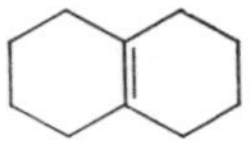	KCN	$H_2SO_4/(n\text{-}C_4H_9)_2O$	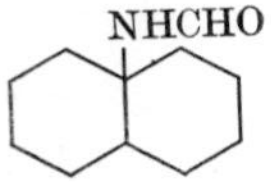 NHCHO	95	6

Note: References 127–149 are on p. 325.

TABLE III. CYCLOALKENES (*Continued*)

Cycloalkene	Nitrile	Reaction Medium	Product(s)	Yield (%)	Refs.
 (Pinene)	HCN	H_2SO_4	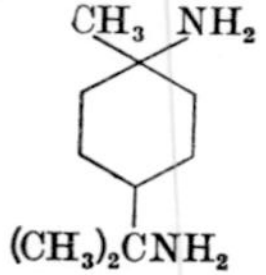 CH_3 NH_2 / $(CH_3)_2CNH_2$	37	38, 39
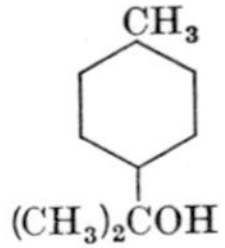 CH_3 / $(CH_3)_2COH$	HCN	H_2SO_4	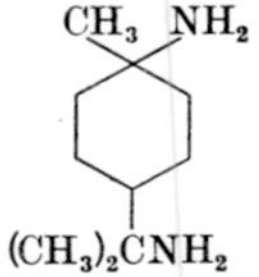 CH_3 NH_2 / $(CH_3)_2CNH_2$	53	38, 39
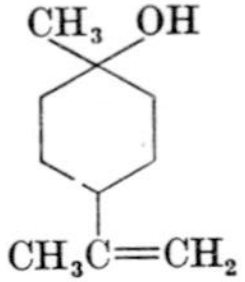 CH_3 OH / $CH_3C{=}CH_2$	HCN	H_2SO_4	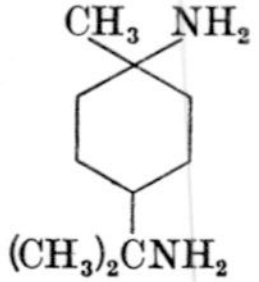 CH_3 NH_2 / $(CH_3)_2CNH_2$	—	38, 39
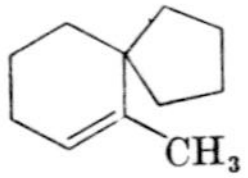 CH_3	KCN	$H_2SO_4/(n\text{-}C_4H_9)_2O$	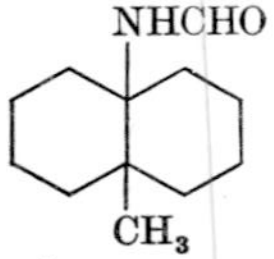 NHCHO / CH_3	Quant.	6
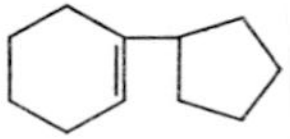	KCN	$H_2SO_4/(n\text{-}C_4H_9)_2O$	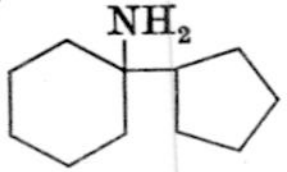 NH_2	40	6
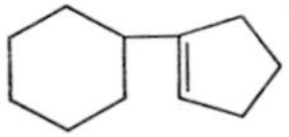	KCN	$H_2SO_4/(n\text{-}C_4H_9)_2O$	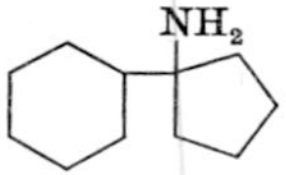 NH_2	25	6
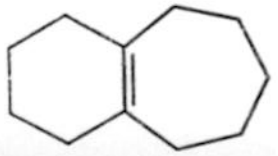	KCN	$H_2SO_4/(n\text{-}C_4H_9)_2O$	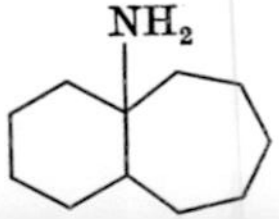 NH_2	40	6

TABLE IV. ALKADIENES

Diene	Nitrile	Reaction Medium	Product	Yield (%)	Ref.
$H_2C{=}C(CH_3)CH_2CH_2C(CH_3){=}CH_2$	HCN	H_2SO_4	$(CH_3)_2C(NH_2)CH_2CH_2C(NH_2)(CH_3)_2$	42	52
CH_3 … $CH_3C{=}CH_2$	HCN	H_2SO_4	CH_3, NH_2 … $(CH_3)_2CNH_2$	37	38, 39
$H_2C{=}C(CH_3)(CH_2)_7CH{=}CHCH_3$	$NC(CH_2)_4CN$	90% HCO_2H	Polyamide	—	53
$H_2C{=}C(CH_3)(CH_2)_8CH{=}CHCH_3$	$NC(CH_2)CN$	90% HCO_2H	Polyamide	—	53
	$NC(CH_2)CN$	72% H_2SO_4	Polyamide	—	53
	$H_2C{=}C(CH_3)CH_2CH_2$-$COCH_2CH_2CN$	72% H_2SO_4	Polyamide	—	53
	$H_2C{=}C(CH_3)CH_2CN$	H_2SO_4	Polyamide	—	53
	$C_6H_4(CN)_2$-1,4	90% HCO_2H	Polyamide	—	53
	$C_6H_4(CH_2CN)_2$-1,4	90% HCO_2H	Polyamide	—	53

Note: References 127–149 are on p. 325.

TABLE V. Primary Alcohols

(Sulfuric acid was used as the reaction medium unless otherwise noted.)

Alcohol	Nitrile	Product	Yield (%)	Refs.
$C_6H_5CH_2OH$	CH_3CN	$C_6H_5CH_2NHCOCH_3$	72	56–58
	$CH_2{=}CHCN$	$C_6H_5CH_2NHCOCH{=}CH_2$	62	56–58
	C_2H_5CN	$C_6H_5CH_2NHCOC_2H_5$	45	58
	C_6H_5CN	$C_6H_5CH_2NHCOC_6H_5$	55	58
	$C_6H_5CH_2CN$	$C_6H_5CH_2NHCOCH_2C_6H_5$	27	58
$4\text{-}CH_3C_6H_4CH_2OH$	CH_3CN	$4\text{-}CH_3C_6H_4CH_2NHCOCH_3$	40	57
$4\text{-}CH_3OC_6H_4CH_2OH$	CH_3CN	$4\text{-}CH_3OC_6H_4CH_2NHCOCH_3$	60	56, 57
$2,4\text{-}(CH_3)_2C_6H_3CH_2OH$	CH_3CN	$2,4\text{-}(CH_3)_2C_6H_3CH_2NHCOCH_3$	87	57
$3,5\text{-}(CH_3)_2C_6H_3CH_2OH$	CH_3CN	$3,5\text{-}(CH_3)_2C_6H_3CH_2NHCOCH_3$	40	56
$C_6H_{11}CH_2OH$ (cyclohexyl-CH_2OH)	KCN*	1-CH_3-1-NH_2-cyclohexane	5	16

Note: References 127–149 are on p. 325.

* Di-*n*-butyl ether was used with the sulfuric acid.

TABLE VI. SECONDARY ALCOHOLS

Alcohol	Nitrile	Reaction Medium	Product	Yield (%)	Refs.
$CH_3CHOHCH_2Cl$	$C_6H_5CH_2CN$	H_2SO_4	$CH_3CH(CH_2Cl)NHCOCH_2C_6H_5$	40	44
$(CH_3)_2CHOH$	$NCCH_2CN$	H_2SO_4/CH_3CO_2H	$(CH_3)_2CHNHCOCH_2CONHCH(CH_3)_2$	40	55
	$CH_2{=}CHCN$	H_2SO_4	$(CH_3)_2CHNHCOCH{=}CH_2$	44	28, 47
	$NCCH{=}CHCN$	H_2SO_4/CH_3CO_2H	$(CH_3)_2CHNHCOCH{=}CHCONHCH(CH_3)_2$	80	55
$CH_3CHOHC_2H_5$	$CH_2{=}CHCN$	H_2SO_4	$CH_3CH(C_2H_5)NHCOCH{=}CH_2$	76	28, 133
$(C_2H_5)_2CHOH$	$CH_2{=}CHCN$	H_2SO_4	$(C_2H_5)_2CHNHCOCH{=}CH_2$	77	133
[cyclobutyl]$CHOHCH_3$	KCN	$H_2SO_4/(n\text{-}C_4H_9)_2O$	[1-methylcyclopentyl] CH_3, NH_2	40	16
[cyclopentyl]$CHOHCH_3$	KCN	$H_2SO_4/(n\text{-}C_4H_9)_2O$	[1-methylcyclohexyl] CH_3, NH_2	45	16
$C_6H_5CHOHCH_2Br$	CH_3CN	H_2SO_4	$C_6H_5CH(CH_2Br)NHCOCH_3$	92	44
$C_6H_5CHOHCH_2Cl$	CH_3CN	H_2SO_4	$C_6H_5CH(CH_2Cl)NHCOCH_3$	78	44
$C_6H_5CHOHCH_3$	KCN	$H_2SO_4/(n\text{-}C_4H_9)_2O$	$C_6H_5CHNH_2CH_3$	60	8
	C_6H_5CN	H_2SO_4	$C_6H_5CH(CH_3)NHCOC_6H_5$	100	8, 135
[cyclopentyl]$CHOHC_2H_5$	KCN	$H_2SO_4/(n\text{-}C_4H_9)_2O$	[1-ethylcyclohexyl] C_2H_5, NH_2	45	16
[cyclopentyl]$CH_2CHOHCH_3$	KCN	$H_2SO_4/(n\text{-}C_4H_9)_2O$	[1-ethylcyclohexyl] C_2H_5, NH_2	54	16
[1-methylcyclopentyl] CH_3, $CHOHCH_3$	KCN	$H_2SO_4/(n\text{-}C_4H_9)_2O$	[cyclohexyl] CH_3, NH_2, CH_3	20	16

Note: References 127–149 are on p. 325.

TABLE VI. SECONDARY ALCOHOLS (*Continued*)

Alcohol	Nitrile	Reaction Medium	Product	Yield (%)	Refs.
 [cyclohexane ring]$CHOHCH_3$	KCN	$H_2SO_4/(n\text{-}C_4H_9)_2O$	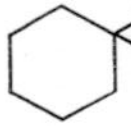 [cyclohexane ring]$(C_2H_5)NH_2$	30	16
$n\text{-}C_6H_{13}CHOHCH_3$	CH_3CN	$H_4Fe(CN)_6$	$n\text{-}C_6H_{13}CH(CH_3)NHCOCH_3$	20	109
$C_6H_5CHOHC_2H_5$	KCN	$H_2SO_4/(n\text{-}C_4H_9)_2O$	$C_6H_5CH(NH_2)C_2H_5$	60	8
	C_6H_5CN	H_2SO_4	$C_6H_5CH(C_2H_5)NHCOC_6H_5$	90	8, 135
$C_6H_5CH_2CHOHCH_3$	KCN	$H_2SO_4/(n\text{-}C_4H_9)_2O$	$C_6H_5CH_2CH(NH_2)CH_3$	60	8
	C_6H_5CN	H_2SO_4	$C_6H_5CH_2CH(CH_3)NHCOC_6H_5$	75	8
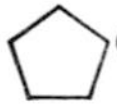 [cyclopentane ring]$CHOHC_3H_7\text{-}n$	KCN	$H_2SO_4/(n\text{-}C_4H_9)_2O$	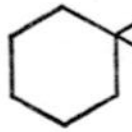 [cyclohexane ring]$(C_3H_7\text{-}n)NH_2$	40	16
 [cyclohexane ring]$CHOHC_2H_5$	KCN	$H_2SO_4/(n\text{-}C_4H_9)_2O$	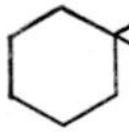 [cyclohexane ring]$(C_3H_7\text{-}n)NH_2$	40	16
 [cyclohexane ring]$CH_2CHOHC_2H_5$	KCN	$H_2SO_4/(n\text{-}C_4H_9)_2O$	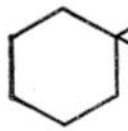 [cyclohexane ring]$(C_3H_7\text{-}n)NH_2$	—	16
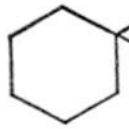 [cyclohexane ring]$(CH_3)CHOHCH_3$	KCN	$H_2SO_4/(n\text{-}C_4H_9)_2O$	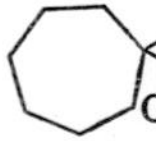 [cycloheptane ring]$(CH_3)NH_2$, CH_3	15	16
 [cycloheptane ring]$CHOHCH_3$	KCN	$H_2SO_4/(n\text{-}C_4H_9)_2O$	 [cycloheptane ring]$(C_2H_5)NH_2$	15	16
$C_6H_5CHOHCH(CH_3)_2$	KCN	$H_2SO_4/(n\text{-}C_4H_9)_2O$	$C_6H_5CHNH_2CH(CH_3)_2$	60	8
	C_6H_5CN	H_2SO_4	$C_6H_5CH(C_3H_7\text{-}i)NHCOC_6H_5$	98	8, 135
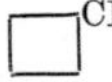 [cyclobutane ring]$CHOHC_6H_5$	KCN	$H_2SO_4/(n\text{-}C_4H_9)_2O$	[cyclobutane ring]$CHNH_2C_6H_5$	40	8
$C_6H_5CHOHC(CH_3)_3$	HCN	$H_2SO_4/(n\text{-}C_4H_9)_2O$	$C_6H_5C(CH_3)(NH_2)C_3H_7\text{-}i$ $C_6H_5CHNH_2C(CH_3)_3$ (9:1)	50	9
	C_6H_5CN	H_2SO_4	$C_6H_5CH(NHCOC_6H_5)C(CH_3)_3$	60	9

$C_6H_5CH_2CHOHCH(CH_3)_2$	KCN	$H_2SO_4/(n\text{-}C_4H_9)_2O$	$C_6H_5CH_2C(CH_3)_2NH_2$	65	8
$C_6H_5CHOHCH_2CH(CH_3)_2$	KCN	$H_2SO_4/(n\text{-}C_4H_9)_2O$	$C_6H_5CHNH_2CH_2CH(CH_3)_2$	40	8
$C_6H_5C(CH_3)_2CHOHCH_3$	HCN	$H_2SO_4/(n\text{-}C_4H_9)_2O$	$C_6H_5C(CH_3)(NH_2)CH(CH_3)_2$	55	9
[cyclopentyl]$CHOHC_6H_5$	KCN	$H_2SO_4/(n\text{-}C_4H_9)_2O$	[cyclopentyl]$CH(NH_2)C_6H_5$	55	8
	C_6H_5CN	H_2SO_4	[cyclopentyl]$CH(C_6H_5)NHCOC_6H_5$	100	8, 135
$C_6H_5CH(CH_3)CHOH$-$CH(CH_3)_2$	HCN	$H_2SO_4/(n\text{-}C_4H_9)_2O$	$C_6H_5CH(CH_3)CH_2C(CH_3)_2NH_2$	30	9
$(C_6H_5)_2CHOH$	HCN	H_2SO_4	$(C_6H_5)_2CHNHCHO$	86	58
	CH_3CN	H_2SO_4	$(C_6H_5)_2CHNHCOCH_3$	85	58
	$CH_2{=}CHCN$	H_2SO_4	$(C_6H_5)_2CHNHCOCH{=}CH_2$	70	58
	$NCCH_2CO_2H$	H_2SO_4	$(C_6H_5)_2CHNHCOCH_2CO_2H$	75	58
	C_2H_5CN	H_2SO_4	$(C_6H_5)_2CHNHCOC_2H_5$	97	58
	$NC(CH_2)_2CN$	H_2SO_4	$(C_6H_5)_2CHNHCO(CH_2)CONHCHN(C_6H_5)_2$	64	58
	$NCCH_2CO_2C_2H_5$	H_2SO_4	$(C_6H_5)_2CHNHCOCH_2CO_2C_2H_5$	71	58
	C_6H_5CN	H_2SO_4	$(C_6H_5)_2CHNHCOC_6H_5$	86	10, 58, 135
[cyclohexyl]$CHOHC_6H_5$	KCN	$H_2SO_4/(n\text{-}C_4H_9)_2O$	[cyclohexyl]$CHNH_2C_6H^5$	65	8
	C_6H_5CN	H_2SO_4	[cyclohexyl]$CH(C_6H_5)NHCOC_6H_5$	90	8, 135
$C_6H_5CH_2CHOHC_6H_5$	C_6H_5CN	H_2SO_4	$C_6H_5CH_2CH(C_6H_5)NHCOC_6H_5$	66	135
[ring]$NCH_2CH_2CHOHC_6H_5$	C_6H_5CN	H_2SO_4	[ring]$NCH_2CH_2CH(C_6H_5)NHCOC_6H_5$	31	136
$(C_6H_5)_2CHCHOHCH_3$	KCN	$H_2SO_4/(n\text{-}C_4H_9)_2O$	$C_6H_5CH(CH_3)CHNH_2C_6H_5$	60	10
$C_6H_5CH(CH_3)CHOHC_6H_5$	KCN	$H_2SO_4/(n\text{-}C_4H_9)_2O$	$C_6H_5CH(CH_3)CHNH_2C_6H_5$	60	10
$(C_6H_5)_2CHCHOHC_6H_5$	HCN	$H_2SO_4/(n\text{-}C_4H_9)_2O$	$(C_6H_5)_2CHCHNH_2C_6H_5$	—	21
	CH_3CN	$H_2SO_4/(n\text{-}C_4H_9)_2O$	$(C_6H_5)_2CHCH(C_6H_5)NHCOCH_3$	60	21*
	C_6H_5CN	$H_2SO_4/(n\text{-}C_4H_9)_2O$	$(C_6H_5)_2CHCH(C_6H_5)NHCOC_6H_5$	60	21*

Note: References 127–149 are on p. 325.

* Ritter reactions of 1,2,2-triphenylethanol-1-^{14}C are also described in this reference.

TABLE VI-A. Secondary Alicyclic Alcohols

Alcohols	Nitrile	Reaction Medium	Product(s)	Yield (%)	Refs.
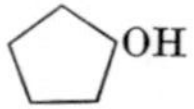 OH	KCN	$H_2SO_4/(n\text{-}C_4H_9)_2O$	 NH_2	28	15, 137
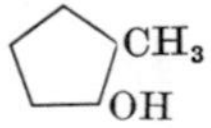 CH_3, OH	KCN	$H_2SO_4/(n\text{-}C_4H_9)_2O$	 CH_3, NH_2	30	16
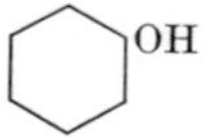 OH	KCN	$H_2SO_4/(n\text{-}C_4H_9)_2O$	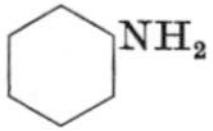 NH_2	40	16, 137
	$CH_2=CHCN$	H_2SO_4	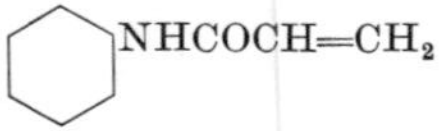 $NHCOCH=CH_2$	72	28, 133
	$NCCH=CHCN$	$H_2SO_4/(n\text{-}C_4H_9)_2O$	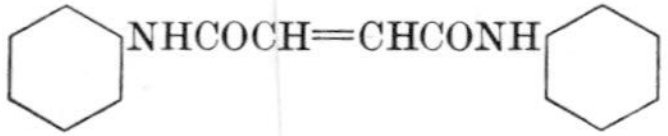 $NHCOCH=CHCONH$	87	55
 CH_3, OH	KCN	$H_2SO_4/(n\text{-}C_4H_9)_2O$	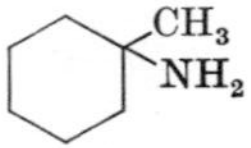 CH_3, NH_2	35	16, 137
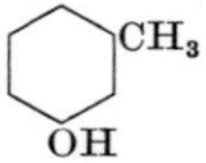 CH_3, OH	KCN	$H_2SO_4/(n\text{-}C_4H_9)_2O$	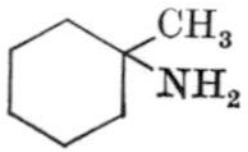 CH_3, NH_2	40	16
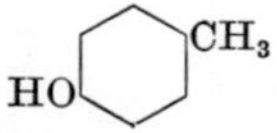 HO, CH_3	KCN	$H_2SO_4/(n\text{-}C_4H_9)_2O$	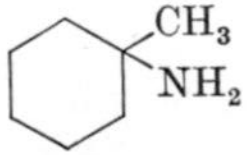 CH_3, NH_2	8	16, 137

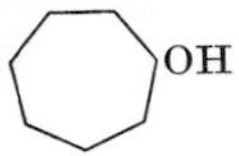 OH	KCN	$H_2SO_4/(n\text{-}C_4H_9)_2O$	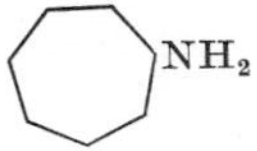 NH_2	35	15
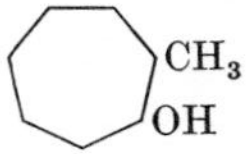 CH_3 OH	KCN	$H_2SO_4/(n\text{-}C_4H_9)_2O$	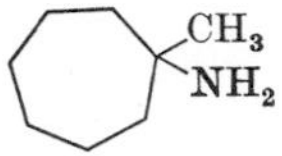 CH_3 NH_2	17	16
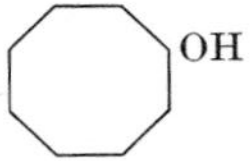 OH	KCN	$H_2SO_4/(n\text{-}C_4H_9)_2O$	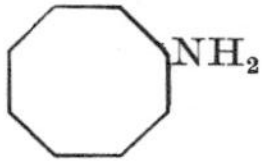 NH_2	20	15
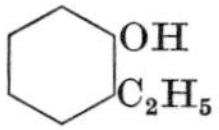 OH C_2H_5	KCN	$H_2SO_4/(n\text{-}C_4H_9)_2O$	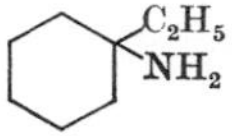 C_2H_5 NH_2	32	16
 CH_3 CH_3 OH	HCN	H_2SO_4	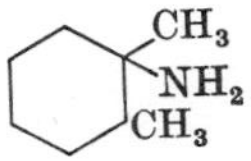 CH_3 NH_2 CH_3	46	13
	HCN	$H_2SO_4/(n\text{-}C_4H_9)_2O$	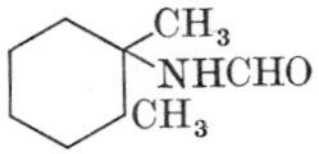 CH_3 NHCHO CH_3	40	60
	KCN	$H_2SO_4/(n\text{-}C_4H_9)_2O$	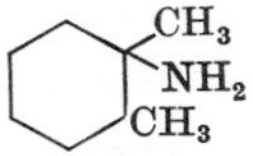 CH_3 NH_2 CH_3	20	16
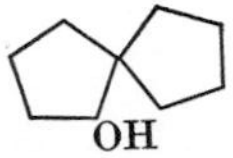 OH	KCN	$H_2SO_4/(n\text{-}C_4H_9)_2O$	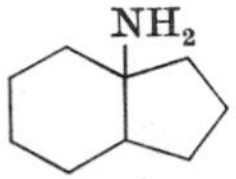 NH_2	35	6, 13

Note: References 127–149 are on p. 325.

TABLE VI-A. SECONDARY ALICYCLIC ALCOHOLS (*Continued*)

Alcohols	Nitrile	Reaction Medium	Product(s)	Yield (%)	Refs.
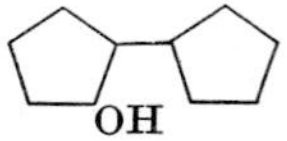 OH	KCN	$H_2SO_4/(n\text{-}C_4H_9)_2O$	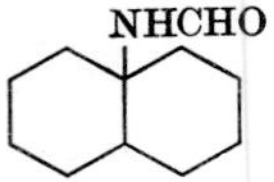 NHCHO	30	6
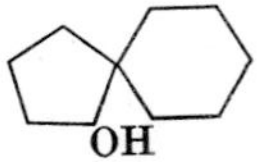 OH	KCN	$H_2SO_4/(n\text{-}C_4H_9)_2O$	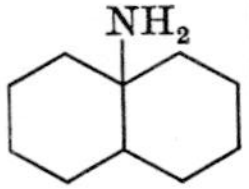 NH_2	10	6
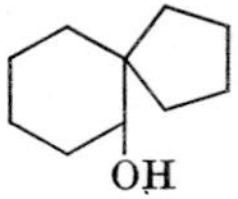 OH	KCN	$H_2SO_4/(n\text{-}C_4H_9)_2O$	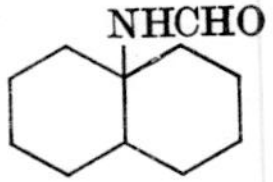 NHCHO	95	6
	CH_3CN	96% H_2SO_4	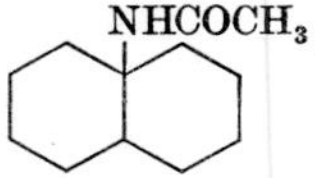 $NHCOCH_3$	90	11
	CH_3CN	96% H_2SO_4/CCl_4	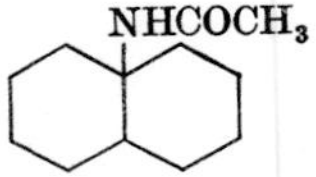 $NHCOCH_3$	15	11
	CH_3CN	100% $H_2SO_4/(n\text{-}C_4H_9)_2O$	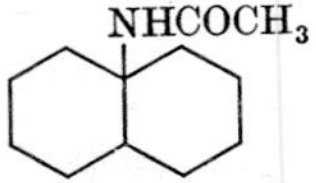 $NHCOCH_3$	53	11

	CH_3CN	100% H_2SO_4/CH_3CN	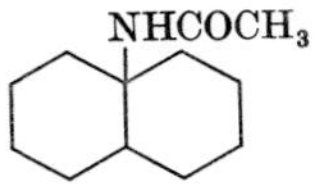 $NHCOCH_3$	65	11
	CH_3CN	96% H_2SO_4/CH_3CN	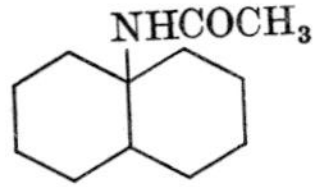 $NHCOCH_3$	70	11
	CH_3CN	96% $H_2SO_4/n\text{-}C_7H_{16}$	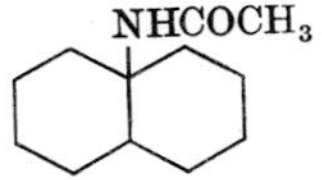 $NHCOCH_3$	10	11
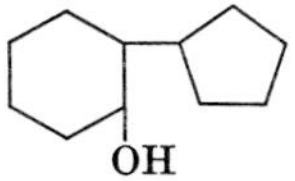 OH	KCN	$H_2SO_4/(n\text{-}C_4H_9)_2O$	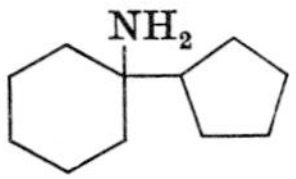 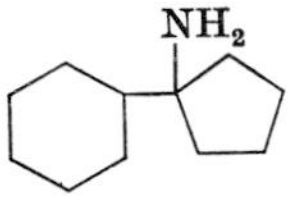 NH_2 NH_2	20	6
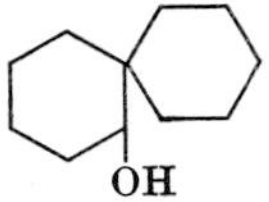 OH	KCN	$H_2SO_4/(n\text{-}C_4H_9)_2O$	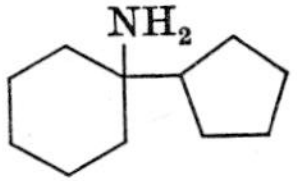 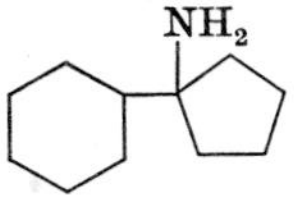 NH_2 NH_2	20	6
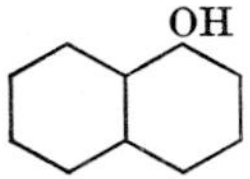 OH	CH_3CN	H_2SO_4	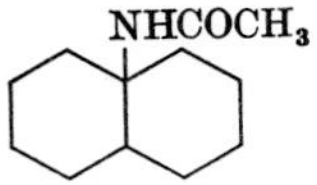 $NHCOCH_3$	90	11
	CH_3CN	H_2SO_4/CCl_4	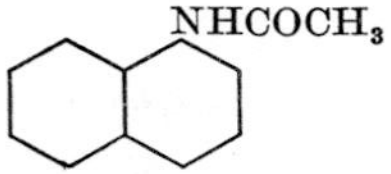 $NHCOCH_3$	25	11

Note: References 127–149 are on p. 325.

TABLE VI-A. Secondary Alicyclic Alcohols (*Continued*)

Alcohols	Nitrile	Reaction Medium	Product(s)	Yield (%)	Refs.
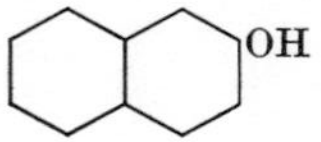 OH	HCN	H_2SO_4	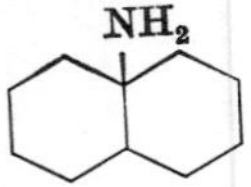 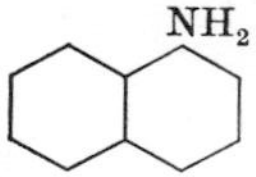 NH_2 ; NH_2	72	11, 25
	KCN	$H_2SO_4/(n\text{-}C_4H_9)_2O$	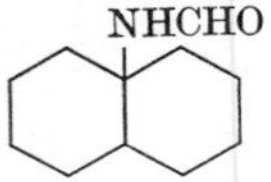 NHCHO	85	6
	NaCN	$H_2SO_4/(n\text{-}C_4H_9)_2O$	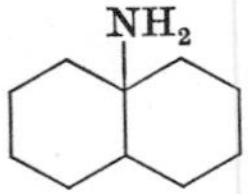 NH_2	22	11
	CH_3CN	H_2SO_4	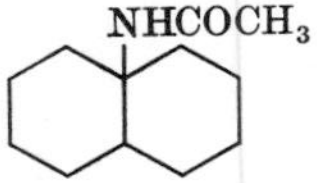 $NHCOCH_3$	20	11
	CH_3CN	H_2SO_4/CCl_4	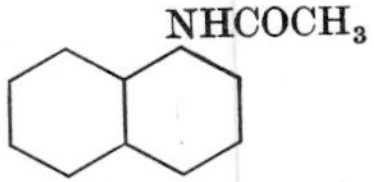 $NHCOCH_3$	2	11
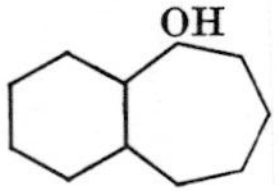 OH	KCN	$H_2SO_4/(n\text{-}C_4H_9)_2O$	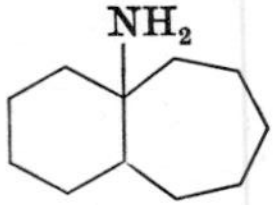 NH_2	20	6
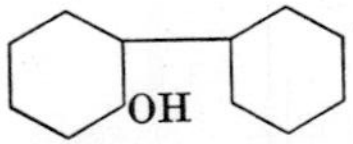 OH	KCN	$H_2SO_4/(n\text{-}C_4H_9)_2O$	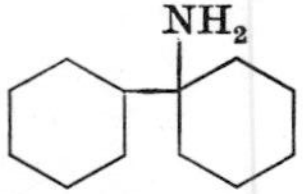 NH_2	40	16

Note: References 127–149 are on p. 325.

TABLE VI-B. HYDROXY ESTERS

Ester	Nitrile	Reaction Medium	Product(s)	Yield (%)	Ref.
$CH_3CHOHCO_2C_2H_5$	C_6H_5CN	H_2SO_4	No reaction	—	63
$(CH_3)_2C(OH)CH_2CO_2C_2H_5$	C_6H_5CN	H_2SO_4	$(CH_3)_2C(NHCOC_6H_5)CH_2CO_2H$	62	63
$C_2H_5C(CH_3)(OH)CH_2CO_2C_2H_5$	C_6H_5CN	H_2SO_4	$C_2H_5C(CH_3)(NHCOC_6H_5)CH_2CO_2H$	76	63
[cyclopentane ring] $-CH_2CO_2C_2H_5$, $-OH$	C_6H_5CN	H_2SO_4	[cyclopentane ring] $-CH_2CO_2H$, $-NHCOC_6H_5$	40	63
n-$C_3H_7C(CH_3)(OH)CH_2CO_2C_2H_5$	C_6H_5CN	H_2SO_4	n-$C_3H_7C(CH_3)(NHCOC_6H_5)CH_2CO_2H$	72	63
$(C_2H_5)_2C(OH)CH_2CO_2H_5$	C_6H_5CN	H_2SO_4	$(C_2H_5)_2C(NHCOC_6H_5)CH_2CO_2H$	78	63
$(CH_3)_2C(OH)C(CH_3)_2CO_2C_2H_5$	C_6H_5CN	H_2SO_4	$(CH_3)_2C(NHCOC_6H_5)C(CH_3)_2CO_2H$	80	63
$C_6H_5C(CH_3)(OH)CH_2CO_2C_2H_5$	CH_3CN	H_2SO_4	$C_6H_5C(CH_3)(NHCOCH_3)CH_2CO_2C_2H_5$	64	62
	$CH_2{=}CHCN$	H_2SO_4	$C_6H_5C(CH_3)(NHCOCH{=}CH_2)CH_2CO_2C_2H_5$	62	62
	C_2H_5CN	H_2SO_4	$C_6H_5C(CH_3)(NHCOC_2H_5)CH_2CO_2C_2H_5$	73	62
[cyclohexane ring] $-CH_2CO_2C_2H_5$, $-OH$	C_6H_5CN	H_2SO_4	[cyclohexane ring] $-CH_2CO_2H$, $-NHCOC_6H_5$	65	63

Note: References 127–149 are on p. 325.

TABLE VI-B. HYDROXY ESTERS (*Continued*)

Ester	Nitrile	Reaction Medium	Product(s)	Yield (%)	Ref.
[cyclopentane ring] $-CH(CH_3)CO_2C_2H_5$, OH	C_6H_5CN	H_2SO_4	[cyclopentane ring] $-CH(CH_3)CO_2H$, $NHCOC_6H_5$	44	63
$(CH_3)_2C(OH)CH(C_3H_7\text{-}i)CO_2C_2H_5$	C_6H_5CN	H_2SO_4	$(CH_3)_2CCH(C_3H_7\text{-}i)CO_2H$	58	63
[cyclohexane ring] $-CH(CH_3)CO_2C_2H_5$, OH	C_6H_5CN	H_2SO_4	[cyclohexane ring] $-CH(CH_3)CO_2H$, $NHCOC_6H_5$	58	63
$n\text{-}C_5H_{11}C(CH_3)(OH)CH_2CO_2C_2H_5$	$4\text{-}CH_3C_6H_4CN$	H_2SO_4	$n\text{-}C_5H_{11}C(CH_3)CH_2CO_2H$ | $NHCOC_6H_4CH_3\text{-}4$	67	63
$4\text{-}ClC_6H_4C(CH_3)(OH)CH_2CO_2C_2H_5$	CH_3CN	H_2SO_4	$4\text{-}ClC_6H_4C(CH_3)CH_2CO_2C_2H_5$ | $NHCOCH_3$	72	62
$C_6H_5C(C_2H_5)(OH)CH_2CO_2C_2H_5$	CH_3CN	H_2SO_4	$C_6H_5C(C_2H_5)CH_2CO_2C_2H_5$ | $NHCOCH_3$	70	62
$C_6H_5CH_2C(CH_3)(OH)CH_2CO_2C_2H_5$	C_6H_5CN	H_2SO_4	$C_6H_5CH_2C(CH_3)CH_2CO_2H$ | $NHCOC_6H_5$	9	63
$(C_6H_5)_2C(OH)CH_2CO_2C_2H_5$	CH_3CN	H_2SO_4	$(C_6H_5)_2CCH_2CO_2C_2H_5$ | $NHCOCH_3$	42	62

Note: References 127–149 are on p. 325.

TABLE VII-A. *t*-Butyl Alcohol

$(CH_3)_3COH + RCN \rightarrow (CH_3)_3CNHCOR$

$(CH_3)_3COH + NC(R)CN \rightarrow (CH_3)_3CNHCO(R)CONHC(CH_3)_3$

R	Reaction Medium	Yield (%)	Refs.
H	H_2SO_4/CH_3CO_2H	9	4
H	$H_4Fe(CN)_6$	10	109
(CH_2)	H_2SO_4/CH_3CO_2H	20	55
$(O_2N)_2CH$	t-C_4H_9OH	9	138
NH_2CO	H_2SO_4/CH_3CO_2H	62	131
$CH_2{=}CH$	H_2SO_4/CH_3CO_2H	88	28, 55, 133
	41–55% H_2SO_4	80	22
$CH_2{=}CH(+Cl_2)$	H_2SO_4	75*	26
CH_3CHOH	H_2SO_4/CH_3CO_2H	40	133
$(CH_2)_2$	H_2SO_4/CH_3CO_2H	25	55
$CH_2{=}C(CH_3)$	H_2SO_4/CH_3CO_2H	55	133
$(CH_3)_2CHCHOH$	H_2SO_4	—	108
$CH_3(CH_2)_2CHOH$	H_2SO_4	—	108
$(CH_3)_2CHCHNHCH_3$	H_2SO_4	—	108
$(CH_3)_3CCHOH$	H_2SO_4	—	108
$(C_2H_5)_2NCH_2$	H_2SO_4/CH_3CO_2H	95	32
C_6H_5	H_2SO_4/CH_3CO_2H	69	139
$(CH_3)_2CHCH[N(CH_3)_2]$	H_2SO_4	—	108
$(CH_3)_2CHCH(NHC_2H_5)$	H_2SO_4	—	108
$(C_2H_5)_2CHCHOH$	H_2SO_4	—	108
$C_6H_5CH_2$	H_2SO_4/CH_3CO_2H	—	139
$(CH_3)_2CHCH[N(CH_3)CH_2CH_2OH]$	H_2SO_4	—	139
$C_6H_5CH{=}CH$	H_2SO_4	70	108
$C_6H_5CH_2CHOH$	H_2SO_4	—	108
n-$C_3H_7CH(-N(CH_2CH_2)_2O)$	H_2SO_4	—	108
i-$C_3H_7CH(-N(CH_2CH_2)_2O)$	H_2SO_4	—	108
i-$C_3H_7CH[N(C_2H_5)_2]$	H_2SO_4	—	108
i-$C_3H_7CH(-N(CH_2CH_2)_2NCOCH_3)$	H_2SO_4	—	108
$(C_2H_5)_2CHCH(-N(CH_2CH_2)_2O)$	H_2SO_4	—	108
$(C_6H_5)_2CH$	H_2SO_4/CH_3CO_2H	100	34
$C_6H_5CH_2CH(-N(CH_2CH_2)_2O)$	H_2SO_4	—	108
$(C_6H_5)_2CHCHOH$	H_2SO_4	—	108

Note: References 127–149 are on p. 325.

* The product is $(CH_3)_3CNHCOCHClCH_2Cl$.

TABLE VII-A. *t*-Butyl Alcohol (*Continued*)

R	Reaction Medium	Yield (%)	Ref.
$(C_6H_5CH_2)_2CHCHOH$	H_2SO_4	—	108
$(C_6H_5CH_2)_2CHCH(NHCH_3)$	H_2SO_4	87	140
$(C_6H_5)_2CHCH$(—N(morpholino)O)	H_2SO_4	—	108
$(C_6H_5CH_2)_2CHCH$(—N(morpholino)O)	H_2SO_4	—	108

Note: References 127–149 are on p. 325.

TABLE VII-B. Tertiary Alcohols

(See also Table VII-A.)

$$\begin{matrix} R \\ \diagdown \\ \end{matrix} C(CH_3)OH + R''CN \rightarrow C(CH_3)NHCOR''$$

$$RR'C(CH_3)OH + R''CN \rightarrow RR'C(CH_3)NHCOR''$$

R	R′	R″	Reaction Medium	Yield (%)	Ref.
CH_3	CH_2Cl	CH_3	H_2SO_4	55	44
CH_3	CH_2Cl	$C_2H_5O_2CCH_2$	H_2SO_4	65	44
CH_3	CH_2Cl	C_6H_5	H_2SO_4	40	44
CH_3	CH_2Cl	$C_6H_5CH_2$	H_2SO_4	74	44
C_2H_5	n-C_3H_7	$CH_2{=}CH$	H_2SO_4/CH_3CO_2H	82	133
C_2H_5	n-C_3H_7	$C_6H_5CH{=}CH$	H_2SO_4/CH_3CO_2H	61	133
C_2H_5	n-C_4H_9	O(morpholino)NCH_2	H_2SO_4/CH_3CO_2H	86	33
C_2H_5	n-C_4H_9	(piperidino)NCH_2	H_2SO_4/CH_3CO_2H	71	33
C_2H_5	i-C_4H_9	$CH_2{=}CH$	H_2SO_4/CH_3CO_2H	57	133
C_2H_5	$C_6H_5CH_2$	H	H_2SO_4	58	129
n-C_3H_7	n-C_4H_9	$C_6H_5CH{=}CH$	H_2SO_4/CH_3CO_2H	74	133
n-C_3H_7	i-C_4H_9	$C_6H_5CH{=}CH$	H_2SO_4/CH_3CO_2H	69	133
n-C_3H_7	n-C_6H_{13}	$C_6H_5CH{=}CH$	H_2SO_4/CH_3CO_2H	79	133
n-C_4H_9	i-C_4H_9	$CH_2{=}CH$	H_2SO_4/CH_3CO_2H	36	133
n-C_4H_9	i-C_4H_9	$C_6H_5CH{=}CH$	H_2SO_4/CH_3CO_2H	67	133
C_6H_5	CH_2Cl	CH_3	H_2SO_4	84	44
C_6H_5	i-C_3H_7	H	$H_2SO_4/(n\text{-}C_4H_9)_2O$	60*	9
C_6H_5	i-C_4H_9	H	$H_2SO_4/(n\text{-}C_4H_9)_2O$	20*	9
C_6H_5	$C_6H_5CH_2$	H	$H_2SO_4/(n\text{-}C_4H_9)_2O$	40*	10

Note: References 127–149 are on p. 325.

* The product is the primary amine.

TABLE VII-C. Tertiary Alcohols

$$RC(CH_3)_2OH + R'CN \rightarrow RC(CH_3)_2NHCOR'$$

R	R′	Reaction Medium	Yield (%)	Refs.
C_2H_5	H	$H_2SO_4/(n\text{-}C_4H_9)_2O$	80*	16
C_2H_5	H	$H_4Fe(CN)_6$	10	109
C_2H_5	$CH_2{=}CH$	H_2SO_4/CH_3CO_2H	80	28, 133
C_2H_5	$CH_3CH(OH)$	H_2SO_4/CH_3CO_2H	35	133
C_2H_5	$CH_2{=}C(CH_3)$	H_2SO_4/CH_3CO_2H	55	133
C_2H_5	O NCH$_2$	H_2SO_4/CH_3CO_2H	85	32
C_2H_5	$(C_2H_5)_2NCH_2$	H_2SO_4/CH_3CO_2H	80	32
C_2H_5	$C_6H_5CH{=}CH$	H_2SO_4/CH_3CO_2H	73	133
C_2H_5	NCH$_2$	H_2SO_4/CH_3CO_2H	80	32
C_2H_5	NCH$_2$CH$_2$	H_2SO_4/CH_3CO_2H	80	32
C_2H_5	$(C_2H_5)_2NCH(C_2H_5)$	H_2SO_4/CH_3CO_2H	50	32
C_2H_5	$(C_2H_5)_2NCH(C_4H_9\text{-}n)$	H_2SO_4/CH_3CO_2H	37	32
C_2H_5	$(C_6H_5)_2CH$	H_2SO_4/CH_3CO_2H	76	34
$n\text{-}C_3H_7$	$CH_2{=}CH$	H_2SO_4/CH_3CO_2H	71	133
$n\text{-}C_3H_7$	$(C_2H_5)_2NCH_2$	H_2SO_4/CH_3CO_2H	46	32
$n\text{-}C_3H_7$	O NCH$_2$	H_2SO_4/CH_3CO_2H	90	32
$n\text{-}C_3H_7$	NCH$_2$	H_2SO_4/CH_3CO_2H	90	32
$n\text{-}C_3H_7$	$C_6H_5CH{=}CH$	H_2SO_4/CH_3CO_2H	69	133
$i\text{-}C_3H_7$	H	$H_2SO_4/(n\text{-}C_4H_9)_2O$	40*	16
$i\text{-}C_3H_7$	$CH_2{=}CH$	H_2SO_4/CH_3CO_2H	71	133
$i\text{-}C_3H_7$	$C_6H_5CH{=}CH$	H_2SO_4/CH_3CO_2H	67	133
$n\text{-}C_4H_9$	$CH_2{=}CH$	H_2SO_4/CH_3CO_2H	70	133
$n\text{-}C_4H_9$	O NCH$_2$	H_2SO_4/CH_3CO_2H	61	33
$n\text{-}C_4H_9$	NCH$_2$	H_2SO_4/CH_3CO_2H	36	33
$n\text{-}C_4H_9$	$C_6H_5CH{=}CH$	H_2SO_4/CH_3CO_2H	96	133
$i\text{-}C_4H_9$	O NCH$_2$	H_2SO_4/CH_3CO_2H	Poor	33
$i\text{-}C_4H_9$	NCH$_2$	H_2SO_4/CH_3CO_2H	Poor	33
	H	$H_2SO_4/(n\text{-}C_4H_9)_2O$	20†	16

Note: References 127–149 are on p. 325.

* The product is the primary amine $RC(CH_3)_2NH_2$.

† The product is NH$_2$ CH$_3$ CH$_3$.

TABLE VII-C. Tertiary Alcohols (*Continued*)

R	R′	Reaction Medium	Yield (%)	Refs.
n-C_5H_{11}	$O(CH_2CH_2)_2NCH_2$	H_2SO_4/CH_3CO_2H	65	33
n-C_5H_{11}	$(CH_2)_5NCH_2$	H_2SO_4/CH_3CO_2H	68	33
i-C_5H_{11}	$O(CH_2CH_2)_2NCH_2$	H_2SO_4/CH_3CO_2H	69	33
i-C_5H_{11}	$(CH_2)_5NCH_2$	H_2SO_4/CH_3CO_2H	75	33
$C_2H_5CH(CH_3)CH_2$	$C_6H_5CH{=}CH$	H_2SO_4/CH_3CO_2H	73	133
C_6H_5	H	$H_2SO_4/(n\text{-}C_4H_9)_2O$	70*	8, 9
C_6H_5	$CH_2{=}CH$	H_2SO_4	20	23
C_6H_5	C_6H_5	H_2SO_4	—	8
cyclo-C_6H_{11}	H	$H_2SO_4/(n\text{-}C_4H_9)_2O$	40‡	16
n-C_6H_{13}	$O(CH_2CH_2)_2NCH_2$	H_2SO_4/CH_3CO_2H	85	33
n-C_6H_{13}	$(CH_2)_5NCH_2$	H_2SO_4/CH_3CO_2H	75	33
$3\text{-}CF_3C_6H_4$	$CH_2{=}CH$	H_2SO_4	72	23
$4\text{-}CF_3C_6H_4$	$CH_2{=}CH$	H_2SO_4	22	23
$2,3\text{-}Cl_2C_6H_3CH_2$	H	H_2SO_4/CH_3CO_2H	95	41
$2,3\text{-}Cl_2C_6H_3CH_2$	H	H_3PO_4	§	41
$3,4\text{-}Cl_2C_6H_3CH_2$	H	H_2SO_4/CH_3CO_2H	95	41
$C_6H_5CH_2$	H	H_2SO_4/CH_3CO_2H	92	4, 129, 141
$C_6H_5CH_2$	$ClCH_2$	H_2SO_4/CH_3CO_2H	87	34
$C_6H_5CH_2$	CH_3	H_2SO_4/CH_3CO_2H	84	34
$C_6H_5CH_2$	$CH_2{=}CH$	H_2SO_4/CH_3CO_2H	90	34
$C_6H_5CH_2$	H_2NOCCH_2	H_2SO_4/CH_3CO_2H	78	34
$C_6H_5CH_2$	C_2H_5	H_2SO_4/CH_3CO_2H	100	34
$C_6H_5CH_2$	n-C_3H_7	H_2SO_4/CH_3CO_2H	94	34
$C_6H_5CH_2$	C_5H_4N(3-pyridyl)	H_2SO_4/CH_3CO_2H	71	34
$C_6H_5CH_2$	C_6H_5	H_2SO_4/CH_3CO_2H	84	34
$C_6H_5CH_2$	$4\text{-}O_2NC_6H_4CH_2$	H_2SO_4/CH_3CO_2H	93	34
$C_6H_5CH_2$	$2\text{-}CH_3C_6H_4$	H_2SO_4/CH_3CO_2H	74	34
$C_6H_5CH_2$	$C_6H_5CH_2$	H_2SO_4/CH_3CO_2H	86	34
$C_6H_5CH_2$	$4\text{-}CH_3OC_6H_4$	$H_2SO_4/(n\text{-}C_4H_9)_2O$	81	34
$C_6H_5CH_2$	2-HO-4,6-$(CH_3)_2$-3-pyridyl	H_2SO_4/CH_3CO_2H	34	34
$C_6H_5CH_2$	$C_6H_5CH_2CH_2$	H_2SO_4/CH_3CO_2H	95	34
$C_6H_5CH_2$	$3,4\text{-}(CH_3O)_2C_6H_3$	$H_2SO_4/(n\text{-}C_4H_9)_2O$	71	34
$C_6H_5CH_2$	$(C_6H_5)_2CH$	H_2SO_4/CH_3CO_2H	50	34

Note: References 127–149 are on p. 325.

* The product is the primary amine $RC(CH_3)_2NH_2$.

‡ The product is 1-NH_2-1,2-$(CH_3)_2$-cycloheptane.

§ The product is $2,3\text{-}Cl_2C_6H_3CH{=}C(CH_3)_2$.

TABLE VII-C. Tertiary Alcohols (*Continued*)

R	R′	Reaction Medium	Yield (%)	Refs.
CH_3–	H	H_2SO_4	53‖	38, 39
CH_2 OH	CH_3	H_2SO_4	57	92
n-C_7H_{15}	O NCH$_2$	H_2SO_4/CH_3CO_2H	56	33
n-C_7H_{15}	$(C_2H_5)_2NCH_2$	H_2SO_4/CH_3CO_2H	70	33, 141
n-C_7H_{15}	NCH$_2$	H_2SO_4/CH_3CO_2H	50	33
n-C_7H_{15}	O NCH$_2$CH$_2$	H_2SO_4/CH_3CO_2H	67	33
n-C_7H_{15}	O NCH(CH$_3$)	H_2SO_4/CH_3CO_2H	25	33
n-C_7H_{15}	$(C_2H_5)_2NCH_2CH_2$	H_2SO_4/CH_3CO_2H	57	33
n-C_7H_{15}	CO$_2$H NCH$_2$	H_2SO_4/CH_3CO_2H	54	33
n-C_7H_{15}	NCH$_2$CH$_2$	H_2SO_4/CH_3CO_2H	64	33
$C_6H_5CH_2CH_2$	H	H_2SO_4	65	129
$C_6H_5CH_2CH_2$	H	$H_2SO_4/(n\text{-}C_4H_9)_2O$	60*	8
$C_6H_5CH(CH_3)$	H	$H_2SO_4/(n\text{-}C_4H_9)_2O$	55*	9
n-C_8H_{17}	O NCH$_2$	H_2SO_4/CH_3CO_2H	95	33
n-C_8H_{17}	NCH$_2$	H_2SO_4/CH_3CO_2H	54	33
$C_6H_5CH_2CH_2CH_2$	H	H_2SO_4	27	129
$C_6H_5CH(CH_3)CH_2$	H	$H_2SO_4/(n\text{-}C_4H_9)_2O$	60*	9
n-C_9H_{19}	O N	H_2SO_4/CH_3CO_2H	77	33
n-C_9H_{19}	NCH$_2$	H_2SO_4/CH_3CO_2H	40	33
n-$C_{11}H_{23}$	O NCH$_2$	H_2SO_4/CH_3CO_2H	95	33
n-$C_{11}H_{23}$	NCH$_2$	H_2SO_4/CH_3CO_2H	71	33

Note: References 127–149 are on p. 325.

* The product is the primary amine $RC(CH_3)_2NH_2$.

‖ The product is CH_3 NH_2 $(CH_3)_2CNH_2$.

TABLE VII-D. Tertiary Alcohols

$$R(R')(R)C\text{—}OH + R''CN \xrightarrow{H_2SO_4} R(R')(R)C\text{—}NHCOR''$$

R	R′	R″	Yield (%)	Refs.
C_2H_5	C_6H_5	H	10*	9
$4\text{-}ClC_6H_4$	CH_3	C_6H_5	—	40
$4\text{-}ClC_6H_4$	CCl_3	CH_3	95	40
$4\text{-}ClC_6H_4$	CF_3	CH_3	90	40
$4\text{-}ClC_6H_4$	CF_3	C_6H_5	96	40
$4\text{-}ClC_6H_4$	C_2F_5	CH_3	72	40
$4\text{-}ClC_6H_4$	$n\text{-}C_3H_7$	CH_3	94	40
$4\text{-}FC_6H_4$	CCl_3	CH_3	93	40
$4\text{-}FC_6H_4$	CF_3	CH_3	90	40
C_6H_5	CH_3	CH_3	—	40
C_6H_5	CH_3	C_6H_5	—	40
C_6H_5	CH_3	CH_2Cl	—	40
C_6H_5	CH_3	H	—	40
C_6H_5	CH_2Cl	CH_3	—	40
C_6H_5	CH_2Cl	C_6H_5	—	40
C_6H_5	$CHCl_2$	C_6H_5	—	40
C_6H_5	CHF_2	C_6H_5	100	40
C_6H_5	CF_3	CH_3	88	40
C_6H_5	CO_2H	CH_3	99	40
C_6H_5	C_6H_5	CH_3	93	58, 76
C_6H_5	C_6H_5	C_2H_5	91	58
C_6H_5	C_6H_5	C_6H_5	68	58
C_6H_5	C_6H_5	$C_6H_5CH_2$	74	58

Note: References 127–149 are on p. 325.

* The product was obtained as the amine $(C_2H_5)_2C(C_6H_5)NH_2$; sulfuric acid and di-*n*-butyl ether constituted the reaction medium.

TABLE VII-E. TERTIARY ALICYCLIC ALCOHOLS

Alcohol	Nitrile	Reaction Medium	Product	Yield (%)	Refs.
1-methylcyclobutanol (CH_3, OH)	KCN	$H_2SO_4/(n\text{-}C_4H_9)_2O$	1-methylcyclobutylamine (CH_3, NH_2)	33	15
1-methylcyclopentanol (CH_3, OH)	NaCN	H_2SO_4	1-methylcyclopentylamine (CH_3, NH_2)	49	142
	KCN	$H_2SO_4/(n\text{-}C_4H_9)_2O$	1-methylcyclopentylamine (CH_3, NH_2)	30	15
1-methylcyclohexanol (CH_3, OH)	HCN	H_2SO_4	1-methylcyclohexylamine (CH_3, NH_2)	60	60
	NaCN	H_2SO_4/CH_3CO_2H	1-methylcyclohexylamine (CH_3, NH_2)	76	142
	KCN	$H_2SO_4/(n\text{-}C_4H_9)_2O$	1-methylcyclohexylamine (CH_3, NH_2)	60	15
	NaCN	90% H_2SO_4	1-methylcyclohexylamine (CH_3, NH_2) : 2-methylcyclohexylamine (CH_3, NH_2) (22:1)	62	25
	CH_3CN	H_2SO_4/CH_3CO_2H	1-methylcyclohexyl (CH_3) $NHCOCH_3$	10–15	139
	morpholino-NCH_2CN	H_2SO_4/CH_3CO_2H	1-methylcyclohexyl (CH_3) $NHCOCH_2N$-morpholine	58	33

Note: References 127–149 are on p. 325.

TABLE VII-E. Tertiary Alicyclic Alcohols (*Continued*)

Alcohol	Nitrile	Reaction Medium	Product	Yield (%)	Refs.
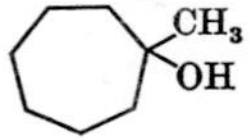 [ring structure] CH_3, OH	NaCN	H_2SO_4/CH_3CO_2H	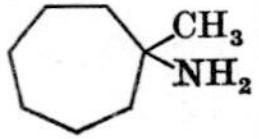 [ring structure] CH_3, NH_2	78	142
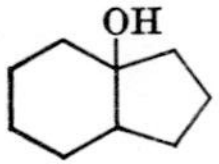 [ring structure] OH (*trans*)	HCN	$H_2SO_4/(n\text{-}C_4H_9)_2O$	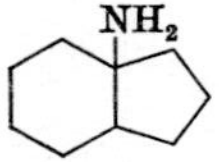 [ring structure] NH_2	34	13
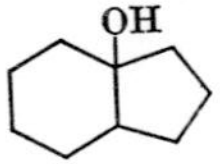 [ring structure] OH (*cis*)	HCN	$H_2SO_4/(n\text{-}C_4H_9)_2O$	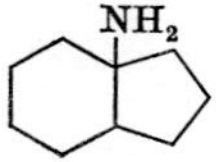 [ring structure] NH_2	Poor	13
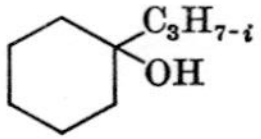 [ring structure] $C_3H_7\text{-}i$, OH	KCN	$H_2SO_4/(n\text{-}C_4H_9)_2O$	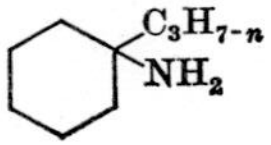 [ring structure] $C_3H_7\text{-}n$, NH_2	50	15
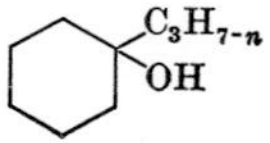 [ring structure] $C_3H_7\text{-}n$, OH	KCN	$H_2SO_4/(n\text{-}C_4H_9)_2O$	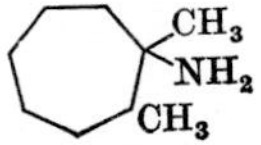 [ring structure] CH_3, NH_2, CH_3	40	16
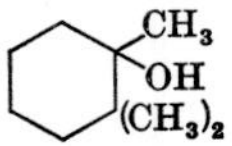 [ring structure] CH_3, OH, $(CH_3)_2$	NaCN	H_2SO_4/CH_3CO_2H	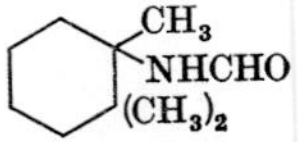 [ring structure] CH_3, $NHCHO$, $(CH_3)_2$	92	37
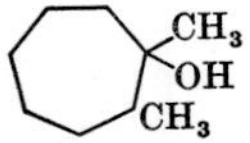 [ring structure] CH_3, OH, CH_3	KCN	$H_2SO_4/(n\text{-}C_4H_9)_2O$	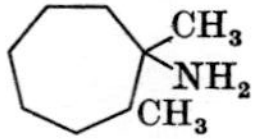 [ring structure] CH_3, NH_2, CH_3	40	16
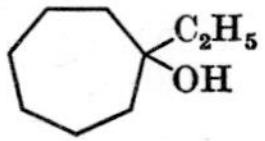 [ring structure] C_2H_5, OH	KCN	$H_2SO_4/(n\text{-}C_4H_9)_2O$	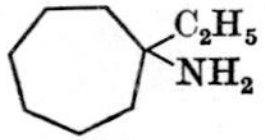 [ring structure] C_2H_5, NH_2	28	15
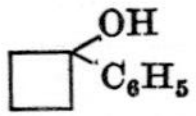 [ring structure] OH, C_6H_5	KCN	$H_2SO_4/(n\text{-}C_4H_9)_2O$	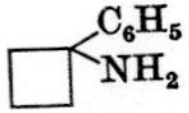 [ring structure] C_6H_5, NH_2	45	8

	C_6H_5CN	H_2SO_4	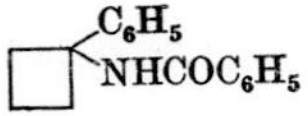 C_6H_5, $NHCOC_6H_5$ (cyclobutane)	—	8
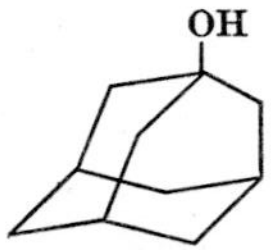 OH (Adamantane)	CH_3CN	H_2SO_4	1-Acetamidoadamantane	61	143, 144
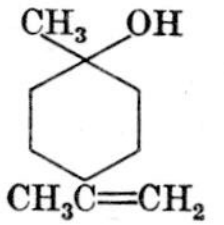 CH_3 OH, $CH_3C{=}CH_2$	HCN	H_2SO_4	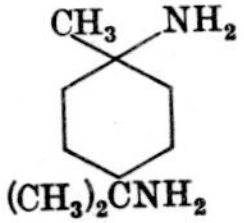 CH_3 NH_2, $(CH_3)_2CNH_2$	—	38, 39
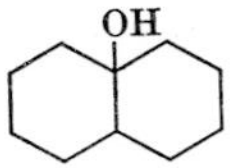 OH *trans*	CH_3CN	96% H_2SO_4	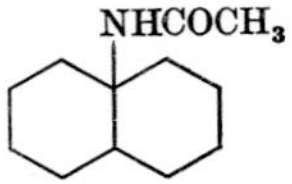 $NHCOCH_3$	90	11
	CH_3CN	H_2SO_4/CCl_4	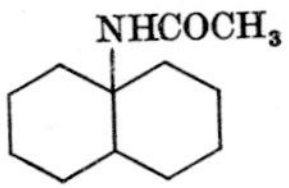 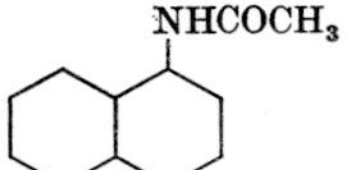 $NHCOCH_3$; $NHCOCH_3$	5	11
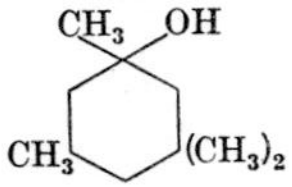 CH_3 OH, CH_3, $(CH_3)_2$	 O NCH₂CN	H_2SO_4/CH_3CO_2H	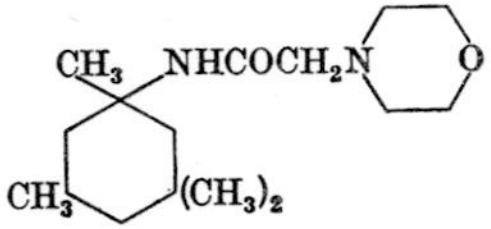 CH_3 $NHCOCH_2N$ O, CH_3, $(CH_3)_2$	Poor	33
	 NCH_2CN	H_2SO_4/CH_3CO_2H	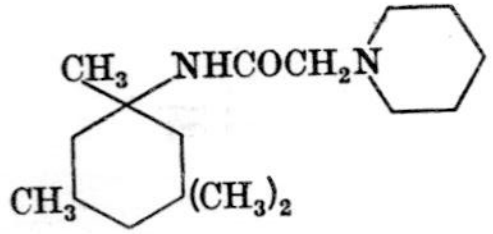 CH_3 $NHCOCH_2N$, CH_3, $(CH_3)_2$	22	33

Note: References 127–149 are on p. 325.

TABLE VII-E. TERTIARY ALICYCLIC ALCOHOLS (*Continued*)

Alcohol	Nitrile	Reaction Medium	Product	Yield (%)	Refs.
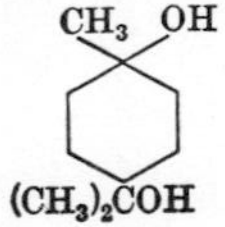 CH_3, OH; $(CH_3)_2COH$	HCN	H_2SO_4	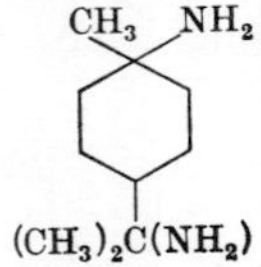 CH_3, NH_2; $(CH_3)_2C(NH_2)$	61	38, 39
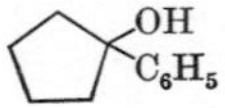 OH, C_6H_5	HCN	$H_2SO_4/(n\text{-}C_4H_9)_2O$	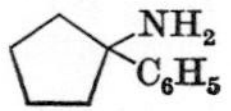 NH_2, C_6H_5	0	9
	KCN	$H_2SO_4/(n\text{-}C_4H_9)_2O$	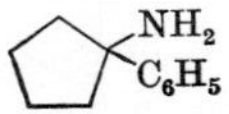 NH_2, C_6H_5	35	8
	C_6H_5CN	H_2SO_4	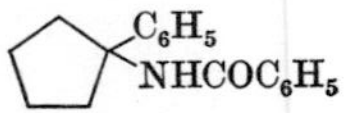 C_6H_5, $NHCOC_6H_5$	50	8
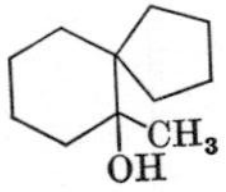 CH_3, OH	KCN	$H_2SO_4/(n\text{-}C_4H_9)_2O$	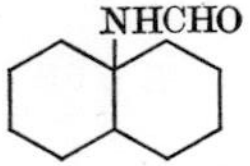 NHCHO	100	6
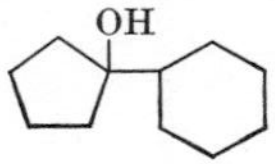 OH	KCN	$H_2SO_4/(n\text{-}C_4H_9)_2O$	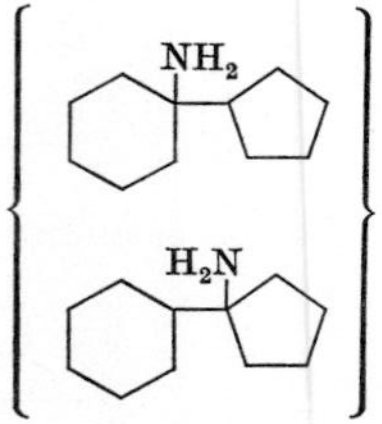 { NH_2; H_2N }	35	6

Reactant	Nitrile	Acid/Solvent	Product	Yield (%)	Refs.
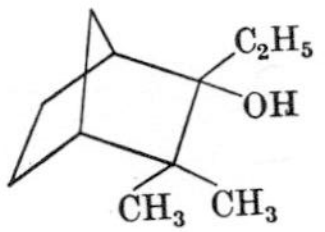 C_2H_5, OH, CH_3, CH_3	NaCN	H_2SO_4/CH_3CO_2H	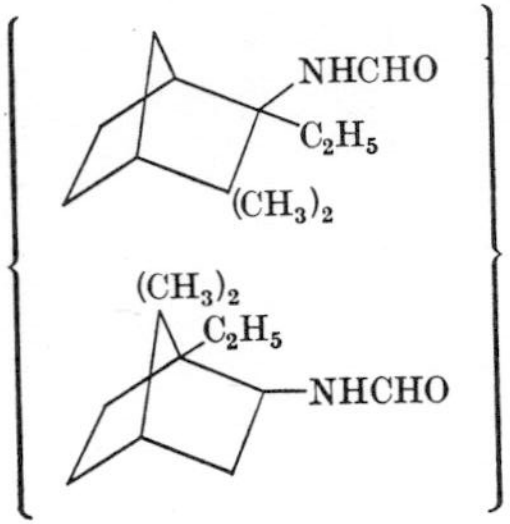 NHCHO, C_2H_5, $(CH_3)_2$; $(CH_3)_2$, C_2H_5, NHCHO	Quant.	37
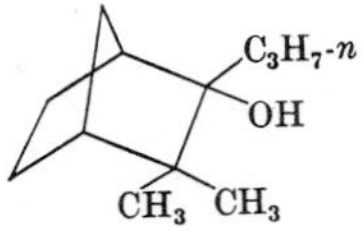 C_3H_7-*n*, OH, CH_3, CH_3	NaCN	H_2SO_4/CH_3CO_2H	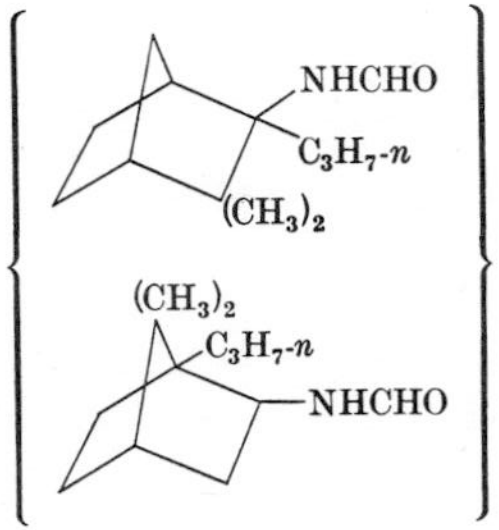 NHCHO, C_3H_7-*n*, $(CH_3)_2$; $(CH_3)_2$, C_3H_7-*n*, NHCHO	83	37
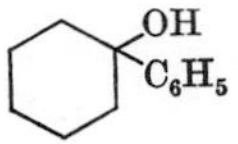 OH, C_6H_5	HCN	$H_2SO_4/(n\text{-}C_4H_9)_2O$	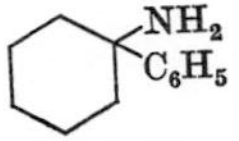 NH_2, C_6H_5	5–10	9
	KCN	$H_2SO_4/(n\text{-}C_4H_9)_2O$	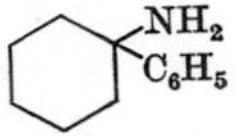 NH_2, C_6H_5	60	8
	C_6H_5CN	H_2SO_4	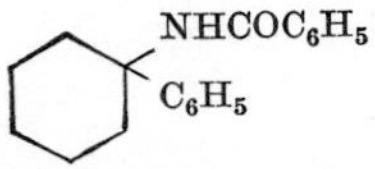 $NHCOC_6H_5$, C_6H_5	65	8

Note: References 127–149 are on p. 325.

TABLE VII-F.[65] Tertiary Heterocyclic Alcohols

Alcohol	Nitrile	Reaction Medium	Product	Yield (%)
1-methyl-4-hydroxy-4-methylpiperidine (CH_3N ring, OH, CH_3)	CH_3CN	H_2SO_4	CH_3N ring, $NHCOCH_3$, CH_3	44
CH_3N ring, OH, C_2H_5	CH_3CN	H_2SO_4	CH_3N ring, $NHCOCH_3$, C_2H_5	60
CH_3N ring, OH, C_4H_9-*n*	CH_3CN	H_2SO_4	CH_3N ring, $NHCOCH_3$, $C_4H_{9\text{-}n}$	39

Note: References 127–149 are on p. 325.

TABLE VIII. GLYCOLS

Glycol	Nitrile	Reaction Medium	Product	Yield (%)	Ref.
CH_2OH (benzene ring) CH_2OH	CH_3CN	H_2SO_4	$CH_2NHCOCH_3$ (benzene ring) $CH_2NHCOCH_3$	62	56
CH_2OH (benzene ring) CH_2OH	CH_3CN	H_2SO_4	$CH_2NHCOCH_3$ (benzene ring) $CH_2NHCOCH_3$	65	57
	$CH_2{=}CHCN$	H_2SO_4	$CH_2NHCOCH{=}CH_2$ (benzene ring) $CH_2NHCOCH{=}CH_2$	64	56
$(CH_3)_2C(OH)CH_2$ | $(CH_3)_2C(OH)CH_2$	$ClCH_2CH_2CN$	H_2SO_4	$(CH_3)_2CNHCOCH_2CH_2Cl$ | $(CH_2)_2$ | $(CH_3)_2CNHCOCH_2CH_2Cl$	63	101
CH_3 CH_3 (benzene ring) CH_2OH CH_2OH	$ClCH_2CN$	H_3PO_4	CH_3 CH_3 (benzene ring) $CH_2NHCOCH_2Cl$ $CH_2NHCOCH_2Cl$	98	57
	CH_3CN	H_2SO_4	CH_3 CH_3 (benzene ring) $CH_2NHCOCH_3$ $CH_2NHCOCH_3$	62	57

Note: References 127–149 are on p. 325.

TABLE VIII. GLYCOLS (*Continued*)

Glycol	Nitrile	Reaction Medium	Product	Yield (%)	Ref.
2-CH_3-4-CH_3-1,3-$(CH_2OH)_2C_6H_2$ (*contd.*)	CH_3CN	H_3PO_4	2-CH_3-4-CH_3-1,3-$(CH_2NHCOCH_3)_2C_6H_2$	74	57
	CH_3CN	H_2SO_4/CH_3CO_2H	2-CH_3-4-CH_3-1,3-$(CH_2NHCOCH_3)_2C_6H_2$	62	57
	$CH_2{=}CHCN$	H_2SO_4/CH_3CO_2H	2-CH_3-4-CH_3-1,3-$(CH_2NHCOCH{=}CH_2)_2C_6H_2$	98	57
	$ClCH_2CH_2CN$	H_3PO_4	2-CH_3-4-CH_3-1,3-$(CH_2NHCOCH_2CH_2Cl)_2C_6H_2$	98	57
	C_6H_5CN	H_3PO_4	2-CH_3-4-CH_3-1,3-$(CH_2NHCOC_6H_5)_2C_6H_2$	98	57
$HOCH(CH_3)(CH_2)_8CH(CH_3)OH$	$C_6H_4(CN)_2$-1,4	H_2SO_4	Polymer	—	145
	$NC(CH_2)_4CN$	H_2SO_4	Polymer	—	145
1,4-$(CH_2CH_2CHOHCH_3)_2C_6H_4$	$NC(CH_2)_4CN$	H_2SO_4	Polymer	—	145

Note: References 127–149 are on p. 325.

TABLE IX. ALKYL CHLORIDES[66]

Alkyl Chloride	Nitrile	Reaction Medium	Product	Yield (%)
$(CH_3)_3CCl$	C_6H_5CN	90% HCO_2H	$(CH_3)_3CNHCOC_6H_5$	39
	C_6H_5CN	80% H_2SO_4	$(CH_3)_3CNHCOC_6H_5$	16
	$C_6H_5CH_2CN$	90% HCO_2H	$(CH_3)_3CNHCOCH_2C_6H_5$	23
	$NC(CH_2)_4CN$	90% HCO_2H	$(CH_3)_3CNHCO(CH_2)_4CONHC(CH_3)_3$	10
$(CH_3)_2C(Cl)(CH_2)_8C(Cl)(CH_3)_2$	$NC(CH_2)_4CN$	90% HCO_2H	Polymer	—

Note: References 127–149 are on p. 325.

TABLE X-A. FORMALDEHYDE

R	Reaction Medium	Yield (%)	Ref.
	$CH_2O + RCN \rightarrow RCONHCH_2NHCOR$		
$CH_2{=}CH$	85% H_2SO_4	86	68
n-C_3H_7	85% H_2SO_4	40	68
$CH_2{=}CHCH_2$	75% H_2SO_4	18	68
$CH_3CH{=}CHCH_2$	90% H_2SO_4	9	68
C_6H_5	85% H_2SO_4	90	68
	90% HCO_2H	22	68
4-$CH_3C_6H_4$	85% H_2SO_4	83	68
	$CH_2O + NCRCN \rightarrow (HNCORCONHCH_2)_n$		
$(CH_2)_4$	H_2SO_4/HCO_2H	95	69
$(CH_2)_5$	H_2SO_4/HCO_2H	65	69
$(CH_2)_7$	H_2SO_4	92	69
$(CH_2)_8$	H_2SO_4/HCO_2H	75	69
 H_2C CH_2	H_2SO_4	80	69
$(CH_2)_3CHCH_3$	H_2SO_4/HCO_2H	13	69
$CH_2CH_2OCH_2CH_2$	H_2SO_4	27	69
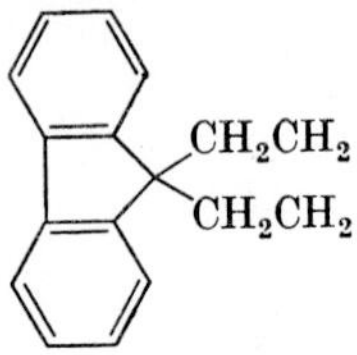 CH_2CH_2 CH_2CH_2	H_2SO_4/HCO_2H	25	69

Note: References 127–149 are on p. 325.

TABLE X-B. ALDEHYDES OTHER THAN FORMALDEHYDE

Aldehyde	Nitrile	Reaction Medium	Product	Yield (%)	Refs.
CCl_3CHO	$CH_2{=}CHCN$	75% H_2SO_4	$CH_2{=}CHCONHCH(CCl_3)NHCOCH{=}CH_2$	18	68
	n-C_3H_7CN	96% H_2SO_4	n-$C_3H_7CONHCH(CCl_3)NHCOC_3H_7$-$n$	67	68
	$NC(CH_2)_4CN$	H_2SO_4	Polyamide	40	66
CH_3CHO	$NC(CH_2)_4CN$	H_2SO_4	Polyamide	13	66
C_2H_5CHO	$NC(CH_2)_4CN$	H_2SO_4	Polyamide	27	66
n-C_3H_7CHO	n-C_3H_7CN	96% H_2SO_4	n-$C_3H_7CONHCH(C_3H_7$-$n)NHCOC_3H_7$-n	19	68
	$NC(CH_2)_4CN$	H_2SO_4	Polyamide	40	66
	$C_6H_4(CN)_2$-1,4	H_2SO_4	Polyamide	—	66
n-C_4H_9CHO	$NC(CH_2)_4CN$	H_2SO_4	Polyamide	—	66
C_6H_5CHO	CH_3CN	H_2SO_4	Benzylidene-*bis*-acetamide	78	70
C_6H_5CHO*	CH_3CN	H_2SO_4	$C_6H_5CHCH(CO_2C_2H_5)_2$ with $NHCOCH_3$ on the benzylic carbon	55	70
$C_6H_{11}CHO$	$NC(CH_2)_4CN$	H_2SO_4	Polyamide	—	66

Note: References 127–149 are on p. 325.

* Diethyl malonate was also a reactant. See p. 233.

TABLE XI. KETONES[70]

Ketone	Nitrile	Reaction Medium	Product	Yield (%)
CH_3COCH_3	CH_3CN	H_2SO_4	$(CH_3)_2C(NHCOCH_3)CH_2COCH_3$	23
	C_6H_5CN	H_2SO_4	$(CH_3)_2C(NHCOC_6H_5)CH_2COCH_3$	62
$C_2H_5COCH_3$	C_6H_5CN	H_2SO_4	$C_2H_5C(CH_3)(NHCOC_6H_5)CH(CH_3)COCH_3$	16
Cyclohexanone	CH_3CN	H_2SO_4	2-(1-$NHCOCH_3$-cyclohexyl)cyclohexanone	72
	C_6H_5CN	H_2SO_4	2-(1-$NHCOC_6H_5$-cyclohexyl)cyclohexanone	51

$(CH_3)_2C(OH)CH_2COCH_3$	C_6H_5CN	H_2SO_4	$(CH_3)_2CCH_2COCH_3$ | $NHCOC_6H_5$	37
$C_6H_5COCH_3$	CH_3CN	H_2SO_4	$C_6H_5C(CH_3)CH_2COC_6H_5$ | $NHCOCH_3$	74
	$CH_2{=}CHCN$	H_2SO_4	$C_6H_5C(CH_3)CH_2COC_6H_5$ | $NHCOCH{=}CH_2$	74
	$C_6H_5CH_2CN$	H_2SO_4	$C_6H_5C(CH_3)CH_2COC_6H_5$ | $NHCOCH_2C_6H_5$	84
$4\text{-}CH_3C_6H_4COCH_3$	CH_3CN	H_2SO_4	$4\text{-}CH_3C_6H_4C(CH_3)CH_2COC_6H_4CH_3\text{-}4$ | $NHCOCH_3$	86
$4\text{-}CH_3OC_6H_4COCH_3$	CH_3CN	H_2SO_4	$4\text{-}CH_3OC_6H_4C(CH_3)CH_2COC_6H_4OCH_3\text{-}4$ | $NHCOCH_3$	67

Note: References 127–149 are on p. 325.

TABLE XII. ETHERS[71]

Ether	Nitrile	Reaction Medium	Product	Yield (%)
sec-$C_4H_9OCH_3$	C_6H_5CN	H_2SO_4	*sec*-$C_4H_9NHCOC_6H_5$	46
t-$C_4H_9OCH_3$	CH_3CN	H_2SO_4	*t*-$C_4H_9NHCOCH_3$	—
	C_6H_5CN	H_2SO_4	*t*-$C_4H_9NHCOC_6H_5$	85
	$NC(CH_2)_4CN$	H_2SO_4/CH_3CO_2H	*t*-$C_4H_9NHCO(CH_2)_4CONHC_4H_9$-*t*	75
i-$C_3H_7OC_3H_7$-*i*	C_6H_5CN	H_2SO_4	*i*-$C_3H_7NHCOC_6H_5$	81
	$NC(CH_2)_4CN$	H_2SO_4	*i*-$C_3H_7NHCO(CH_2)_4CONHC_3H_7$-*i*	63
$CH_3OC(CH_3)_2—(CH_2)_8—C(CH_3)_2—OCH_3$	$NC(CH_2)_4CN$	H_2SO_4/CH_3CO_2H	$[—C(CH_3)_2—(CH_2)_8—C(CH_3)_2—NHCO(CH_2)_4CONH—]_n$	—
	NCC_6H_4CN-1,4	H_2SO_4/CH_3CO_2H	$[—C(CH_3)_2—(CH_2)_8—C(CH_3)_2—NHCO—C_6H_4—CONH—]_n$	—

Note: References 127–149 are on p. 325.

TABLE XIII. α,β-UNSATURATED CARBONYL COMPOUNDS

Amides	Nitrile	Product	Yield (%)	Refs.
$C_6H_5C(CH_3)=CHCONH_2$	CH_3CN	$C_6H_5C(CH_3)(NHCOCH_3)CH_2CONH_2$	67	62
$C_6H_5C(C_2H_5)=CHCONH_2$	CH_3CN	$C_6H_5C(C_2H_5)(NHCOCH_3)CH_2CONH_2$	22	62
Esters				
$CH_2=CHCO_2CH_3$	C_6H_5CN	No reaction		63
$CH_3CH=CHCO_2CH_3$	C_6H_5CN	No reaction		63
$(CH_3)_2C=C(C_2H_5)CO_2C_2H_5$	C_6H_5CN	$(CH_3)_2C(NHCOC_6H_5)CH(C_2H_5)CO_2H$	62	63
$C_6H_5CH=CHCO_2C_2H_5$	C_6H_5CN	$C_6H_5CH(NHCOC_6H_5)CH_2CO_2H$	26	63
Ketones				
$(CH_3)_2C=CHCOCH_3$	CH_3CN	$(CH_3)_2C(NHCOCH_3)CH_2COCH_3$	25	72
	C_6H_5CN	$(CH_3)_2C(NHCOC_6H_5)CH_2COCH_3$	78	70, 72
$C_6H_5CH=CHCOC_6H_5$	CH_3CN	$C_6H_5CH(NHCOCH_3)CH_2COC_6H_5$	Poor	72
	C_6H_5CN	$C_6H_5CH(NHCOC_6H_5)CH_2COC_6H_5$	Poor	72
Acids				
$HCCO_2H \parallel HCCO_2H$	C_6H_5CN	No reaction		63
$CH_3CCO_2H \parallel HCCO_2H$	C_6H_5CN	No reaction		63
$(CH_3)_2C=CHCO_2H$	C_6H_5CN	$(CH_3)_2C(NHCOC_6H_5)CH_2CO_2H$	72	63

Note: References 127–149 are on p. 325.

TABLE XIV. METHYLOLAMIDES

Methylolamide	Nitrile	Reaction Medium	Product	Yield (%)	Ref.
$C_6H_5CONHCH_2OH$	$CH_2{=}CHCN$	H_2SO_4/CH_3CO_2H	$C_6H_5CONHCH_2NHCOCH{=}CH_2$	83	74
$C_6H_4(C{=}O)_2NCH_2OH$	CH_3CN	H_2SO_4	$C_6H_4(C{=}O)_2NCH_2NHCOCH_3$	83–93	73
	$CH_2{=}CHCN$	H_2SO_4/CH_3CO_2H	$C_6H_4(C{=}O)_2NCH_2NHCOCH{=}CH_2$	95	74
	$CH_2{=}C(CH_3)CN$	H_2SO_4/CH_3CO_2H	$C_6H_4(C{=}O)_2NCH_2NHCOC(CH_3){=}CH_2$	89	74
$C_6H_4(C{=}O)(SO_2)NCH_2OH$	$CH_2{=}CHCN$	H_2SO_4/CH_3CO_2H	$C_6H_4(C{=}O)(SO_2)NCH_2NHCOCH{=}CH_2$	—	74

Note: References 127–149 are on p. 325.

TABLE XV. CARBOXYLIC ACIDS

(The reaction medium was sulfuric acid of unspecified concentration unless otherwise indicated.)

Acid	Nitrile	Product	Yield (%)	Ref.
Tertiary				
$(CH_3)_3CCO_2H$	NaCN	$(CH_3)_3CNH_2$	68	25
$C_2H_5C(CH_3)_2CO_2H$	NaCN	2-Amino-2-methylbutane (95%) 3-Amino-2-methylbutane (5%)	58	25
n-$C_3H_7C(CH_3)_2CO_2H$	NaCN	2-Amino-2-methylpentane (73%) 3-Amino-3-methylpentane (11%) 3-Amino-2-methylpentane (13%) 2-Amino-3-methylpentane (2%)	53	25
$(C_2H_5)_2(CH_3)CO_2H$	NaCN	2-Amino-2-methylpentane (39%) 3-Amino-3-methylpentane (40%) 3-Amino-2-methylpentane (7%) 2-Amino-3-methylpentane (14%)	51	25
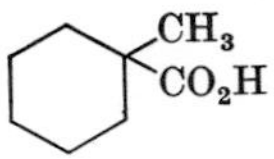 CH_3, CO_2H	NaCN	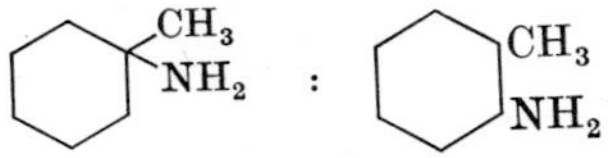 CH_3, NH_2 : CH_3, NH_2 (65:35)	66	25
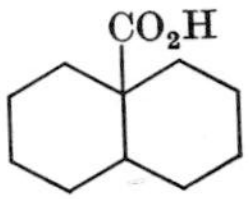 CO_2H	NaCN	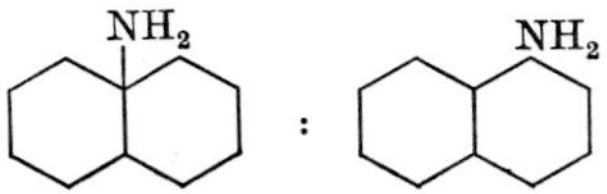 NH_2 : NH_2 (25:75)	67	25
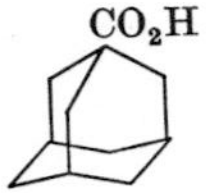 CO_2H	NaCN	1-Formamidoadamantane	56	25
	CH_3CN	1-Acetamidoadamantane	51	25

Note: References 127 and 149 are on p. 325.

TABLE XV. CARBOXYLIC ACIDS (*Continued*)

(The reaction medium was sulfuric acid of unspecified concentration unless otherwise indicated.)

Acid	Nitrile	Product	Yield (%)	Ref.
α-Hydroxy				
$(C_6H_5)_2C(OH)CO_2H$	CH_2ClCN	$(C_6H_5)_2C(NHCOCH_2Cl)CO_2H$	82*	64
	CH_3CN	$(C_6H_5)_2C(NHCOCH_3)CO_2H$	82	64
	$NC(CH_2)_4CN$	$(CH_2)_4[CONHC(C_6H_5)_2CO_2H]_2$	41*	64
	C_6H_5CN	2-C_6H_5-4,4-$(C_6H_5)_2$-oxazolidin-5-one ($O=$ ring with O, C_6H_5, NH, $(C_6H_5)_2$)	90	61, 105
Unsaturated (see also Table XIII)				
$CH_2{=}CHCH_2CO_2H$	C_6H_5CN	No reaction		63
$CH_2{=}CHCH_2CH_2CO_2H$	C_6H_5CN	No reaction		63
$CH_2{=}CH(CH_2)_8CO_2H$	HCN	$CH_3CH(NHCHO)(CH_2)_8CO_2H$	80†	42
$CH_3(CH_2)_7CH{=}CH(CH_2)_7CO_2H$	HCN	$CH_3(CH_2)_nCH(NHCHO)(CH_2)_mCO_2H$‡	96†	42
	CH_3CN	$CH_3(CH_2)_nCH(NHCOCH_3)(CH_2)_mCO_2H$‡	99	19
	$NCCH_2CN$	{ $CH_3(CH_2)_nCH(NHCOCH_2CN)(CH_2)_mCO_2H$‡ ; $CH_3(CH_2)_nCH(CH_2)_mCO_2H$‡ with NHCOCH$_2$CONH– linked to $CH_3(CH_2)_nCH(CH_2)_mCO_2H$ }	—	19

Nitrile	Product	Yield (%)	Ref.
$CH_2{=}CHCN$	$CH_3(CH_2)_nCH(CH_2)_mCO_2H$‡ (N-substituent: $NHCOCH{=}CH_2$)	89	19
$NCCH_2CO_2H$	$CH_3(CH_2)_nCH(CH_2)_mCO_2H$‡ (N-substituent: $NHCOCH_2CO_2H$)	91	19
C_2H_5CN	$CH_3(CH_2)_nCH(CH_2)_mCO_2H$‡ (N-substituent: $NHCOC_2H_5$)	94	19
$NC(CH_2)_2CN$	{ $CH_3(CH_2)_nCH(CH_2)_mCO_2H$‡ (N-substituent: $NHCOCH_2CH_2CN$); $CH_3(CH_2)_nCH(CH_2)_mCO_2H$‡ (N-substituent: $NHCO(CH_2)_2CONH$–$CH(CH_2)_mCO_2H$ with $CH_3(CH_2)_n$) }	—	19
n-C_3H_7CN	$CH_3(CH_2)_nCH(CH_2)_mCO_2H$‡ (N-substituent: $NHCOC_3H_7$-n)	—	19
C_6H_5	$CH_3(CH_2)_nCH(CH_2)_mCO_2H$‡ (N-substituent: $NHCOC_6H_5$)	91	19
$NC(CH_2)_4CN$	$CH_3(CH_2)_nCH(CH_2)_mCO_2H$‡ (N-substituent: $NHCO(CH_2)_4CONH$–$CH_3(CH_2)_nCH(CH_2)_mCO_2H$)	87	43
$NC(CH_2)_7CN$	$CH_3(CH_2)_nCH(CH_2)_mCO_2H$‡ (N-substituent: $NHCO(CH_2)_7CONH$–$CH_3(CH_2)_nCH(CH_2)_mCO_2H$)	72	43
$NC(CH_2)_8CN$	$CH_3(CH_2)_nCH(CH_2)_mCO_2H$ (N-substituent: $NHCO(CH_2)_8CONH$–$CH_3(CH_2)_nCH(CH_2)_mCO_2H$)	81	43

Note: References 127–149 are on p. 325.

* The reaction medium was H_2SO_4 and $(C_2H_5)_2O$.
† The reaction medium was 85% H_2SO_4.
‡ The position of the addition was not determined; therefore $m = 8$, $n = 7$, or vice versa.

TABLE XV. CARBOXYLIC ACIDS (*Continued*)

(The reaction medium was sulfuric acid of unspecified concentration unless otherwise indicated.)

Acid	Nitrile	Product	Yield (%)	Ref. i
$CH_3(CH_2)_{10}CH{=}CH(CH_2)_4CO_2H$	NaCN	$CH_3(CH_2)_aCH(NHCHO)(CH_2)_bCO_2H$§	82	27
	CH_3CN	$CH_3(CH_2)_aCH(NHCOCH_3)(CH_2)_bCO_2H$§	94	27
	$CH_2{=}CHCN$	$CH_3(CH_2)_aCH(NHCOCH{=}CH_2)(CH_2)_bCO_2H$§	100	27
	C_2H_5CN	$CH_3(CH_2)_aCH(NHCOC_2H_5)(CH_2)_bCO_2H$§	99	27
	C_6H_5CN	$CH_3(CH_2)_aCH(NHCOC_6H_5)(CH_2)_bCO_2H$§	—	27
$CH_3(CH_2)_5CHOHCH_2CH{=}CH(CH_2)_7CO_2H$	HCN	$CH_3(CH_2)_5CHOH(CH_2)_cCH(NHCHO)(CH_2)_dCO_2H$‖	84	42

Note: References 127–149 are on p. 325.

§ The position of the addition was not determined; therefore $a = 11$, $b = 4$, or $a = 10$, $b = 5$.
‖ The position of the addition was not determined; therefore $c = 1$, $d = 8$, or $c = 2$, $d = 7$.

TABLE XVI. Carboxylic Esters
(See also Table XIII.)

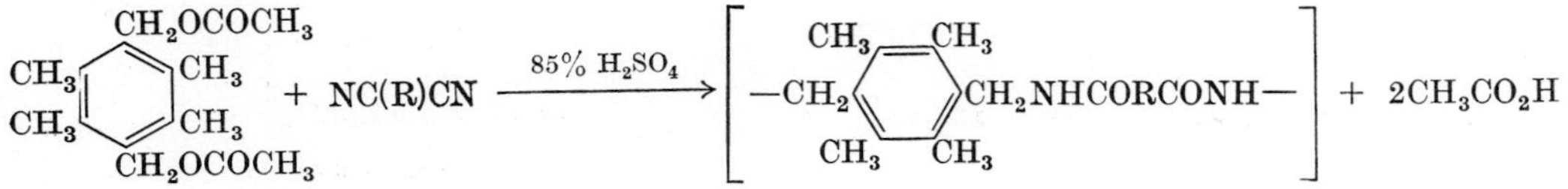

$$CH_3, CH_3, CH_3, CH_3\text{-}C_6(CH_2OCOCH_3)_2 + NC(R)CN \xrightarrow{85\%\ H_2SO_4} \left[-CH_2\text{-}C_6(CH_3)_4\text{-}CH_2NHCORCONH- \right] + 2CH_3CO_2H \quad \text{(Ref. 77)}$$

R	Yield (%)	R	Yield (%)
$(CH_2)_3$	100	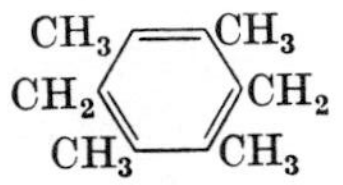 CH_3, CH_3, CH_2, CH_2, CH_3, CH_3 (benzene ring: $-CH_2C_6(CH_3)_4CH_2-$)	86
$(CH_2)_4$	99		
$CH_2CH_2OCH_2CH_2$	95		
$(CH_2)_5$	—		
1,3-C_6H_4	80		
1,4-C_6H_4	98		
$(CH_2)_7/(CH_2)_4$ (1:1)*	94	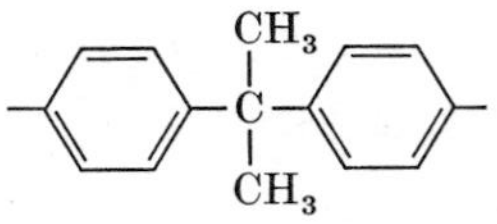 $-C_6H_4-C(CH_3)_2-C_6H_4-$	89
$(CH_2)_8$	99		
i-$(CH_2)_8$†	97		

Ester	Nitrile	Solvent	Product	Yield (%)	Ref.
$(C_6H_5)_3COCHO$	CH_3CN	98% H_2SO_4	$(C_6H_5)_3CNHCOCH_3$	Quant.	76
$CH_3CHOCOCH_3-(CH_2)_8-CH_3CHOCOCH_3$	$NC(CH_2)_4CN$	H_2SO_4	Polymer	—	145
	$C_6H_4(CN)_2$-1,4	H_2SO_4	Polymer	—	145

Note: References 127–149 are on p. 325.

* A 1:1 mixture of $NC(CH_2)_7CN$ and $NC(CH_2)_4CN$ was used in this reaction.
† Isosebaconitrile was used in this reaction.

TABLE XVII-A. Dihydro-1,3-oxazines

$$(CH_3)_2C(OH)CH_2CHOHCH_3 + RCN \longrightarrow$$ [4,4,6-trimethyl-5,6-dihydro-1,3-oxazine ring: $(CH_3)_2$, N, R, O, CH_3]

R	Yield (%)	Refs.
CH_3	44	83
$CH_2{=}CH$	47	87, 88
C_2H_5	59	59, 87
$CH_2{=}C(CH_3)$	53	88
C_6H_5	47	83
$C_6H_5CH_2$	26	83
CH_2CH_2CN	40	85
2-$CH_3C_6H_4$	—	87
4-$CH_3C_6H_4$	—	87
4-$H_2NC_6H_4$	—	87

Note: References 127–149 are on p. 325.

TABLE XVII-B. 2-Substituted 4,4-Dimethyl-5,6-dihydro-1,3-thiazines

$$(CH_3)_2C(OH)CH_2CH_2SH + RCN \xrightarrow{H_2SO_4}$$ [thiazine ring: S, R, N, $(CH_3)_2$]

R	Yield (%)	Refs.
H	42	59, 95
CH_3	41	59, 87
C_2H_5	46	59, 87
$CH_2{=}CH$	51	59
C_6H_5	48, 75	59, 95*,†
2-$CH_3C_6H_4$	45	59
4-$CH_3C_6H_4$	50	59
4-$H_2NC_6H_4$	53	59
$CH_2CO_2C_2H_5$	45, 59†	96
$CH_2CO_2CH_3$	57†	96

Note: References 129–149 are on p. 325.

* The thio ether $(CH_3)_2C(OH)CH_2CH_2SCH_2C_6H_5$ was used in this reaction.

† Boron fluoride etherate was employed instead of sulfuric acid.

TABLE XVII-C. Δ^2-THIAZOLINES

Method A: $CH_2{=}C(CH_3)CH_2SH$ $\xrightarrow[RCN]{H_2SO_4}$ 2-R-4,4-$(CH_3)_2$-Δ^2-thiazoline $\xleftarrow[RCN]{H_2SO_4}$ $CH_3C(CH_3)(OH)CH_2SH$ (Refs. 59, 90)

R	Yield (%) of Δ^2-Thiazoline from Methallyl Mercaptan	Yield (%) of Δ^2-Thiazoline from 2-Methyl-2-hydroxypropanethiol
CH_3	23	50
$CH_2{=}CH$	22	47
C_6H_5	24	51
4-$H_2NC_6H_4$	—	55

Method B: R_1CN + thiirane (R_2, R_5, R_4, R_3) $\xrightarrow{Acid}$ 2-R_1-Δ^2-thiazoline (R_2, R_3, R_4, R_5)

R_1	R_2	R_3	R_4	R_5	Acid	Yield (%)	Refs.
CH_3	H	H	H	H	H_2SO_4	Trace	94
CH_3	CH_3	H	H	H	H_2SO_4	17	93
CH_3 (*cis*)*	CH_3	CH_3	H	H	H_2SO_4	79, 75	93, 94
CH_3 (*cis*)*	CH_3	CH_3	H	H	$HClO_4$	9	94
CH_3 (*cis*)*	CH_3	CH_3	H	H	H_2SO_4	38	94
CH_3 (*trans*)*	CH_3	CH_3	H	H	$HClO_4$	6	94
$(CH_3)_3C$ (*trans*)*	CH_3	CH_3	H	H	H_2SO_4	31	94
CH_3		Cyclohexene sulfide			H_2SO_4	36	94
C_6H_5		Cyclohexene sulfide			H_2SO_4	52	93
C_6H_5 (*cis*)*	CH_3	CH_3	H	H	$HClO_4$	37	94
C_6H_5 (*trans*)*	CH_3	CH_3	H	H	$HClO_4$	16	94
C_6H_5	CH_3	CH_3	CH_3	CH_3	H_2SO_4	73	94
C_6H_5	CH_3	CH_3	CH_3	CH_3	$HClO_4$	44	94
$C_6H_5CH_2$ (*cis*)*	CH_3	CH_3	H	H	H_2SO_4	76	94
4-$CH_3OC_6H_4$ (*trans*)*	CH_3	CH_3	H	H	H_2SO_4†	11	94

Note: References 127–149 are on p. 325.

* The parenthesized *cis* or *trans* denotes the configuration of the product.
† 1,2-Dimethoxyethane was used as the solvent.

TABLE XVII-D. α,ω-*Bis*(HETEROCYCLYL)ALKANES[104]

(Sulfuric acid was used as the reaction medium.)

Alcohol	Nitrile	R	Product	Yield (%)
$(CH_3)_2C(OH)CH_2CH_2C(CH_3)_2OH$	NCRCN	$—(CH_2)_2—$	$[(CH_3)_2, (CH_3)_2, N]_2R$	72
		$—(CH_2)_3—$		76
		$—(CH_2)_4—$		74
$CH_3CH(OH)CH_2C(CH_3)_2OH$	NCRCN	$—(CH_2)_2O(CH_2)_2—$	$[CH_3, O, (CH_3)_2, N]_2R$	64
		$—(CH_2)_2—$		53
		$—(CH_2)_4—$		—
$HSCH_2CH_2(CH_3)_2OH$	NCRCN	$—(CH_2)_2—$	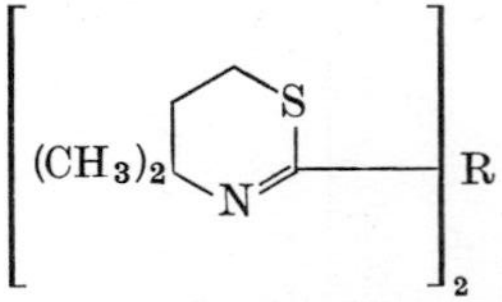 $[S, (CH_3)_2, N]_2R$	45

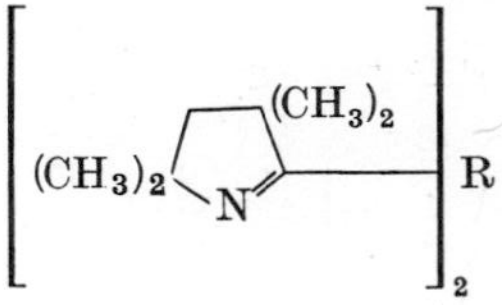

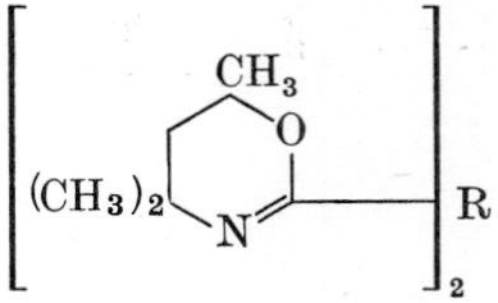

Note: References 127–149 are on p. 325.

TABLE XVII-E. 1-PYRROLINES

$$(CH_3)_2C(OH)CH_2CH_2C(OH)(CH_3)_2 + R^1CN \xrightarrow{H_2SO_4}$$ (1-pyrroline product: ring of $RCCH_3$, CH_2, N, $RC{=}C{-}CR'$ with CH_3)

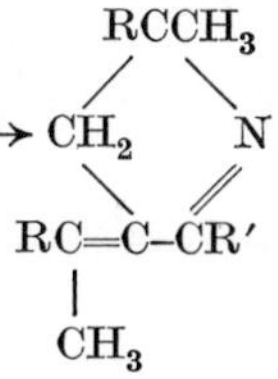

Nitrile	R	R^1	Yield (%)	Ref.
CH_3CN	CH_3	CH_3	80	90
$CH_2{=}CHCN$	CH_3	$CH_2{=}CH$	78	90
C_6H_5CN	CH_3	C_6H_5	72	90
$4\text{-}O_2NC_6H_4CN$	CH_3	$4\text{-}O_2NC_6H_4$	78	90
$4\text{-}CH_3OC_6H_4CN$	CH_3	$4\text{-}CH_3OC_6H_4$	71	90
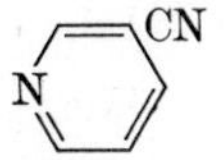 3-cyanopyridine (pyridine ring with CN)	CH_3	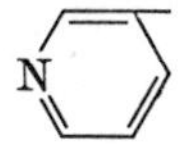 3-pyridyl	62	90
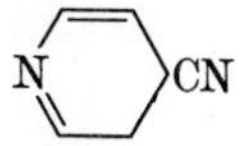 4-cyanopyridine (pyridine ring with CN)	CH_3	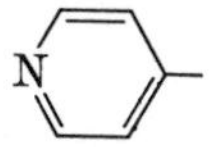 4-pyridyl	55	90
CH_3CN	C_2H_5*	CH_3	56	90
$CH_2{=}C(CH_3)CN$	CH_3	$CH_2{=}C(CH_3)$	—	91
HCN	CH_3	H	—	91
$(NCCH_2CH_2)_2O$ / $(NCCH_2CH_2)_2S$	CH_3	$CH_2{=}CH$	—	86
$2\text{-}CH_3C_6H_4CN$	CH_3	$2\text{-}CH_3C_6H_4$	—	91
$3\text{-}CH_3C_6H_4CN$	CH_3	$3\text{-}CH_3C_6H_4$	—	91
$4\text{-}CH_3C_6H_4CN$	CH_3	$4\text{-}CH_3C_6H_4$	—	91
$CH_2{=}CHCN$	C_2H_5*	$CH_2{=}CH$	—	91
$CH_2{=}C(CH_3)CN$	C_2H_5*	$CH_2{=}C(CH_3)$	—	91

Note: References 127–149 are on p. 325.

* 2,6-Dimethyloctane-2,6-diol was used in this reaction.

TABLE XVII-F. 5,6-DIHYDROPYRIDINES

$$(CH_3)_2C(OH)CH_2C(OH)(CH_3)_2 + RCN \xrightarrow{H_2SO_4}$$ (ring of $C(CH_3)_2$, CH_2, N, CH_3C, CR, CH) (Ref. 90)

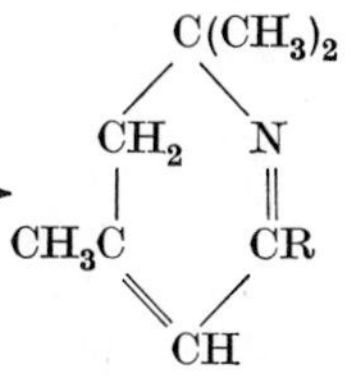

Nitrile	R	Yield (%)
CH_3CN	CH_3	23
$CH_2{=}CHCN$	$CH_2{=}CH$	20
C_6H_5CN	C_6H_5	21

Note: References 127–149 are on p. 325.

TABLE XVII-G. ISOQUINOLINES[92]

Alcohol	Nitrile	Reaction Medium	Product	Yield (%)
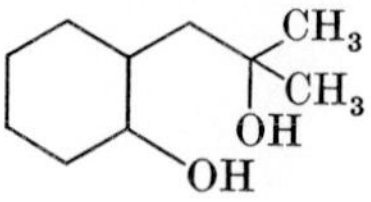 [structure]	CH_3CN	96% H_2SO_4	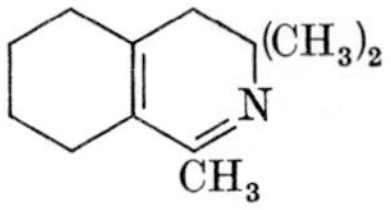 [structure]	30
	CH_3CN	98% H_2SO_4	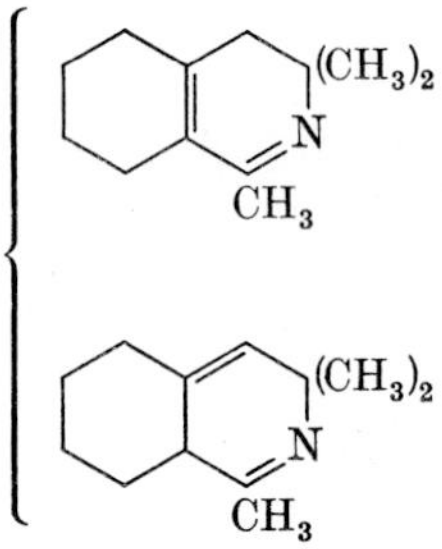 [structures]	20
	CH_3CN	$H_2SO_4/NaBH_4$	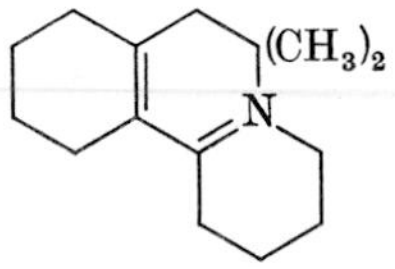 [structure]	17

Note: References 127–149 are on p. 325.

TABLE XVII-H. 3,4-DIHYDROISOQUINOLINES

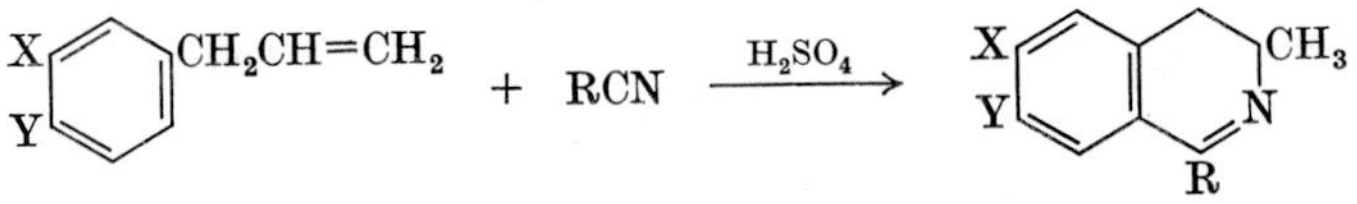

$$X,Y\text{-}C_6H_3CH_2CH{=}CH_2 + RCN \xrightarrow{H_2SO_4} \text{3,4-dihydroisoquinoline (X, Y; 3-}CH_3\text{; 1-R)}$$

X	Y	R	Yield (%)	Refs.
CH_3O	CH_3O	CH_3	—	54*
CH_3O	CH_3O	C_6H_5	15	34, 54*
CH_3O	CH_3O	$C_6H_5CH_2$	—	54*
CH_3O	CH_3O	4-$CH_3OC_6H_4$	16	34
CH_3O	CH_3O	3,4-$(CH_3O)_2C_6H_3$	53	34
CH_3O	CH_3O	3,4-$(C_2H_5O)_2C_6H_3$	Poor	34
—OCH_2O—		C_6H_5	—	54*
—OCH_2O—		$C_6H_5CH_2$	—	54*

Note: References 127–149 are on p. 325.

* Isosafrol and methylisoeugenol were used.

TABLE XVII-I. 2-PYRIDONES[97]
(The reaction medium was 96% sulfuric acid.)

Reactant	Product	Yield (%)
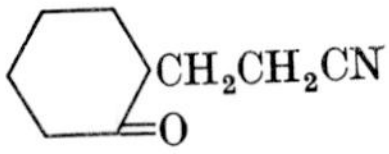 CH_2CH_2CN, =O	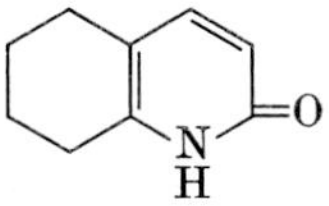 =O, N, H	54
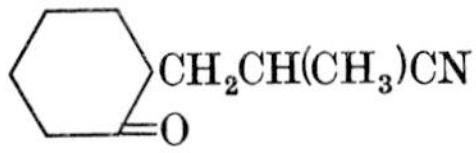 $CH_2CH(CH_3)CN$, =O	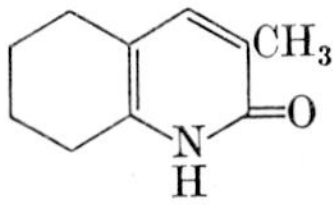 CH_3, =O, N, H	54
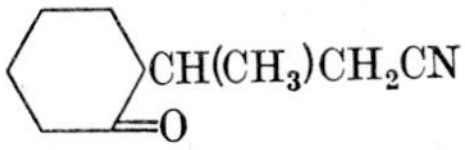 $CH(CH_3)CH_2CN$, =O	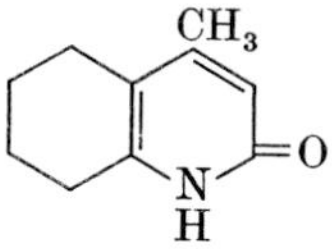 CH_3, =O, N, H	40
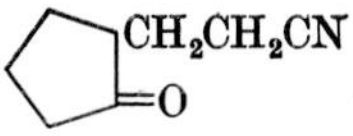 CH_2CH_2CN, =O	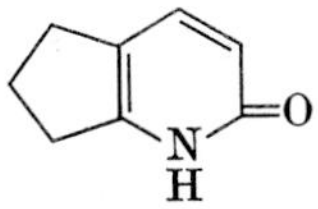 =O, N, H	71
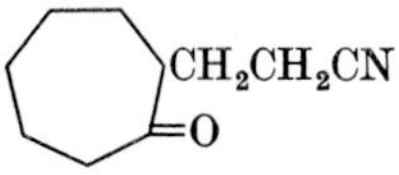 CH_2CH_2CN, =O	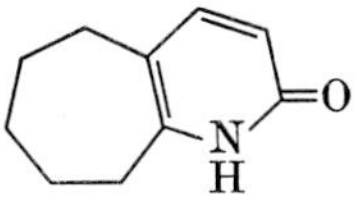 =O, N, H	45

Note: References 127–149 are on p. 325.

TABLE XVII-J. Dihydropyridones

(The reaction medium was sulfuric acid.)

Reactant	Product	Yield (%)	Ref.
$C_6H_5CH(CN)CH_2CH(CH_3)COCH_3$	3-C_6H_5, 5-CH_3, 6-CH_3-3,4-dihydro-2(1H)-pyridone	48	98
$C_6H_5CH(CN)C(CH_3)_2CH_2COCH_3$	3-C_6H_5, 4-$(CH_3)_2$, 6-CH_3-3,4-dihydro-2(1H)-pyridone	74	98
$C_6H_5CH(CN)CH_2CH(CH_3)COC_6H_5$	3-C_6H_5, 5-CH_3, 6-C_6H_5-3,4-dihydro-2(1H)-pyridone	79	98
$C_6H_5CH(CN)CH_2CH(C_6H_5)COCH_3$	3-C_6H_5, 5-C_6H_5, 6-CH_3-3,4-dihydro-2(1H)-pyridone	12	98
$C_6H_5CH(CN)CH_2CH(C_6H_5)COC_6H_5$	3-C_6H_5, 5-C_6H_5, 6-C_6H_5-3,4-dihydro-2(1H)-pyridone	70	98
$C_6H_5COCH(C_6H_5)CH_2CH_2CN$	5-C_6H_5, 6-C_6H_5-3,4-dihydro-2(1H)-pyridone	65	99

Note: References 127–149 are on p. 325.

TABLE XVII-K. 1-AZABICYCLOALKANES

(The reaction medium was sulfuric acid; the Ritter reaction was followed by reduction with sodium borohydride.)

Glycol	Nitrile	Product	Yield (%)	Refs.
$(CH_3)_2C(OH)CH_2CH_2C(OH)(CH_3)_2$	$ClCH_2CH_2CN$	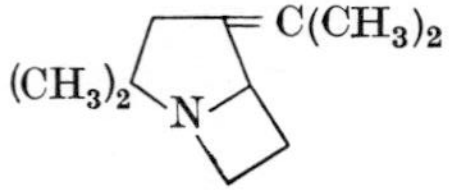 $(CH_3)_2$ N $C(CH_3)_2$	30	103, 106
	$ClCH_2CH_2CH_2CN$	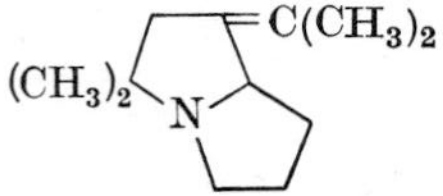 $(CH_3)_2$ N $C(CH_3)_2$	61	101
	$Cl(CH_2)_4CN$	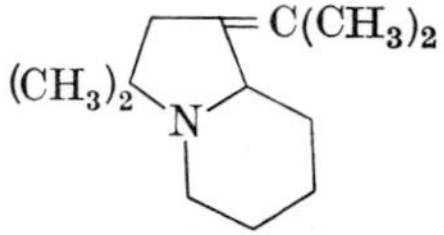 $(CH_3)_2$ N $C(CH_3)_2$	65	101
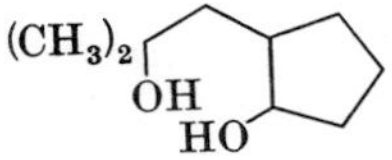 $(CH_3)_2$ OH HO	$ClCH_2CH_2CN$	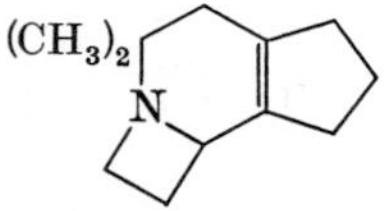 $(CH_3)_2$ N	50	102
	$Cl(CH_2)_3CN$	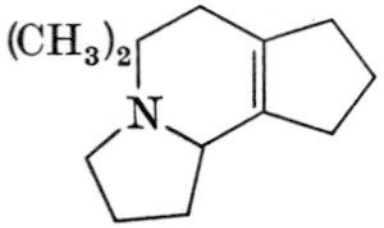 $(CH_3)_2$ N	54	102
	$Cl(CH_2)_4CN$	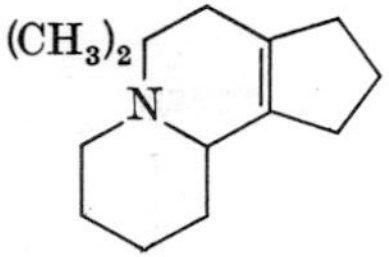 $(CH_3)_2$ N	46	102

Note: References 127–149 are on p. 325.

TABLE XVII-L. 1,4,5,6-TETRAHYDRO-*as*-TRIAZINES

$$R^2-\underset{OH}{\overset{R^3}{C}}-\underset{NCH_3(NH_2)}{\overset{R^1}{CH}} + RCN \xrightarrow{H_2SO_4} \text{1,4,5,6-tetrahydro-}as\text{-triazine (5-}R^1\text{, 6-}R^2\text{, 6-}R^3\text{, 4-NCH}_3\text{... 3-R)}$$

(Ref. 100)

R	R^1	R^2	R^3	Yield (%)
CH_3	H	H	C_6H_5	67
CH_3	H	CH_3	CH_3	19
$CH_2{=}CH$	H	H	C_6H_5	16
3-NC_5H_4	CH_3	H	C_6H_5	15
$(C_6H_5)_2NCH_2$	H	H	C_6H_5	9
2-ClC_6H_4	H	H	C_6H_5	29
2-ClC_6H_4	H	CH_3	CH_3	13
2-ClC_6H_4	CH_3	H	C_6H_5	26
4-ClC_6H_4	H	H	C_6H_5	17
4-ClC_6H_4	CH_3	H	C_6H_5	23
4-FC_6H_4	CH_3	H	C_6H_5	14
C_6H_5	H	H	C_6H_5	52
C_6H_5	H	CH_3	CH_3	12
C_6H_5	CH_3	H	C_6H_5	35
4-$CF_3C_6H_4$	H	H	C_6H_5	6
4-$CF_3C_6H_4$	CH_3	H	C_6H_5	7
2-$CH_3C_6H_4$	CH_3	H	C_6H_5	15
3-$CH_3C_6H_4$	CH_3	H	C_6H_5	30
4-$CH_3C_6H_4$	CH_3	H	C_6H_5	34
4-$CH_3OC_6H_4$	H	H	C_6H_5	8
4-$CH_3OC_6H_4$	CH_3	H	C_6H_5	10

References 127–149 are on p. 325.

TABLE XVIII. REACTIONS OF 5-METHYL- AND 5-PHENYL-ISOXAZOLE[110]

Substituent in Isoxazole	Addend	Product	Yield (%)
5-Methyl	$(CH_3)_2C{=}CHCOCH_3$	CH_3 CH_3 NH CH_3 =O $COCH_3$	47
5-Phenyl	$(CH_3)_2C{=}CHCOCH_3$	CH_3 CH_3 NH CH_3 =O COC_6H_5	—
5-Methyl	$HC{\equiv}CCH_2C(CH_3)_2OH$	$HC{\equiv}CCH_2C(CH_3)_2NHCOCH_2COCH_3$	—
	$(CH_3)_3COH$	$(CH_3)_3CNHCOCH_2COCH_3$	—
	$(CH_3)_2C{=}CHCO_2H$	$(CH_3)_3CNHCOCH_2COCH_3$	66
	$(CH_3)_2C{=}CHCO_2C_2H_5$	$(CH_3)_2CNHCOCH_2COCH_3$ $CH_2CO_2C_2H_5$	56

Note: References 127–149 are on p. 325.

TABLE XIX. TETRACYCLINES

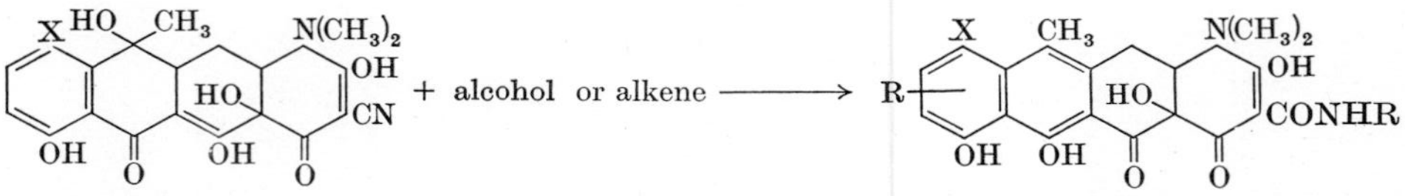

+ alcohol or alkene ⟶

Reactant	X	R	Reaction Medium	Ref.
$(CH_3)_2C{=}CH_2$	H	$(CH_3)_3C$	H_2SO_4/CH_3CO_2H	146
	Cl	$(CH_3)_3C$	H_2SO_4/CH_3CO_2H	146
$(CH_3)_2C{=}CHCH_3$	H	$(CH_3)_2C(C_2H_5)$	$H_2SO_4/(n\text{-}C_4H_9)_2O$	146
$C_6H_5CH_2OH$	Cl	$C_6H_5CH_2$	H_2SO_4/CH_3CO_2H	146
$C_6H_5CHOHCH_3$	H	$CH_3CH(C_6H_5)$	H_2SO_4/CH_3CO_2H	146
$n\text{-}C_6H_{13}CHOHCH_3$	H	$n\text{-}C_6H_{13}CHCH_3$	H_2SO_4/CH_3CO_2H	146

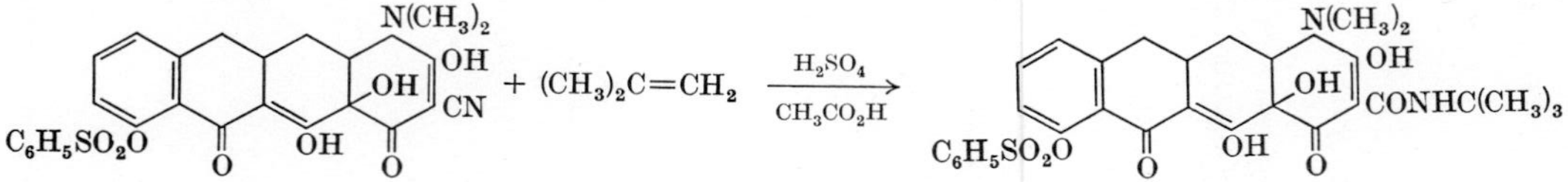

Reaction	Ref.
($C_6H_5SO_2O$-substituted nitrile) + $(CH_3)_2C{=}CH_2$ $\xrightarrow[CH_3CO_2H]{H_2SO_4}$ ($C_6H_5SO_2O$-substituted $CONHC(CH_3)_3$)	128
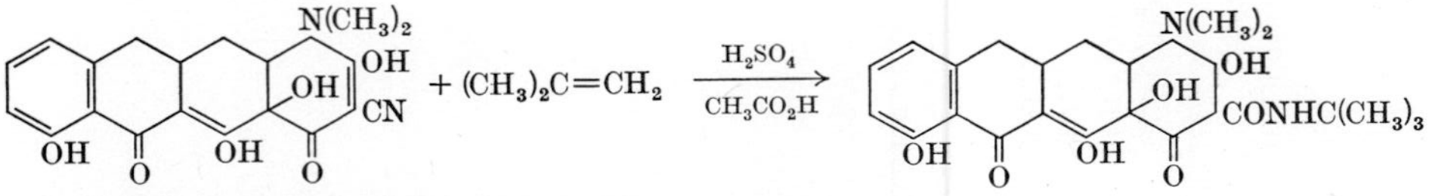 (nitrile) + $(CH_3)_2C{=}CH_2$ $\xrightarrow[CH_3CO_2H]{H_2SO_4}$ ($CONHC(CH_3)_3$)	128

Note: References 127–149 are on p. 325.

TABLE XX. MISCELLANEOUS RITTER REACTIONS

Reactant	Nitrile	R in Product	Solvent	Product	Yield (%)	Refs.
$CH_2{=}C(CH_3)C(CH_3){=}CH_2$	CH_3CN	—	H_2SO_4 (fuming)	$(CH_3)_2C{=}C(CH_3)CH_2N$ (ring with CH_3, CH_3; CH_3, $NHCOCH_3$)	—	54
(cyclopentenyl)$CH_2C(CH_3)_2OH$	CH_3CN	—	H_2SO_4	(bicyclic) CH_3, N, $(CH_3)_2$	19	86
(1-hydroxycyclopentyl)$CH_2C(CH_3)_2OH$ (OH)	CH_3CN	—	H_2SO_4	(bicyclic) CH_3, N, $(CH_3)_2$	76	86
					a / b	
(cyclopentenyl, a, b)$CH_2C(CH_3)_2OH$	CH_3CN	CH_3	H_2SO_4	(spiro) O, R, N, $(CH_3)_2$	63 / 51	86, 147
	$CH_2{=}CHCN$	$CH_2{=}CH$	H_2SO_4		58 / 38	86
	C_2H_5CN	C_2H_5	H_2SO_4		57 / 29	86
	C_6H_5CN	C_6H_5	H_2SO_4		72 / 39	86
(tetrahydroquinoline) CH_3, $(CH_3)_2$, NH	CH_3CN	—	H_2SO_4	No reaction		148
	C_6H_5CN	—	H_2SO_4	No reaction		148
$CH_2{=}CH(CH_2)_8CO_2CH_3$	C_6H_5CN	—	—	$CH_2(CH_2)_9CO_2H$ / $NHCOC_6H_5$	16	63
Turpentine	$(CH_3)_3CCH_2CH(CH_3)$-$CH_2CHOHCN$	—	$H_2SO_4/(n\text{-}C_4H_9)_2O$	N-Dihydroterpenyl-2-hydroxydecanamide*	—	134
	(bicyclic) H_3C, CH_3, CH_3, CHOHCN	—	$H_2SO_4/(n\text{-}C_4H_9)_2O$	(bicyclic) H_3C, CH_3, CH_3, CHOHCONH-dihydroterpenyl*		134

Note: References 127–149 are on p. 325.

*** The product corresponded in composition to this formula; its structure was not established.**

TABLE XXI. MISCELLANEOUS INTRAMOLECULAR RITTER REACTIONS

Reactant	Reaction Medium	Product	Yield (%)	Ref.
$CH_2{=}C(CH_3)CH_2CH_2CN$	Polyphosphoric acid	No reaction	—	46
$CH(CO_2H){=}C(CH_3)CH{=}C(CH_3)CN$	Alkaline hydrolysis	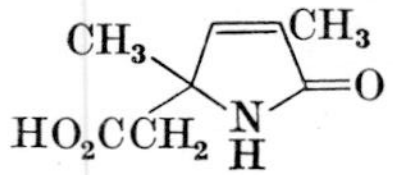 CH_3, CH_3, HO_2CCH_2, N H, =O	—	149
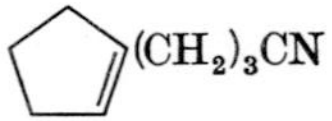 $(CH_2)_3CN$	Polyphosphoric acid	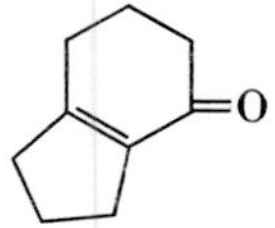 =O	—	45
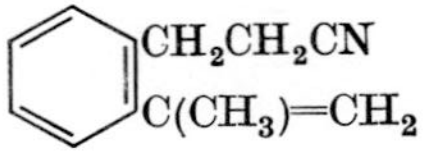 CH_2CH_2CN, $C(CH_3){=}CH_2$	Polyphosphoric acid	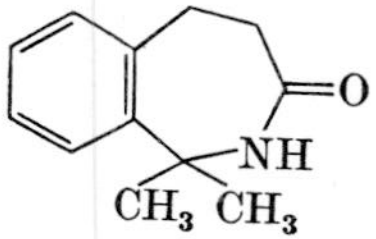 =O, NH, CH_3 CH_3	24	50
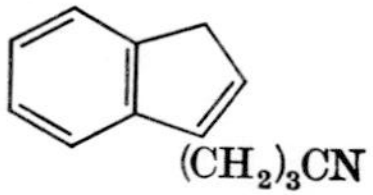 $(CH_2)_3CN$	H_2SO_4/C_2H_5OH	No reaction	—	47
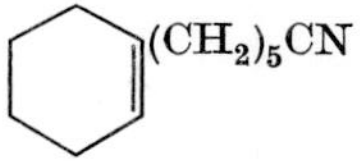 $(CH_2)_5CN$	H_2SO_4	No reaction	—	48
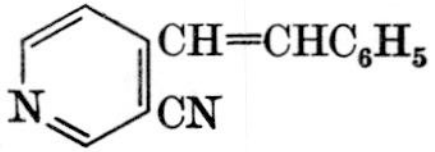 $CH{=}CHC_6H_5$, CN, N	Polyphosphoric acid	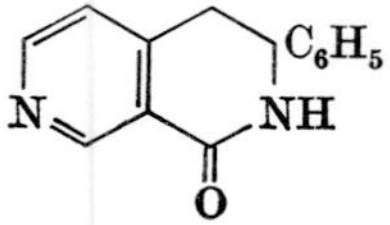 C_6H_5, N, NH, O	90	51
$(C_6H_5)_2C{=}CH(CH_2)_4CN$	Polyphosphoric acid	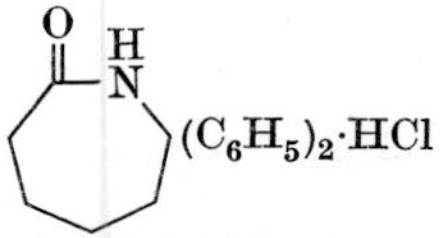 O, H N, $(C_6H_5)_2\cdot HCl$	85	49

REFERENCES TO TABLES

[127] J. R. Boissier, C. Malen, and C. Dumont, *Compt. Rend.*, **242**, 1086 (1956).
[128] C. R. Stephens, C. R. Beereboom, H. H. Rennhard, P. N. Gordon, K. Murai, R. K. Blackwood, and M. S. von Wittenau, *J. Am. Chem. Soc.*, **85**, 2643 (1963).
[129] H. D. Moed, J. van Dyk, and H. Niewind, *Rec. Trav. Chim.*, **74**, 919 (1955).
[130] A. E. Clarke and M. F. Grundon, *J. Chem. Soc.*, 4190 (1964).
[131] R. P. Welcher, M. E. Castellion, and V. P. Wystrach, *J. Am. Chem. Soc.*, **81**, 2541 (1959).
[132] J. J. Ritter, *J. Am. Chem. Soc.*, **70**, 4253 (1948).
[133] H. Plaut and J. J. Ritter, *J. Am. Chem. Soc.*, **73**, 4076 (1951).
[134] P. L. de Benneville, U.S. pat. 2,632,766 [*C.A.* **47**, 6686c (1953)].
[135] M. Mousseron, H. Christol, and A. Laurent, *Compt. Rend.*, **248**, 1904 (1958).
[136] A. Pohland, U.S. pat. 2,778,835 [*C.A.*, **52**, 12930b (1959)].
[137] M. Mousseron, R. Jacquier, and H. Christol, *Compt. Rend.*, **235**, 57 (1952).
[138] C. D. Parker, W. D. Emmons, A. S. Pagano, H. A. Rolewicz, and K. S. McCallum, *Tetrahedron*, **17**, 96 (1962).
[139] S. Chiavarelli, E. F. Rogers, and G. B. Mariui-Bettolo, *Gazz. Chim. Ital.*, **83**, 347 (1953).
[140] J. Anatol, *Compt. Rend.*, **235**, 249 (1952).
[141] B. V. Shetty, *J. Org. Chem.*, **26**, 3002 (1961).
[142] H. J. Barber and E. Lunt, *J. Chem. Soc.*, 1187 (1960).
[143] H. Stetter, M. Schwarz, and A. Hirschhorn, *Angew. Chem.*, **71**, 429 (1959).
[144] H. Stetter, M. Schwarz, and A. Hirschhorn, *Chem. Ber.*, **92**, 1629 (1959).
[145] E. E. Magat, U.S. pat. 2,628,216 [*C.A.*, **47**, 5129e (1953)].
[146] C. R. Stephens, U.S. pat. 3,028,409 [*C.A.*, **57**, 9769g (1962)].
[147] S. Hünig, W. Lücke, V. Meuer, and W. Grässmann, *Angew. Chem.*, *Intern. Ed. Engl.*, **2**, 213 (1963).
[148] J. P. Brown, *J. Chem. Soc.*, 3012 (1964).
[149] A. T. Balaban, T. H. Crawford, and R. H. Wiley, *J. Org. Chem.*, **30**, 879 (1965).

AUTHOR INDEX, VOLUMES 1–17

CHAPTER AND TOPIC INDEX, VOLUMES 1–17

Many chapters contain brief discussions of reactions and comparisons of alternative synthetic methods which are related to the reaction that is the subject of the chapter. These related reactions and alternative methods are not usually listed in this index.

In this index the volume number is in BOLDFACE, the chapter number in ordinary type.

SUBJECT INDEX, VOLUME 17

Since the table of contents provides a quite complete index, only those items not readily found from the contents pages are listed here.

Numbers in BOLDFACE type refer to experimental procedures.